Illustrated Experiments in Fluid Mechanics

The NCFMF Book of Film Notes

The MIT Press
Cambridge, Massachusetts, and
London, England

Illustrated Experiments in Fluid Mechanics

The NCFMF Book of Film Notes

Based on material in films
produced under direction of the
National Committee
for Fluid Mechanics Films

This book was printed on White Semline Offset
by Semline, Inc.
and bound by Semline, Inc.
in the United States of America.

Library of Congress Cataloging in Publication Data

National Committee for Fluid Mechanics Films.
 Illustrated experiments in fluid mechanics.

 "Based on material in films produced under direction
of the National Committee for Fluid Mechanics Films."
 1. Fluid mechanics. I. Title.
QC145.2.N37   532'.0028   72-5264
ISBN 0-262-14014-4
ISBN 0-262-64012-0 (pbk)

Contents

Preface

"Since things in motion sooner catch the eye than what not stirs." *Troilus and Cressida*

This volume contains text and photographic material related to the sound films* prepared under the direction of the National Committee for Fluid Mechanics Films (NCFMF). The films, and the related text material herein, cover nearly all of the fundamental phenomena of fluid motions.

Each chapter of this volume is based on the script of a particular film. Each is heavily illustrated with experimental scenes from the film. The many photographs of interesting and significant phenomena in fluid mechanics are in themselves highly educative to both beginning and advanced students. In addition, descriptions of experimental phenomena help to develop in students that valuable faculty: a physical intuition, a "feel" for the diverse ways in which fluids behave. With the "film in print," in the form of a chapter of this book, the reader who has seen a film can relive the film, as it were: he can recapture the phenomena, the motion, and the vividness of the film, but he can do so at his own speed, without being pressured by the pace of the film. A student can reflect on what he has seen, and relate it to his classroom and textbook learning. Furthermore, a reading of the appropriate chapter before viewing a film is most helpful in getting the maximum satisfaction and benefit from the film.

Formed in 1961, the NCFMF functions in collaboration with Education Development Center, Inc. (EDC) in Newton, Massachusetts, formerly Educational Services Incorporated. The NCFMF, a self-constituted group of eleven university faculty members interested in education and active in various branches of fluid mechanics, was responsible for the general structure of the film program, for the substantive content, and for the creation of this volume. Education Development Center was responsible for film production, for a part of the editing and composition of this book, and also provided administrative and laboratory services. Funding was provided mainly by the National Science Foundation, with some contribution by the Office of Naval Research. The program was completed in 1969.

The concept that brought the NCFMF into existence was the conviction that films would provide

*The films are currently distributed by Encyclopaedia Britannica Educational Corp., 425 N. Michigan Ave., Chicago, Ill. 60611, from whom a brochure is available on request.

a convenient, economic, and powerful way of filling a serious educational gap. On the one hand are the concrete, experimental, perceptual phenomena of the real world of fluid motion. On the other hand is the abstract, mathematical, conceptual treatment of the subject in textbooks and classroom lectures.

With the objective of bridging this gap, we shaped the film program along certain policy guidelines that are of course reflected in the chapters of this book.

1. First priority was given to the presentation and elucidation of *experimental* phenomena. The usual kinds of lecture and text material, and of mathematics, are notably absent. The films and the chapters of this book were not designed as substitutes for lectures (although they have been so used!), but rather as teaching aids. Our aim was to use the power of graphical illustration to support the weakest point of the lecture/textbook system.

2. Out of the extremely large number of experiments and topics available, only those of fundamental and broad nature, and of permanent character, were selected. In each case, an "introductory" treatment was used: one aimed at the student's first study of a particular topic, whether the topic were elementary or advanced.

3. Each film, and the related chapter in this book, was designed to show the broad ramifications of fluid mechanics by drawing on examples from many fields. Parochial viewpoints and unnecessary jargon were avoided. As a consequence, we expect them to have a broadening influence, and, further, that they will be useful to other professions than those principally occupied with fluid mechanics, for instance, medicine, oceanography, acoustics, and architecture.

4. No attempt was made to organize the films, or these related chapters, in an ordered sequence or curriculum. There is too much variation in courses on fluid mechanics for such to be satisfactory. Moreover, the capability of films to range over wide areas in a short time is a stimulating contrast to the slower unfolding of topics in a long series of lectures.

Out of these governing views grew a virtually new component of scientific and engineering education, certainly one previously untried on a large scale: demonstration-experiment films, and related written material, for a difficult subject having a strong mathematical flavor. In 1961 it was admittedly a speculation whether such films could be effective, and, even if they were, whether they would be widely used. In the intervening years both these speculations have become resounding positive statements. The films have won their share of awards at film festivals, to be sure. But more important, they have been widely adopted and used inside and outside the classroom. The remarks of a generation of students, who now take these films for granted as part of their educational experience, confirm that they produce a deep imprint connected with the other parts of their learning experience in fluid mechanics. Groups attempting similar programs in other scientific and engineering fields have emulated the NCFMF. Journals that regularly review books have been stimulated by the NCFMF product to review films as well. In its issue of July 1967, the distinguished *Journal of Fluid Mechanics* inaugurated its reviews of films with the statement:

The work of the U.S. National Committee for Fluid Mechanics Films is now well known. . . . Films on fluid mechanics have been made before now by other people, but the scale of this programme, the quality of the Committee's work, and the wide distribution of the products make these films a notable development in the academic world. Already the impact on the teaching of fluid mechanics in colleges and universities has been considerable, and it is likely that films are here to stay as a regular teaching aid. Fluid mechanics *is* a photogenic subject; and the Committee have opened our eyes to what can be done by an effective combination of thought and money.

We hope that the related written material presented in this volume will be received with equal pleasure by those already familiar with the films, and, further, that the book will augment the film audience.

Many different ways of using the films have evolved on university campuses. The sound films are sometimes shown in class hours. More often each film is shown several times each term, in the afternoon or evening, at well-advertised times, open to the entire community. Where a well-equipped audiovisual facility exists, students can schedule a film at their convenience. Sometimes the library will serve a similar function. Complete courses have been built around the films and the film notes as central exhibits. There are obvious ways in which this present volume may be worked into the related program of film showings.

For the convenience of interested readers, the

volume also contains brief descriptions of the 133 four-minute silent films produced by the NCFMF. These are *single topic* films that show a particular experiment or illustrate a particular phenomenon. They are ideal for illustrating a point that arises during the course of a lecture. They can be operated in corridor showcases, or left in the laboratory to be viewed at any time. Many schools have viewing rooms for cartridge projection in the library or audiovisual center. Students can study the cartridge loops at will on recommendation of the instructor, or they may be assigned as homework.

While the cost of the program, some $3 million, was substantial, no less important were the intellectual and professional contributions so generously made by so many individuals. Mindful that we were acting for a national and international scholarly community, and that a similar effort could not be undertaken for at least a generation, we felt especially conscious of a responsibility for high quality and broad acceptance. An important mechanism for this—and a highly successful one—was the Advisory Committee for each film, which was always encouraged to be aggressive and outspoken in establishing and maintaining the highest of standards. As one principal put it, the development of his film was a "chastening experience." The chapters of this volume reflect these standards of quality.

To the members of the Advisory Committees, and most particularly to the principals, the NCFMF offers its thanks. Acknowledgment is made also of the dedicated efforts of the professional staff at EDC, especially to Charles R. Conn, II, Bruce Egan, and Benjamin T. Richards for their contributions to the editing of the material of this book.

Its work completed, the NCFMF remains as an entity only to keep a paternal eye on its product, and to ensure that it survives its youth. While not all we produced was of even quality, the best should survive without growing old.

Ascher H. Shapiro,
for the National Committee for
Fluid Mechanics Films
Cambridge, Massachusetts
February 1972

Part I

16-mm Sound Films

Each of the following 21 chapters has the same title as the motion picture on which it is based.*

The references at the end of each chapter provide recommended sources of further study.

The chapters are arranged so that the early ones are more general and introductory in character, while the later ones are more specialized and more advanced.

*In lieu of a chapter in this volume, the film, "The Fluid Dynamics of Drag," is represented in written form by the paperback *Shape and Flow: The Fluid Dynamics of Drag,* by Ascher H. Shapiro, Doubleday and Company, Inc., New York (1961).

Eulerian and Lagrangian Descriptions in Fluid Mechanics

John L. Lumley
PENNSYLVANIA STATE UNIVERSITY

Introduction

In order to calculate forces exerted by moving fluids and to calculate other effects of flows, such as transport, we must be able to describe the dynamics of flow mathematically. To discuss the dynamics, we have to be able to describe the motion itself. The description of motion is called kinematics. We are interested in the kinematics of continuous media, that is, in describing the motion of deformable stuff that fills a region. Specifically, we are interested in describing the displacement, velocity, and acceleration of material points in two kinds of reference frames commonly used in fluid mechanics. We will show how these two descriptions are related to one another.

In addition to moving from place to place, an elementary piece of fluid (a piece small compared to the flow field) is usually distorted and rotated as it goes. Here we focus our attention on the translation.*

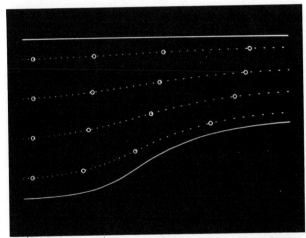

1. Computer simulation of steady flow in a contracting channel. The open circles mark moving material points. The dashed lines are the pathlines of the particles.

Figure 1 shows a computer simulation of the flow of water through a contraction. Note that a few typical material points are identified by open circles; we adopt this convention throughout.

* Deformation is dealt with in the NCFMF film DEFORMATION OF CONTINUOUS MEDIA.

Lagrangian Description

In elementary mechanics, we are accustomed to describing the position of a material point as a function of time, using a vector drawn from some arbitrary location to indicate the displacement. We will use

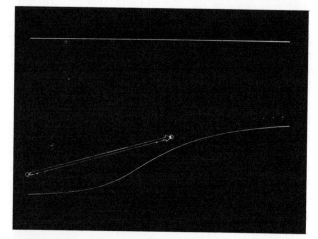

2. The open vector indicates *displacement* of a material point from its initial location.

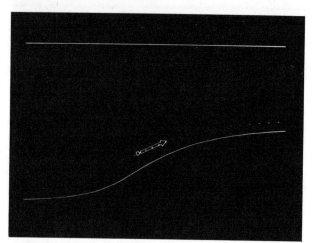

3. The velocity of a moving material point is indicated by an open vector attached to it.

open vectors, (Fig. 2), to indicate velocity and displacement relating to the material points. In a given motion, we can compute velocity and acceleration of such a point at each instant. In Fig. 3, we indicate the velocity by a vector attached to the point. In a continuous fluid, of course, we have an infinity of mass points and we have to find some way of tagging them for identification. A convenient way, though not the only one, is to pick some arbitrary reference time (which we will call the initial time) and identify the material point by its location at the time. Mathematically, we would say that the velocity is a function of initial position and time. To accord with this description, in Fig. 4 the vector is shown attached to the initial position. We could show the vector attached to the moving point, or use both, if we were displaying

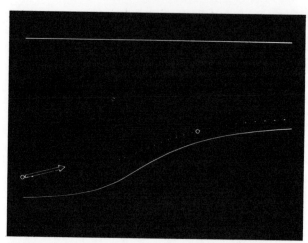

4. The velocity of a moving material point is here indicated by an open vector attached to its initial position.

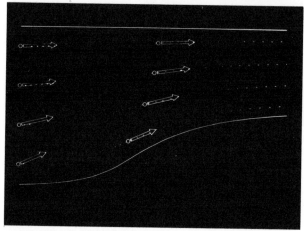

5. Velocity vectors displayed both at the moving points and at their initial locations.

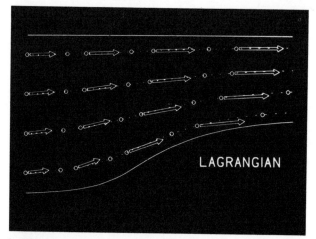

6. The velocities of material points distributed throughout the flow field are displayed at the respective initial positions of the points.

the motion of a group of points whose vectors do not interfere with one another (Fig. 5). To display the whole motion, and in more complicated situations, we avoid interference by showing the vector only at the initial location, as in Fig. 6. To describe the whole motion, we would have to give the velocity of all the pieces of matter in the flow as a function of time and initial position.

Such a description, in terms of material points, is called a *Lagrangian* description of the flow. The identifying co-ordinates are called *Lagrangian,* or sometimes *material,* co-ordinates. Given the Lagrangian velocity field, we can easily calculate the Lagrangian displacement by integration in time, and the acceleration field by partial differentiation with respect to time.

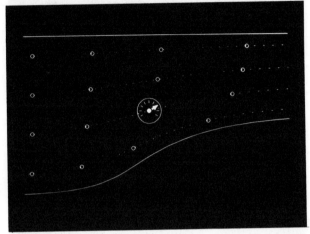

7. A pressure gauge is attached to one of the moving points.

To make what we might call a Lagrangian measurement, we can imagine attaching an instrument like a pressure gauge to a fluid material point (Fig. 7). This sort of measurement is attempted in the atmosphere with balloons of neutral buoyancy. If the balloon does indeed move faithfully with the air, it gives the Lagrangian displacement, i.e. the displacement of an identified fluid "element." Such Lagrangian measurements are actually very difficult, particularly in the laboratory. We usually prefer to make measurements at points fixed in laboratory co-ordinates; it is relatively easy to hold an instrument at a fixed location.

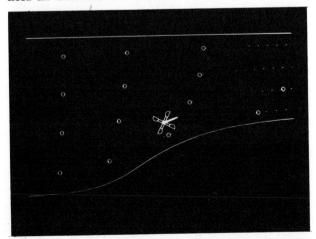

8. An anemometer is placed at a fixed position in the flow field.

Eulerian Description

Classically, the idea of a field, such as an electric, magnetic, or temperature field, is defined by how the response of a test body or probe, like the anemometer in Fig. 8, varies with time at each point in some spatial co-ordinate system. In Fig. 8 the fixed anemometer probes in laboratory co-ordinates. We will always use

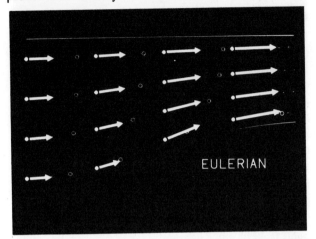

9. The solid vectors indicate the velocities at fixed points in the contraction.

solid points and solid arrows to indicate such probing positions, fixed in our laboratory, and the velocities measured there.

In Fig. 9 we have a grid of points fixed in space with an arrow at each to indicate the velocity at each point. A description like this which gives the spatial velocity distribution in laboratory co-ordinates is called an *Eulerian* description of the flow.

Relation Between Eulerian and Lagrangian Frames

Although the physical field is the same, the Eulerian and Lagrangian representations are not the same, be-

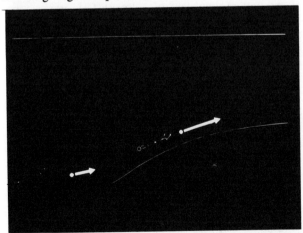

10. The open vectors indicate the Lagrangian velocity of the moving material point. The solid vectors indicate the Eulerian velocities at two fixed points. When the moving point coincides with a fixed point, the Eulerian and Lagrangian velocities coincide.

cause the velocity at a point in laboratory co-ordinates does not always refer to the same piece of matter. Different material points are continually streaming through the same laboratory point. The velocity that a fixed probe indicates is the velocity of the material point that is passing through the laboratory point (probe location) at that instant (Fig. 10).

Change of Reference Frame

A possible advantage of laboratory co-ordinates (and their Galilean transformations, which are also Eulerian) is that the Eulerian field may be steady in one of these frames. This is illustrated by the case of a simple surface wave. Figure 11 is a computer simulation of the flow under a free-surface gravity wave. To make things clearer, the wave amplitude has been rather exaggerated. Figure 12 is a close-up of the same flow, showing moving material points, and their pathlines. The Lagrangian velocities of the moving points are indicated by arrows attached to the points. In any flow, the Lagrangian field can only be steady if each material point always experiences the same

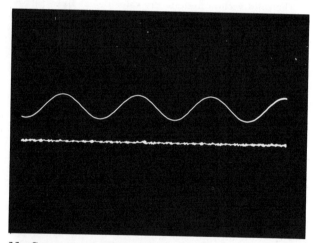

11. Computer simulation of a free-surface gravity wave which moves from left to right.

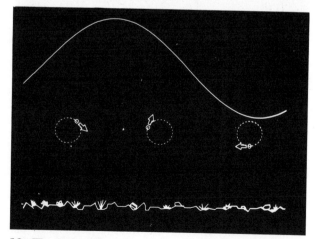

12. The dashed lines show the pathlines of material points under a moving wave. The open vectors indicate the Lagrangian velocities.

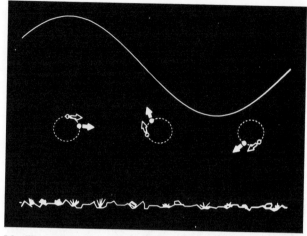

13. The solid arrows show the Eulerian velocities at points fixed in the laboratory frame of reference. (The Eulerian velocity vectors rotate about the fixed points as the waves pass.)

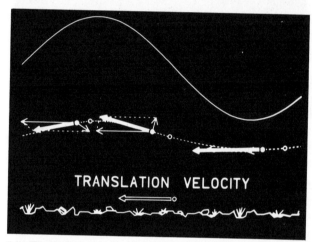

TRANSLATION VELOCITY

14. The Eulerian velocities (solid arrows) as seen from a laboratory reference frame moving to the right at the wave speed. In this frame, the fluid has an equal velocity directed to the left, which adds with the material-point velocity in the original frame to produce the resultant Eulerian velocities. The dashed line is the pathline of the particles.

velocity. This degenerate case happens only in a steady parallel flow.

Figure 13 shows the Eulerian description. In this wave motion neither the Eulerian nor the Lagrangian description is steady. In fact, they have an identical appearance. However, in this flow, if the laboratory frame is moved with the wave speed, the Eulerian pattern will become stationary. Figure 14 illustrates the result; the translation velocity is indicated by an arrow at the bottom. The velocities have been resolved into components: one component is the velocity with which the laboratory frame is translating; the other component is the material point velocity in the original frame of reference.

The pathlines are also streamlines in this frame of reference, since the flow is steady. The pathlines resemble the form of the free surface. As the material point passes through each laboratory point, its velocity is instantaneously the same as that of the laboratory point. It is partly this possibility of eliminating time as a variable that makes the Eulerian representation attractive.

Most laws of nature are more simply stated in terms of properties associated with material elements, that is, quantities described in Lagrangian frames. But it is nearly always much easier mathematically, when describing a continuum, to deal with these laws in laboratory co-ordinates. Thus, to write the conservation equations of fluid mechanics, one must be able to transform from one set of co-ordinates to the other. We will discuss first the relation between time derivatives of a scalar field in a flowing fluid, in the two types of co-ordinates.

Material Derivative in a Scalar Field

Let us imagine a river in which a radioactive tracer is suddenly and uniformly distributed. Since derivatives measure local changes, let us look at an infinitesimal part of this river. The dots in Fig. 15 symbolize

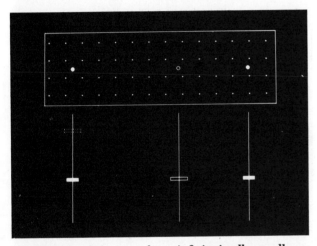

15. An expanded view of an infinitesimally small rectangular area in a river. The dots represent decaying radioactive tracer particles which are uniformly distributed there. The solid circles are laboratory points; the solid bars on the counters below indicate the level of radioactivity at these fixed points. The open circle is moving on a streamline from one laboratory point to the other; the open bar below indicates the level of radioactivity experienced by the moving point.

the tracer which is gradually decaying everywhere. The filled-in circles represent two fixed ("laboratory") points which are infinitesimally close together on the same streamline: they only appear to be far apart as a result of our expanded view. Since in this case the tracer was distributed uniformly, the radioactivities at the two laboratory points are the same, but are changing with time. Radiation counters are indicated at the laboratory points. The solid bars on these Eulerian radiation counters show the levels of radioactivity at the two laboratory points. The level experienced by a material point traveling from one laboratory point to

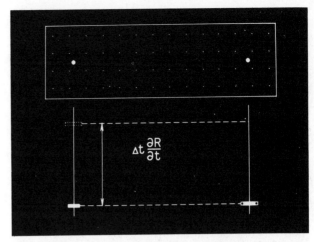

16. Situation at the instant the moving material point coincides with the right-hand laboratory point. $\Delta t \frac{\delta R}{\delta t}$ **is the total change since it coincided with the left-hand point.**

the other is monitored by watching the open bar on the Lagrangian counter carried by it. The dashed bar represents the value recorded by the Lagrangian counter as the material point passed through the left-hand laboratory point. From the before and after values of the Lagrangian counter (Fig. 16) it is evident that the traveling point sees just the same change that each of the laboratory points sees. This can be written as the time difference multiplied by the rate of change with time, as indicated on the figure.

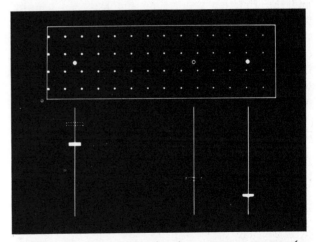

17. Now the tracer intensity is greater upstream (as shown by larger dots there) and falls off downstream.

If the tracer is not uniformly distributed, but instead has greater intensity upstream (Fig. 17), both intensities decrease with time as before. Just as before, the only change experienced by a material point is due to decay. The change seen at a fixed laboratory point is not, however, since new material of originally higher intensity is being swept past. To express the change experienced by a material point, but in Eulerian variables, we need two terms (Fig. 18): (1) the change of

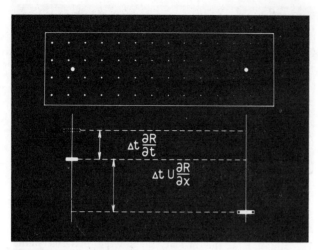

18. The total change experienced by a material point as it travels from one laboratory point to the other is the sum of (1) the change with time at either laboratory point (the upper expression), and (2) the intensity difference between the laboratory points at a fixed time (the lower expression).

intensity with time at a fixed point, and (2) the intensity difference between neighboring laboratory points at a fixed time. The total change when the material point has reached the right-hand laboratory point is given by the difference in level between the dashed counter on the left and the Lagrangian counter. The change with time experienced by either laboratory point (they have only infinitesimal separation) is given by the difference in level between the dashed counter and the Eulerian counter on the left, and can be written, as before, as the time difference multiplied by the spatially local rate of change with time.

The change due to the intensity difference between the laboratory points at any time is indicated by the difference in level between the two Eulerian counters, and can be written as the distance traveled multiplied by the spatial gradient in the direction traveled. The distance traveled can be written as the time difference

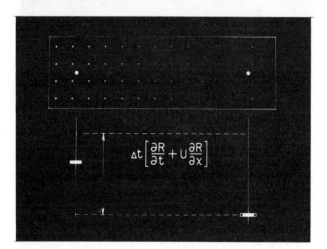

19. The total change is the sum of the two expressions in Fig. 18. The expression in brackets is the *material derivative*.

multiplied by the magnitude of the velocity. The total change (Fig. 19) is the sum of the two changes described. *Material*, or *substantial*, derivative is the name given to the expression multiplying the time difference in Fig. 19. This is simply the time rate of change experienced by the material point as it passes the laboratory point, *expressed in laboratory co-ordinates*. The importance of this point cannot be over-emphasized. Since this derivative operator occurs in every Eulerian conservation equation, we often give it a special symbol in fluid mechanics:

$$\frac{DR}{Dt} = \frac{\delta R}{\delta t} + U \frac{\delta R}{\delta x}$$

In vector notation, the velocity times the gradient in its direction can be written as the scalar product of velocity and the gradient vector:

$$\frac{DR}{Dt} = \frac{\delta R}{\delta t} + (\mathbf{U} \cdot \bigtriangledown)R.$$

Material Derivative in a Steady Vector Field

We are also interested in the material derivative of a vector field, such as the velocity, particularly because the material derivative of the velocity expresses the acceleration in a form which we need for the momentum equation in an Eulerian frame.

The expression deduced for the material derivative of a scalar field is correct for each component of a

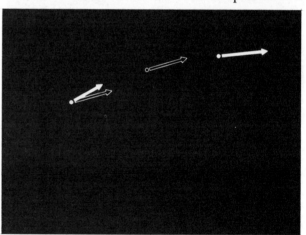

20. The solid circles are fixed laboratory points infinitesimally close together in a magnified view of an arbitrary steady flow. The solid vectors indicate velocities at the laboratory points. The open vectors indicate the velocity of the moving material point (the open circle).

vector field, but we can also operate on the vector field directly. The two laboratory points in Fig. 20 are an infinitesimal distance apart on the same pathline. The material point travels from one to the other. The material point velocity is indicated by open arrows attached to it, and to its "initial" location, the left-hand laboratory point. Although the flow is steady in the laboratory frame, the moving material point ex-

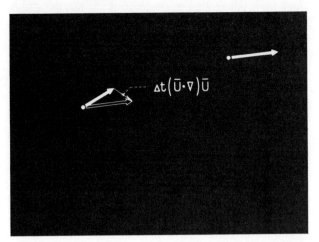

$$\Delta t \left(\bar{U} \cdot \nabla \right) \bar{U}$$

21. When the material point arrives at the right-hand laboratory point, the change in velocity it has experienced is the difference between the Eulerian (solid) vector and the Lagrangian vector at the left-hand laboratory point.

periences change as it travels through regions where the steady velocity is different. The total change is simply the difference between the velocities at the two laboratory points, indicated by the solid Eulerian vectors. The difference between the Eulerian vector and the Lagrangian vector at the left-hand point gives at each instant the change that the material point has experienced. The total change (Fig. 21) when it arrives at the right-hand point, a vector distance Δr away, after a time Δt, is the vector distance traveled times the gradient of the velocity. The distance traveled is just the time difference times the velocity.

Material Derivative in an Unsteady Vector Field

If the velocity of the entire flow changes with time, the Eulerian vectors (at fixed laboratory points) also change with time (Fig. 22). For clarity, we include as

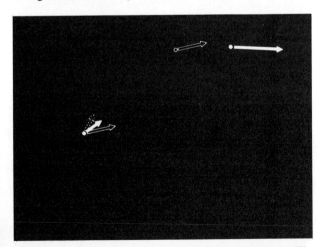

22. The velocity field is here changing with time. The dashed vector shows the initial value at the left-hand laboratory point; the solid vectors show the values at the fixed laboratory points; the open vectors show the velocity of the moving material point.

a dashed vector the initial value of the left-hand Eulerian vector, in addition to placing the Lagrangian vector at the left-hand point. When the material point arrives at the right-hand laboratory point (Fig. 23) the total change it has experienced is the difference be-

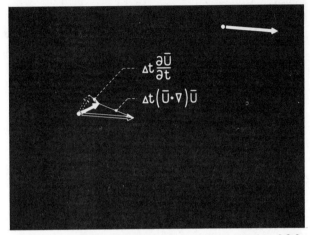

$$\Delta t \frac{\partial \bar{U}}{\partial t}$$

$$\Delta t \left(\bar{U} \cdot \nabla \right) \bar{U}$$

23. The material point has arrived at the right-hand laboratory point. The total change has two parts. $\Delta t \dfrac{\delta U}{\delta t}$ is the temporal velocity difference; $\Delta t \, (U \cdot \nabla) \, U$ is the spatial velocity difference.

tween the dashed vector and the Lagrangian vector. But this can be broken into two parts: the difference between the velocities at the left and right-hand laboratory points at this instant is given by the difference between the Eulerian and Lagrangian vectors on the left. The change each laboratory point has undergone during this time is given by the difference between the dashed and the Eulerian vectors on the left. The spatial velocity difference can be written as before as

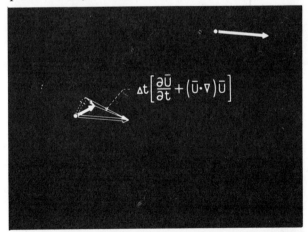

$$\Delta t \left[\frac{\partial \bar{U}}{\partial t} + \left(\bar{U} \cdot \nabla \right) \bar{U} \right]$$

24. The total change is the vector sum of the two components in Fig. 23. The *material derivative* is the expression multiplying the time difference.

the time difference times the velocity times the gradient of the velocity. The temporal velocity difference can be written as the time difference times the rate of change with time at a laboratory point. The total change is the vector sum of the two effects (Fig. 24).

The *material* (or *substantial*) derivative is just the expression multiplying the time difference. This is the rate of change seen by the material point as it passes the laboratory point, written in laboratory co-ordinates. The acceleration, more simply written in a Lagrangian frame, has been expressed in an Eulerian frame.

Summary

To summarize: we can "tag" the material points in a flow by using their locations at some reference time, and then give their displacements, velocities, and accelerations as functions of time and initial positions. This is called a *Lagrangian* description. Alternatively, we can choose a "laboratory" co-ordinate system arbitrarily, and probe to find the displacement, velocity, and acceleration at points fixed in that system. This is called an *Eulerian* description, and it has the advantage that some fields are steady in a correctly-chosen frame of this type. In many problems, Eulerian frames are mathematically enormously more convenient, so we nearly always write the conservation equations for a continuum in this system. It has the disadvantage that we are not always referring to the same material point. We can, however, transform between Eulerian and Lagrangian systems by using the fact that displacement and velocity at a laboratory point are the displacement and velocity of the material point that happens to be there.

To express in Eulerian field variables the change experienced by a moving material point, we must take into account not only the change with time of properties at a fixed point, but also the change of properties with position at a fixed time.

Reference

Almost any classical text in fluid mechanics has a brief discussion of this subject. Of these, perhaps the most extensive (though by no means generous) is that in:

Prandtl, L. and O. G. Tietjens, *Fundamentals of Hydro- and Aero-mechanics*, New York, Dover, 1957, 270 pp. S374 Paperbound $1.85.

Deformation of Continuous Media

John L. Lumley

PENNSYLVANIA STATE UNIVERSITY

Introduction

The motion of a rigid body can be decomposed into translation of one point in the body and rotation. The motion of a deformable medium can be decomposed into translation of one material point, local rotation, and local distortion of shape, usually called the strain. Local rotation and distortion, together called deformation, is the subject of this discussion. First the deformation that takes place during a finite time will be examined; later the rate of deformation will be discussed. The discussion will be limited to flows at constant density and for the most part to two-dimensional flows (although the ideas carry over with little change to compressible and three-dimensional flows).

Understanding deformation is important to the understanding of the general kinematics of motion. In addition, from a dynamical point of view, deformation and its time derivatives determine the stresses in most continuous media.

Since the deformation of a typical small element of fluid is of primary interest in this film, a reference point at the center of the element will be selected and the relative motion of surrounding points then examined. The time rate of change of position of a line of these points can be represented by velocity profiles, and these will usually have length scales associated with them — related to their curvatures. If a small-enough segment is examined, however, the profile can be represented by straight lines; looking at a still smaller region then gives the same picture. By "local" is meant the largest neighborhood in which the profiles can be represented by straight lines. A local velocity profile will look something like that in Fig. 1 to an observer who himself moves with the reference point.

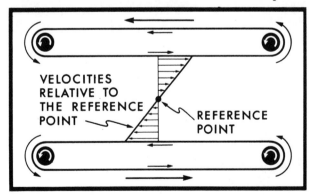

Figure 1

Characterization of the Flow

The film, DEFORMATION OF CONTINUOUS MEDIA, shows an apparatus consisting of two endless parallel belts immersed in a rectangular tank of glycerine. The belts move in opposite directions at equal speeds to produce a flow like that in Fig. 1. Glycerine is used because its high viscosity establishes the flow quickly, and in a short distance from the ends of the parallel section.

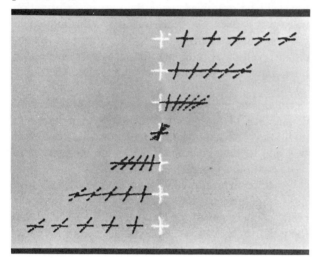

Figure 2

The surface of this flow can be marked with powder; the powder patterns move with the fluid. Figure 2 is a multiple-exposure (at equal time intervals) photograph of an initially vertical row of equally spaced crosses. Note that the point in the middle does not move — it can be taken as the reference point. The distances between successive images of the same point are equal, and the paths are straight lines; this is a steady flow with straight parallel streamlines. Since displacement is proportional to velocity in a steady flow, the velocity profile is evidently a straight line. Such a flow is called a steady rectilinear shear flow, or a steady homogeneous shear. Because straight lines remain straight, this flow evidently has no length scale — similar patterns will distort similarly, no matter what their size. This flow can be used as a model of the local flow.

The Strain Ellipse

To examine systematically the distortion of the fluid element whose center is at the reference point, a circle of neighboring points can be marked, equally distant from the reference and equally spaced (Fig. 3). The lines joining the fluid point at the center to each of the marked fluid points are fluid lines.

Figure 4 shows the pattern after 25 seconds; it is not difficult to show that this shape is an ellipse. It is called the *strain ellipse*.

Every other point has been numbered like the

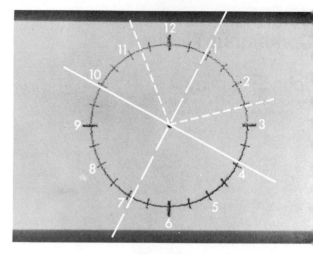

Figure 3

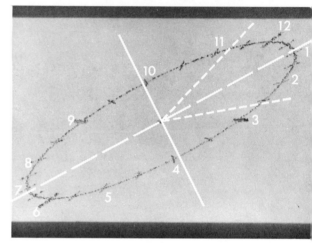

Figure 4

hours on a clock to help analyze the deformation. We will now examine what happens to the different fluid lines (or rays). Comparing Figures 3 and 4, it is evident that there has been both distortion and rotation of the rays, but of greatly variable amount. It is not obvious how to characterize the rotation or the stretching as a whole. For instance, different rays have rotated different amounts: twelve o'clock has rotated clockwise, while three o'clock has not rotated at all.

The Principal Axes

The axes of symmetry of the ellipse represent the directions of extreme strain (these are indicated on Fig. 4, with their initial locations on Fig. 3). That is, the point at the end of the major axis (dashed) has evidently moved farthest from the reference point, while that at the end of the minor axis (solid) has moved closest. These axes appear to be (Fig. 3) at right angles initially, in addition to being at right angles after the deformation.* Therefore they rotate equal

---

* This can be shown to be mathematically precise for the instants chosen. However, these particular fluid lines are not at right angles at intermediate times.

amounts. No other pair of lines originally perpendicular to each other remains so. These so-called *principal axes* can be used as a basis for the entire analysis of deformation.

Measurement of Rotation

In Figures 3 and 4 the initial and final locations of a pair of dotted lines on either side of the major axis are shown. Initially they are symmetrically placed with respect to the major axis. The line on the right evidently does not rotate as much as the major axis, while the one on the left rotates more. Since the lines are symmetrically placed with respect to the major axis after the rotation, the *average* of the rotation of these two lines must be the same as that of the major axis. All the points on the ellipse can be paired this way; hence, the principal axes rotate an amount that is the average of the amounts rotated by all the fluid rays emanating from the reference point. Specifying the angle through which the principal axes have rotated is thus a convenient way of specifying the average rotation of the deforming fluid element.

The Reciprocal Strain Ellipse

The numbered marks on the periphery of the circle are a somewhat artificial way of finding the initial

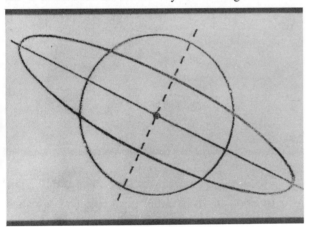

Figure 5

Figure 6

position of the principal axes. It would be more satisfying to find an intrinsic way of marking the initial position. Such a way can be found by asking the question "What initial pattern will turn into a circle after the deformation?" Since running this flow backward is very much like running it forward, it can be anticipated that probably an ellipse with a backward orientation will turn into a circle.

In Fig. 5 is shown a pattern consisting of the old circle plus a backward ellipse oriented about the initial position of the principal axes. Figure 6 shows what happens after deformation. The circle, of course, turns into the strain ellipse, and the backward ellipse turns into a circle. Notice that the axes of the strain ellipse have grown out of the axes of the backward ellipse. That is not surprising, since the minor axis of the strain ellipse marks the point that *has* moved closest, while the major axis of the backward ellipse marks the point that *will* move closest, and vice-versa. The backward ellipse is called the *reciprocal strain ellipse* and the initial principal axes are its principal axes. Thus, to discover the original principal axes of any strain ellipse, we merely flip it over into the reciprocal strain ellipse, and look at the principal axes of the latter. The angle between the minor axis of the reciprocal strain ellipse and the major axis of the strain ellipse gives the average angle of rotation during the deformation.

Properties of the Principal Axes

Two characteristics of the principal axes have already been noted. First, *they each rotate an amount equal to the average rotation of the fluid;* and second, *they lie in the directions of extreme strain.* There are two more important properties associated with deformation in the absence of rotation. Such deformation without rotation is called pure strain. To study it, a pattern based on lines parallel to the axes of the reciprocal strain ellipse is convenient.

Figure 7 shows such a pattern, and Fig. 8 a picture of the same pattern after deformation. During the deformation, the lines that were parallel remained parallel but the spacing changed. The angle between the two sets of lines also changed. While the deformation progressed, these sets of lines were not at right angles to each other. However, after the deformation the lines are at right angles again. It is evident that *the displacement of a point in the direction of one axis is not a function of position along the other axis.* This is a third important property of the principal axes. A line initially parallel to a principal axis is parallel to the same principal axis after deformation; note, however, that the distance between any two points on such a line changes during the deformation.

The principal axes can also be defined as those lines which undergo no shear deformation, since they are

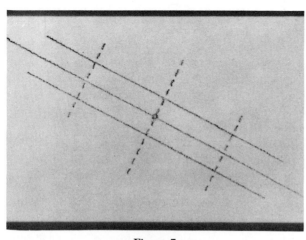

Figure 7

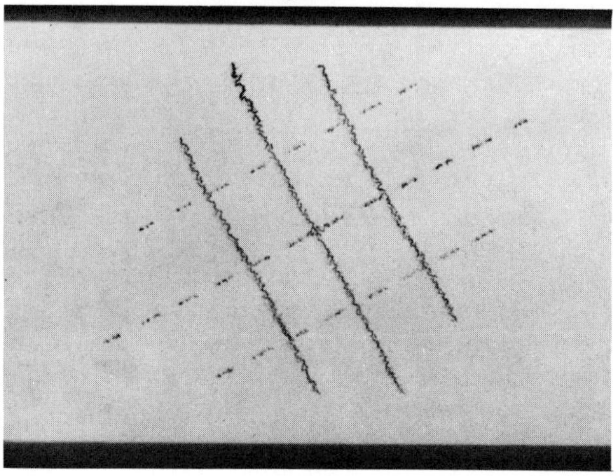

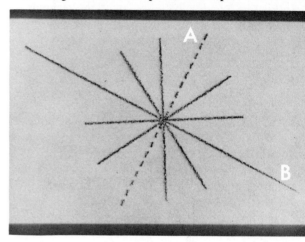

Figure 9

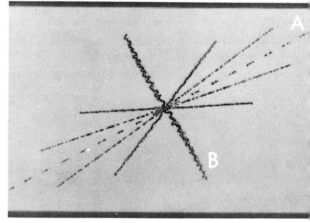

Figure 10

Figure 8

that the principal axes are made up of different material points at every instant.

To describe the deformation as a function of time it is necessary to give the rotation, and the strain, at every time and for every neighborhood in the fluid. According to the description developed above, this

mutually perpendicular both before and after the deformation. In Fig. 9, three sets of mutually perpendicular lines are shown, the principal axes (*A, B*) and two other crosses. During the deformation all the crosses were sheared, but after deformation (Fig. 10) the principal axes (*A, B*) are again at right angles, while the other pairs of lines, which also started out at right angles, are no longer mutualy perpendicular. This is the fourth major characteristic of the principal axes: *principal axes are the only pair of lines that are mutually perpendicular before and after the deformation.*

Figure 11 is a multiple-exposure photograph taken of the patterns at equally spaced times. The reciprocal strain ellipses corresponding to each of these instants can be formed in this flow by reversing Fig. 11. Such a mirror-image reversal has been superimposed on Fig. 11 to make Fig. 12. (The major axes of the successive strain ellipses and the minor axes of the corresponding reciprocal strain ellipses have been indicated for future reference.) It is evident that as the strain ellipse rotates one way, the reciprocal strain ellipse rotates the other way, so that the initial location of material points on the principal axes is different at every instant. Put another way, this means

would be in terms of the present location of the points as a function of their initial position. This is called a *Lagrangian* specification. The Lagrangian specification is of great interest as a description of deformation in a solid, and it is also sometimes of interest in a fluid. In a solid the material is tied together and neighboring material points can never get very far away from each other. In a fluid, however, there is no primeval undeformed state to serve as a reference; in an unsteady inhomogeneous flow, the relation between the present state and the initial state can be extremely complicated. For that reason the deformation in a fluid is usually described in terms of the rates of deformation at a particular place, at a particular time, using the present state as a reference; that is, the rate of change from the present state, the rate of change of the rate of change, and so on. Such a specification, involving only laboratory coordinates, is called an *Eulerian* specification. The number of deriv-

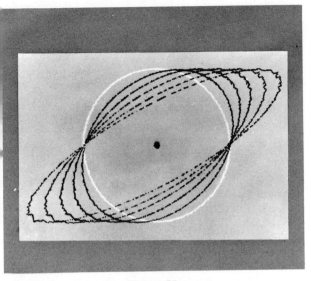

Figure 11

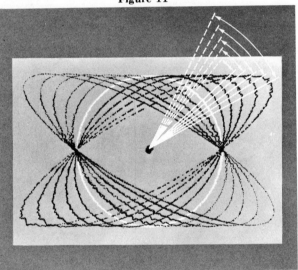

Figure 12

atives that must be determined depends on how complete a specification is needed, and that depends on the use to which it will be put. Incidentally, a very important reason for describing deformation is because stress depends on it. In many fluids stress is quite accurately described by a function only of the present rate of deformation, as for example in those fluids obeying the Navier-Stokes equations (i.e. Newtonian fluids). Our discussion will be limited to *deformation rate.*

Rate of Rotation

To analyze the present rate of deformation, "the initial instant" of our previous discussion will always be taken as the *present,* that is, the instant of special interest.

The position of the principal axes at the initial instant in this flow can be determined by an examination of Fig. 12; as time is successively closer to zero, the major axis of the strain ellipse and the minor axis

of the reciprocal strain ellipse not only approach each other, but both approach 45 degrees. Forty-five degrees and 135 degrees are evidently the initial positions of the principal axes.

Since the amount of rotation at any instant can be measured by the angle between the principal axes of the strain ellipse and those of the reciprocal strain ellipse, a picture like Fig. 12 could be used to infer the rotation rate at the initial instant; it would be necessary to measure the angle between the major axis of the strain ellipse and the minor axis of the reciprocal strain ellipse as a function of time from the initial instant, then to differentiate to obtain the *rate* at the initial instant.

Since the principal axes have at each instant an angular displacement that is the average of that of all the other fluid lines, they have an angular velocity which is the average of that of all the other fluid lines. Thus their rotation rate is related to the moment of momentum, or angular momentum, in the fluid. The *vorticity* in the fluid is exactly twice the rotation rate of the principal axes. The averaging property suggests a simple experiment for exhibiting the average rotation in the fluid.

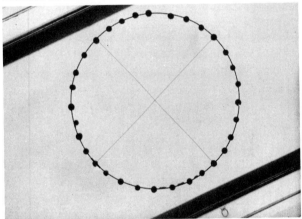

Figure 13

The floats shown in Fig. 13 support a rigid wire ring just above the surface; they are small enough and far enough apart not to seriously affect the flow. The ring may be expected to rotate at the average angular velocity of the rays to each of the floats, because in this situation the clockwise and counter-clockwise drags due to the relative motion will be equal. This rotation can be removed by rotating the frame of reference — that is, by rotating the camera — so that the crosshairs on the circle are stationary. The walls of the channel then appear to revolve in the opposite direction.

The Deformation Relative to a Rotating Frame

Having thus picked the correct average angular velocity of the fluid by the foregoing experiment, the wire ring can be removed and a pattern placed on the flow that consists just of the lines that will become the principal axes at the initial instant. Viewed in the rotating framework, Figures 14 and 15 show the pattern slightly before and slightly after the initial instant. At first the figure's major axis is in the 135° direction. The ellipse contracts in this direction, reaches a circle, and expands to an ellipse with major axis in the 45° direction. We can turn the pattern

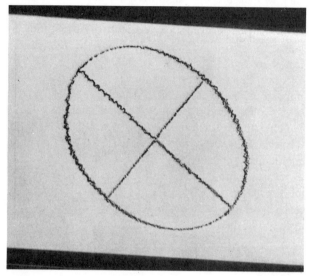

Figure 14

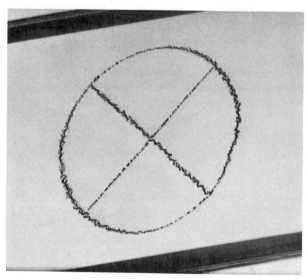

Figure 15

so that the 135° and 45° lines — the principal axes at the initial instant — can be used as coordinate axes. It is only necessary to start the rotation at a different time, relative to the time at which the fluid is marked, and to view the motion in the rotating frame previously established by the wire-ring experiment. Figure 16 (slightly before the initial instant) shows the bulge coming in slightly to the right of the new ordinate, but approaching the ordinate as the figure approaches a circle. As the figure passes out through the circle

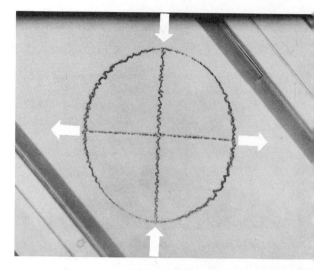

Figure 16

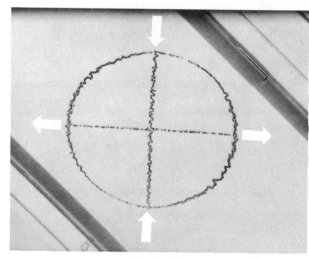

Figure 17

(Fig. 17 — slightly after the initial instant) the bulge moves out along the abscissa.

At the instant that the figure passes through the circle, the velocity is inward along the ordinate and outward along the abscissa. Since in the rotating frame of reference the velocity *is* precisely inward along one axis and outward along the other, it is obvious that *the principal axes are the only pair of lines that are not changing direction (relative to one another) at the instant of interest;* this is why this pair of lines can be used to measure the rate of rotation of the fluid region. In addition, at the instant of interest, it is evident that these axes are in the direction of the maximum rates of stretching and shrinking. These axes are called the *principal axes of strain rate.*

Properties of the Principal Axes of Strain Rate

In this rotating reference frame three of the properties of the principal axes have already been seen. At the instant of interest, (1) *they mark the directions of extreme strain rate;* (2) *they rotate with the aver-*

age angular velocity of the fluid; and (3) they are not only mutually perpendicular instantaneously, but they are the only pair of lines whose included angle is not changing, and therefore they are the only axes for which the shear strain rate is zero. By analogy with finite strain, a fourth property of these axes may be

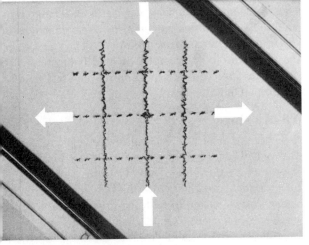

Figure 18

expected. This was seen before by putting lines parallel to the axes. Figure 18 shows such a pattern at the instant of interest. The lines remain parallel as they move toward or away from the axes. This means that the *velocity parallel to one principal axis is not a function of position relative to the other principal axis.*

Conclusions

The analysis of deformation rate is similar to the analysis of deformation itself. They can both be resolved into a rotation and a strain. In both there are mutually-perpendicular principal axes that serve as *average representatives of the angular velocity,* and that *lie in the directions of the extreme stretch,* and *these directions are not changed by the stretch.* In

addition, *the motion in the direction of one axis is not a function of position relative to the other axis.* There are differences, of course. The principal axes of strain continually change direction as the deformation progresses. The principal axes of initial strain rate only occupy one position. Generally speaking, deformation rate is easier to describe mathematically than deformation, but that is because an Eulerian specification can be used. The fact that it is easier is not surprising, because it deals only with the first derivative of the deformation at the initial instant. Fortunately, to describe the stress in Newtonian fluids, deformation rate is all that is needed.

To extend these observations to three-dimensional situations, few changes are necessary. Direct and reciprocal strain ellipses become ellipsoids. The velocity profile in a very small region of an incompressible fluid is a plane instead of a line. There are three mutually-perpendicular principal axes for which the strain is purely lineal; two of these axes have the extrema of lineal strain. The reciprocal strain ellipsoid cannot be obtained by simply turning over the strain ellipsoid, but it is defined in an analogous way, i.e. as the surface that turns into a sphere. Deformation can still be analyzed into rotation and strain; deformation rate into angular velocity and strain rate. The principal axes still represent the average rotation. They still are in the directions that are not changed by the stretching. Strain rate in one principal-axis direction is still not a function of position along the other two axes.

No fundamental changes are required to include compressibility; it is necessary only to allow the area (or volume) of the pattern to change with time. The conclusions are the same.

Reference

Long, Robert R., *Mechanics of Solids and Fluids*, Prentice-Hall, Englewood Cliffs, N.J., 1961.

Rheological Behavior of Fluids

Hershel Markovitz
MELLON INSTITUTE

Introduction

Rheology in its broadest sense is the study of the relation between stress and deformation in continuous media. For the *inviscid fluid* model this relation is simply that the *shear stresses are always zero* — or equivalently that the stress is always isotropic. In the *Newtonian fluid* model, the stress is isotropic only when the fluid is stationary; when it is in relative motion, the *stresses are assumed to be linear functions of the instantaneous velocity gradients*. The purpose of the film is to show examples of fluid behavior which cannot be approximated by either the inviscid fluid or Newtonian fluid models.

Yield Value

There are materials, particularly suspensions, which flow only when the shear stress exceeds a critical value, known as a yield value. The clay suspension in Fig. 1, for example, does not flow through the tube either under its own weight or with the driving pressure of a small added weight. Only when a sufficiently large weight is added does it flow. If the large weight is removed, the flow immediately stops. Such substances can support a shear stress in a state of equilibrium if the shear stress is less than a critical yield value.

The Memory Fluid Model

Some materials, such as molten plastics, protein solutions (e.g., egg white), and rubber cement, flow under the slightest shear stress, but are still not Newtonian. For these the *memory fluid* model provides a good description. In this model, the stress is isotropic only when the fluid has been stationary for a long time; otherwise *the stress is assumed to be a nonlinear function of the deformation-gradient history*.

It is instructive to consider the effects of non-linearity and history independently, although they are really not separable. The dependence on history affects the nonlinear phenomena even in steady flow, and there are non-linearity effects with large deformation in the memory experiments.

Dependence on History

The current value of the stress in a memory fluid cannot be determined from a knowledge of the current state of deformation alone, since it depends on the

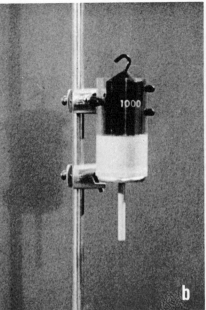

1. A clay suspension in a cylinder does not flow out through the open exit tube under its own weight (a) or even when a small weight is placed on the light piston (b). With a larger weight the suspension does flow (c). The shear stress must exceed a critical yield value for flow to occur.

deformation which the material has experienced previously. Conversely, the current stress does not determine the current state of deformation. As a consequence, memory fluids show elastic-like behavior (hence the term *viscoelastic* is frequently used for these fluids). However, unlike the elastic solid, the memory fluid does not have a preferred configuration to which it invariably returns when all stresses are removed.

In Fig. 2 the silicone-putty ball flows into a puddle even under such a small force as that of gravity, if given sufficient time. But during the short time of a rapid impact (Fig. 3), it behaves like an elastic solid because it does not have a chance to "forget" its previous spherical shape. The significant relaxation times for stresses in silicone putty are much less than the time for settling due to gravity, but much greater than the time of rapid impact. Thus the putty may exhibit a range of behavior from elastic to viscous, depending on how the characteristic time of the experiment compares with the significant relaxation times.

2. Ball of silicone putty relaxes into a puddle in an hour under the force of its own weight.

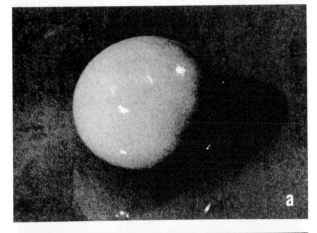

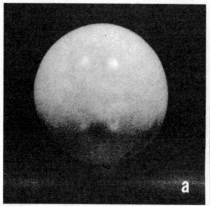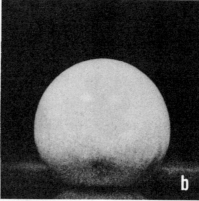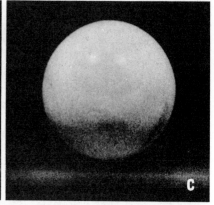

3. When bounced on a hard surface, the silicone-putty ball deforms on impact, but recovers its original shape immediately on the rebound, like an elastic solid.

A similar phenomenon can be observed when a shear is applied to a polymer solution between coaxial cylinders (Fig. 4a). When the inner cylinder is quickly rotated and then immediately released, almost half the deformation is recovered (Fig. 4b). When the inner cylinder is held for a while before being released, the deformation recovered is much less (Fig. 4d), because of the fading memory of the fluid.

The deformation history can be followed more precisely in the coaxial cylinder apparatus of Fig. 5. A constant torque is applied to the inner cylinder, and

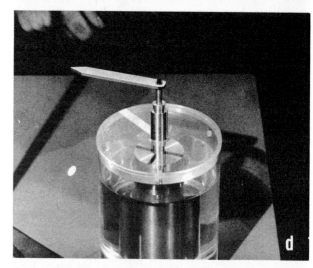

4. A polymer solution is in the annulus between coaxial cylinders. The white stripe on the top marks the starting positions. When the inner cylinder is quickly rotated clockwise through 360° (a) and released immediately, the fluid pulls it back almost half the way to its original position (b). When, instead, the inner cylinder is rotated clockwise but held for some time in a 360° position (c), the deformation recovery upon release is much less (d).

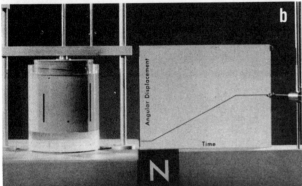

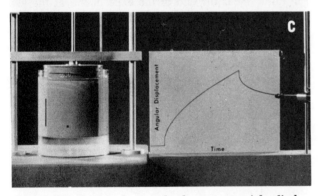

5. (a) The fluid in the annulus between coaxial cylinders is stressed by suddenly subjecting the rotatable inner cylinder to a constant torque through the pulley-and-weight system. The torque is removed when the weight reaches the platform and disengages from the string by means of a slip toggle. The pen is moved vertically by a string attached to the driving drum on the inner cylinder. With the chart moving at uniform speed to the left, angular position of the inner cylinder vs. time is recorded. (b) Record for a Newtonian fluid ("N"). (c) Record for a polymer solution.

later removed. Its angular rotation as a function of time is recorded on the chart. With the Newtonian fluid (Fig. 5b), a constant speed of rotation is obtained almost as soon as the torque is applied, and the motion ceases when the weight is removed. With the polymer solution in the apparatus (Fig. 5c), it takes considerable time for the inner cylinder to achieve a constant rate of rotation. It is clear that the motion at any instant is not determined simply by the value of the instantaneous stress; although the torque is constant, the rate of rotation is changing. Furthermore, when the torque is removed, the material deforms although

no stress is being applied. The reversal in direction also indicates that not all the energy used to produce the flow was dissipated; some was stored and then recovered.

Non-linearity and Shear Stresses

The non-linearity in the stress-deformation relation of the memory fluid affects both the shear stresses and the normal stresses in simple steady flow situations.

The effect on shear stresses is seen in experiments of the type usually used to measure viscosity. For the Newtonian fluid, when the pressure head driving the fluid through a tube is doubled (Fig. 6a), the rate of flow is also doubled (Fig. 6b). With some fluids, such as some polymer solutions, doubling the head more than doubles the rate of flow (Fig. 6c). This behavior is called *pseudoplastic* or *shear-thinning*. With some suspensions, a contrary situation arises (Fig. 6d). This is called *dilatant* or *shear-thickening* behavior.

To predict the flow behavior of an incompressible Newtonian fluid, only the density and the viscosity are needed. For an incompressible memory fluid, the rate of flow through tubes in steady laminar flow is governed by the density and a *viscosity function*.* This function also determines the velocity profile, which is parabolic for the Newtonian fluid (Fig. 7a), but can be quite different for a more complex fluid (Fig. 7b).

*The viscosity function also governs the relation between torque and angular velocity in steady laminar shearing between rotating coaxial cylinders, and in some other viscometers.

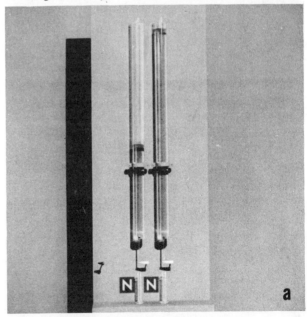

6. Non-linearity in pipe flow. The same fluid is contained in two identical large-diameter reservoirs which are provided with identical outlet tubes and flapper valves which open and close simultaneously. The heads (measured from the bottom of the tubes to the free surfaces) are in the ratio of two to one, and remain nearly so during the experiment. The experiment is performed with three different fluids in succession.

In *shear-thinning* behavior the viscosity function has lower values at higher rates of shear. In *shear-thickening* behavior, the opposite holds (Fig. 8).

In the "race experiment" illustrated in Fig. 9, the flow rate under a varying head of a fluid with shear-thinning behavior was compared to that of a Newtonian fluid. The conditions were deliberately selected so that at the higher rates of shear associated with the initially larger pressure head the viscosity of the poly-

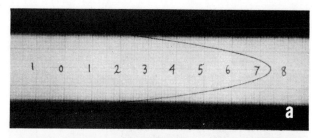

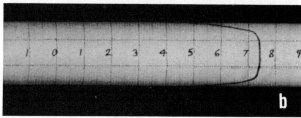

7. The position of an initially transverse straight line of marked fluid in a pipe of circular cross-section is shown shortly after impulsive start of flow. With the viscous Newtonian fluid of (a), the deformation of the line indicates a parabolic velocity profile. With the polymer solution of (b), the indicated velocity profile is not parabolic.* (Courtesy A. G. Frederickson and N. N. Kapoor, University of Minnesota)

mer solution was much lower than that of the Newtonian fluid. Near the end of the experiment, at the low rates of shear associated with the low pressure head, the value of the viscosity for the polymer solution was much greater. Thus, although the shear-thinning fluid achieved an early lead in the race to empty the reservoirs, it was overtaken near the end by the Newtonian fluid.

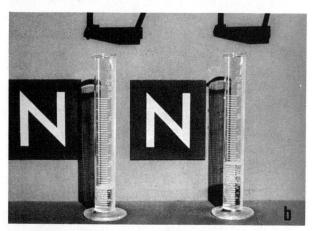

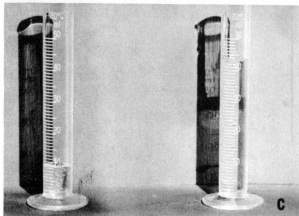

(Fig. 6 continued) With the Newtonian fluid of (b), the flow rates are linearly related to the driving heads. With the polymer solution of (c), doubling the head more than doubles the flow rate (shear-thinning behavior). With the suspension of (d), the rate of flow at twice the head is less than doubled (shear-thickening behavior).

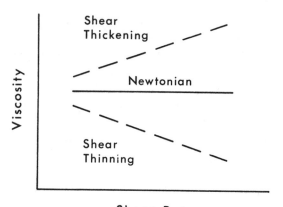

8. Viscosity functions.

*The marked line of fluid of Fig. 7b retains approximately the same shape as the flow proceeds, with most of the velocity gradient concentrated near the walls of the tube. This indicates that non-linearity in the viscosity function is responsible for the shape of the profile and that it is not simply an early stage in the development of a parabolic profile.

9. **The race experiment. (a)** Indentical reservoirs with identical outlet tubes are filled to the same level with a Newtonian fluid (left) and a "shear-thinning" fluid (right). The flapper valves are opened simultaneously. At first, the non-Newtonian fluid leads in the race to empty **(b)**, but as the heads get smaller, the non-Newtonian fluid slows down relative to the Newtonian, which ultimately wins the race **(c)**.

Non-linearity and Normal Stresses

Non-linearity in the stress-deformation relation affects not only the shear stresses but also the normal stresses. In steady simple shearing flow between infinite parallel plates, the normal stresses on an infinitesimal volume element in the three directions indicated in Fig. 10a are all equal for the Newtonian fluid, but not for the memory fluid. The same is true for the normal

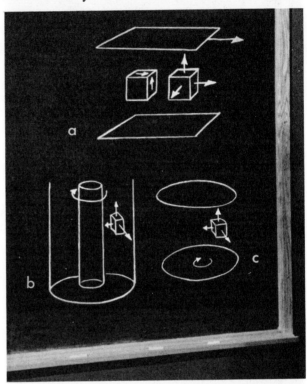

10. **Normal stresses on volume element in (a) simple shear, (b) circular Couette flow, (c) torsional shear.**

stresses in the flow between rotating cylinders (Fig. 10b), in the flow between rotating parallel discs (Fig. 10c), and in some other simple laminar flows. As a result, the *distribution* of normal stresses can be quite different for a memory fluid from what it is for the Newtonian fluid.

For example, in the flow of a Newtonian fluid between rotating cylinders (Fig. 11a), a higher normal stress is exerted on the outer cylinder wall than on the inner wall. For the polymer solution just the opposite is observed* (Fig. 11b). The shape of the surface of the fluid in the annulus is also quite different in the two cases.

The normal stress distribution is also quite different for Newtonian and memory fluids in the flow between

*This experiment is sufficient to indicate the inapplicability of the theory of the viscoinelastic fluid (sometimes called the Reiner-Rivlin fluid) whose force-deformation relation is based on the assumption that the fluid stress is a non-linear function of the velocity gradients. Although such a fluid exhibits normal stress effects in other flows, it would not in this experiment.

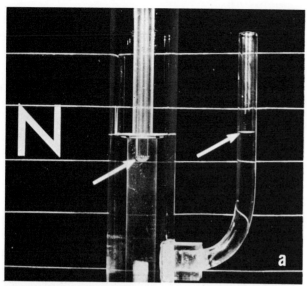

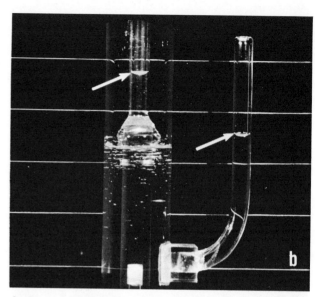

11. Couette flow in the annulus between an outer cylinder and a coaxial rotating shaft. The shaft is a hollow tube, with a hole communicating to the annulus; the height of fluid in the tube indicates the normal stress at the hole. There is a corresponding hole and a side-arm manometer tube to indicate the stress normal to the outer cylinder. Arrows point to the liquid levels in the manometers. With the Newtonian fluid of (a), the normal stress is greater at the greater radius, owing to centrifugal forces. With the polymer solution (b), the non-linear effects overwhelm the centrifugal pressure gradient; the normal stress is substantially greater at the inner cylinder. The polymer solution also climbs the rotating shaft.

12. Flow between rotating parallel discs. The upper disc is stationary, and is instrumented with pressure taps and manometers across a diameter. The outer cylinder rotates, so that fluid is sheared between the upper disc and the floor of the container. The distribution of normal stress on the upper disc is shown for a Newtonian fluid (a) and for a particular non-Newtonian fluid in (b).

rotating parallel discs. For a Newtonian fluid the normal stress distribution is governed by centrifugal force, and the *lowest* stress occurs at the center of the plates (Fig. 12a). With the polymer solution, the *highest* normal stress is exerted at the center (Fig. 12b). A similar situation arises in the steady laminar flow between a rotating shallow cone and a flat disc (Fig. 13). For this geometry the stress normal to the stationary disc is proportional to the logarithm of the distance from the axis of rotation.

13. Experiment similar to that of Fig. 12b except that a shallow 2-degree cone (with apex just below the center of the upper disc) constitutes the floor of the outer container. The normal stress is proportional to the logarithm of the distance from the axis of rotation.

Other Flows

Another example where there is a dramatic difference between the behavior of Newtonian and memory fluids is in the flow through an orifice (Fig. 14). The jet of polymer solution has a considerable greater

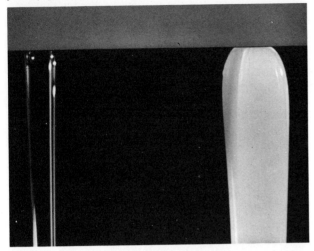

14. The shapes of the jets driven by the same pressure and issuing from two identical orifices in a thick plate are compared for a Newtonian fluid (left) and a polymer solution (right).

diameter than the orifice through which it has been forced, whereas the jet of glycerin (a Newtonian fluid) has a smaller diameter than the orifice. When plastic articles are to be made by the extrusion process, this effect must often be considered and the die designed smaller than the dimensions of the finished product.

One further example that shows sharply contrasting behavior is the flow pattern around a rotating sphere. With the sphere rotating in a high polymer solution the flow spirals inward at the equator and out at the poles (Fig. 15). Just the opposite happens for a sphere rotating in a Newtonian fluid.

Conclusion

Not all non-Newtonian fluids show effects as dramatic as those chosen for demonstration in this film. However, when dealing with fluids containing long-chain polymers and suspensions, one should not be surprised to find flow patterns and stress distributions which are quite different from those predicted for a Newtonian fluid.

References

1. H. Markovitz in *Rheology: Theory and Applications* (F. R. Eirich, ed.), Vol. 4, Chapter 6, Academic Press, New York, 1967

2. B. D. Coleman, H. Markovitz and W. Noll, *The Viscometric Flows of Non-Newtonian Fluids,* Springer, New York, 1966

3. A. G. Fredericksen, *Principles and Applications of Rheology,* Prentice-Hall, Englewood Cliffs, New Jersey, 1964

4. H. Giesekus, *Sekundärströmungen in elastoviskosen Flüssigkeitten,* a 16-mm film available from Institut für den Wissenschaftlichen Film, 34 Göttingen Nonnenstieg 72, Germany

15. A sphere rotating in a non-Newtonian fluid. Dye issuing from a tube at the left shows that the flow spirals radially inward in the equatorial plane and leaves the neighborhood of the sphere in the axial direction at the poles. The secondary flow induced by a sphere rotating in a Newtonian viscous fluid would be the opposite: toward the sphere at the poles and away from it at the equator. (Courtesy H. Giesekus, Farbenfabriken Bayer AG. See Reference 4).

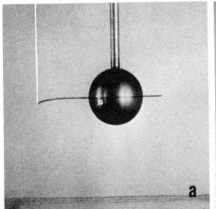
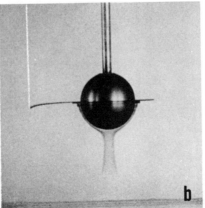
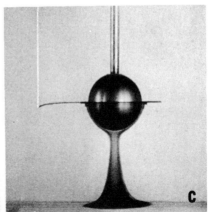

Surface Tension in Fluid Mechanics

Lloyd Trefethen
TUFTS UNIVERSITY

Introduction

Liquid surfaces act as if they are in tension (Fig. 1). The thread is pulled into circular arcs. This pull, called surface tension, exists with all liquid surfaces and is uniform in all directions. Surface-tension forces affect the shapes and the motions of liquids that have open surfaces.

The film in Fig. 1 would contract if free to do so. It also will enlarge if the string is pulled down, as in Fig. 1b. The string performs work, pulling more molecules onto the surface to make it larger.

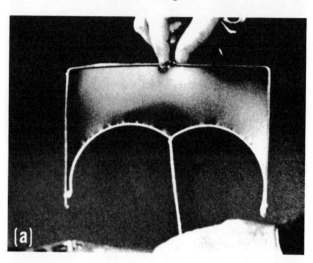

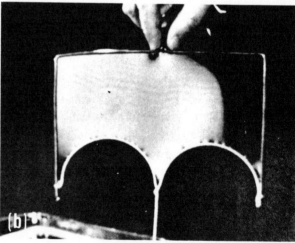

1. Soap film on wire frame pulls thread into circular arcs. The film can be enlarged by doing work, by drawing the string down. This pulls more molecules onto the surface to make it larger.

This tension of surfaces can be understood in terms of forces between molecules. Molecules repel each other when close, attract each other when farther away, and at some in-between distance neither repel nor attract. Inside a liquid, the molecules on the average usually repel each other slightly, just enough to counteract the pressure applied by the surroundings to the liquid. But molecules at the surface are farther apart than the neutral distance, and attract each other. This attraction, or surface tension, is necessary if surface molecules are to be kept from moving from the surface into the bulk liquid.

Since all phenomena in mechanics can be explained in terms of either forces or energies, the predictions one makes about surfaces are the same whether one starts with the concept of surface tension or surface energy. Surface energy results from surface molecules' having more energy than molecules in the bulk liquid; they have had, in effect, half their neighbors removed from them. Dimensionally, surface tension, often expressed as dynes-per-centimeter, is the same as surface energy, which is often expressed as ergs per square centimeter (erg/cm² = dyne cm/cm² = dyne/cm).

Magnitude of Surface Tension

The surface tension of a film could be calculated by measuring the pull on the string in Fig. 1. A soap film would have a surface tension of about 20 or 30 dynes/cm on each of its two sides.

Reference 2 describes better techniques for measuring surface tension. Representative values obtained are given in Table I. It is useful to remember that:

Water has a higher surface tension than most other liquids. (Liquid metals are an exception.)

Other molecules in water usually lower the surface tension. (Some salts are an exception.)

Increased temperature lowers surface tension. Electrical conditions at interfaces can raise or lower surface tension.

TABLE I

Approximate surface-tension values for several liquid-air interfaces. (Values are given in dynes/cm.; to convert to lbf./ft., divide by 14,590.)

Liquid	Surface Tension
Water	73 dynes/cm.
Salt water	75
Soapy water	20-30
Ether	17
Alcohol	23
Carbon tetrachloride	27
Lubricating oil	25-35
Mercury	480

Surface Tension Boundary Conditions

Since surfaces exert forces, they force liquids into shapes and motions that differ from what they would be in the absence of surface tension. The effects can be summarized by three general rules:

Contact angles. Where surfaces meet, the contact angle is determined by the energies of the interfaces.

Pressures caused by curved surfaces. A curved liquid surface has a higher pressure on the concave side:

$$\Delta p = \sigma \left(\frac{1}{R_1} + \frac{1}{R_2} \right)$$

Shear forces caused by surface tension gradients. Variation of surface tension along a surface is balanced by shear forces in the bounding materials:

$$\nabla \sigma = \tau_A + \tau_B$$

σ is surface tension, and R_1 and R_2 are the two radii of curvature necessary to specify curvature at a point on a surface.

These three statements describe the boundary conditions that surface tension imposes on liquids and, together with the equations for fluid mechanics, are sufficient to determine the shapes and motions. All but a few simple cases, however, are difficult to solve mathematically. Free-surface boundaries introduce mathematical complexity even into cases that look as though they should be very simple, such as a static hanging drop.

Contact Angles

When soap films meet, they must meet symmetrically, at 120° angles (Fig. 2). Usually, however, liquid surfaces terminate on solid surfaces, and the contact angle can vary widely, as illustrated in Figs. 3 and 4.

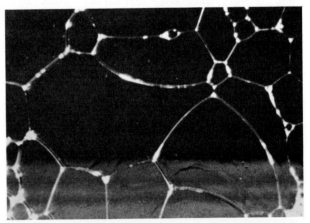

2. Intersecting soap films, between two closely spaced glass plates. Only three-film intersections are observed, because intersections of four or more films are unstable. All angles are 120°, as required by symmetry.

3. A drop of water placed on wax by means of a hypodermic needle demonstrates a liquid angle greater than 90°. Water does not wet wax.

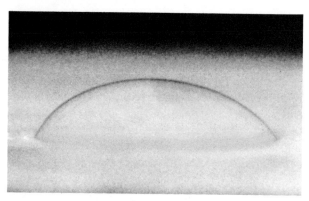

4. Water drop on a bar of soap. The angle included by the water at the edge is less than 90°.

Contact angles are usually "explained" by an equation called Young's equation:

$$\sigma_{\text{solid/gas}} = \sigma_{\text{solid/liq.}} + \sigma_{\text{gas/liq.}} \cos (\text{contact angle})$$

which is a statement that the three surface tensions pulling at the intersection of a gas, a liquid, and a flat solid should balance along that surface. Alternately, the σ's can be regarded as the free energies of the three interfaces, and the equation then amounts to a statement that any perturbation of a stable intersection increases the total free energy of the surfaces.

Young's equation is not very useful in practice. The surface energies of real solids are often not known. The energy of a surface over which liquid is advancing is often different from that of the same surface if the liquid has retreated. (If it were not for such a hysteresis effect, water drops could not stand still on windowpanes.) The thermodynamically correct statement would be that the contact angle will be that angle which leads to the minimum available energy of the total system, a principle even more difficult to apply than Young's equation. Hence, since contact angles typically cannot be predicted, they are usually measured.

Pressure Caused by Curved Surfaces

If an interface is curved, surface tension causes a pressure difference between the fluids on each side of the interface. The pressure causes the bulge, and is therefore always higher on the concave side. A force balance for a curved surface shows the pressure difference to be:

$$\Delta p = \sigma \left(\frac{1}{R_1} + \frac{1}{R_2}\right).$$

Such a pressure difference causes liquid to be sucked into a crevice or a hole if it wets the wall material. In a small tube, if the liquid wets the wall, as in Fig. 5, liquid is sucked up until its hydrostatic head, $\rho g h$, balances the pressure difference, $2\sigma \cos \theta / r$, where θ is the contact angle. The capillary rise can therefore be calculated to be: $h = 2\sigma \cos \theta / \rho g r$. Between the two glass plates, Fig. 6, the liquid rises least where the glass surfaces are farthest apart. Liquid will suck itself into holes only if $\theta < 90°$, that is, if it wets the material.

5. The capillary rise of water (in the tube on the right) is greater than of alcohol, showing that water has greater surface tension than alcohol.

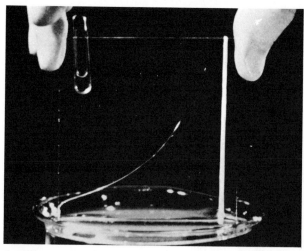

6. Water rises in hyperbolic curve between two glass plates slightly separated at the left by paper clip.

7. Water drops on waterproofed cloth. Because the contact angle of the water with the treated fibers is greater than 90°, the water will not penetrate the holes in the fabric.

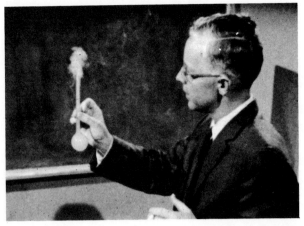

9. The higher pressure inside a soap bubble on the end of a glass tube expels the smoke used to blow it up, converting surface energy of the film into kinetic energy of the smoke jet.

For that reason, the water in Fig. 7 does not penetrate the quite porous fabric.

Applied to a cylindrical jet of liquid of radius r, the equation for pressure drop indicates that the pressure inside is σ/r higher than outside. As r gets smaller, the pressure gets higher. Irregularities in a jet therefore amplify themselves; the higher pressure in any smaller region forces liquid away, into adjacent bigger regions. Since there are always irregularities, a jet is always unstable, and pinches itself into drops. Lord Rayleigh calculated that the most unstable disturbance wavelength is about four and a half times the diameter of the jet, which is about what one sees in Fig. 8.

Other examples of increased pressures are the interiors of bubbles or drops, where $R_1 = R_2 = r$, and the pressure excess is $2\sigma/r$. A soap bubble, with both an inner and an outer surface, has an excess pressure of $4\sigma/r$. In Fig. 9, it is that excess pressure that blows the smoke out of the soap bubble.

Bubble Nucleation

Until recently, people have found it difficult to understand how bubbles can be generated in boiling water and other bubble nucleation systems. New bubbles, to start from a zero size, would require an initially infinite internal pressure ($p = 2\sigma/r_{r=0} = \infty$). Even an assumption that bubbles start from statistical voids, of sizes near the intermolecular distance, would require that initial vapor pressures be thousands of pounds per square inch, a situation likely only under such conditions as those accompanying collisions of high-energy particles; nuclear bubble chambers do provide new bubbles. The more common bubble sources, such as boiling and electrolysis, remained a puzzle until about 1955, when people began to adopt the idea that new bubbles are simply pinched-off enlargements of old bubbles remaining in cracks on solid surfaces. Subsequent research has supported this now generally accepted idea. Bubbles in boiling come from sites where

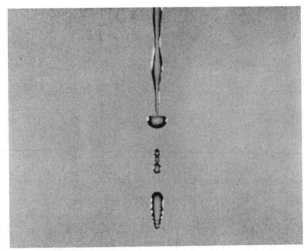

8. A jet of water from a faucet is unstable, and pinches itself into separate drops.

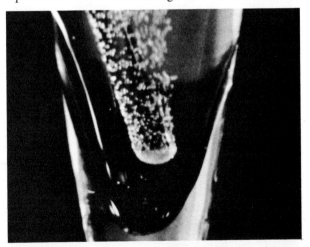

10. Bubbles in beer do not form on smooth glass, but do form at the numerous vapor-filled cavities on the roughened end of a glass rod.

vapor cavities already exist. A roughened end of glass rod has many such sites, and nucleates many new bubbles in the glass of beer shown in Fig. 10.

Break-up of Sheets of Liquids

A free edge of a liquid sheet moves into the sheet, pulled by surface tension. An example is a soap bubble touched by a pencil that has been dipped in alcohol (Fig. 11). A hole appears and rapidly grows larger, consuming the bubble. The alcohol ruptures the liquid sheet by reducing the surface tension of a small portion of the surface. The higher surface tension of the surrounding surface stretches the weaker region until its thickness reduces to about that of the distance between molecules, and a hole appears. Its rolling-up edge is pulled by surface tension at a velocity determined by the rate at which the momentum flux of liquid into the edge equals the pull of surface tension.

If a water jet is deflected by the end of a cylinder,

11. Soap bubble a moment after rupture by a pencil dipped in alcohol.

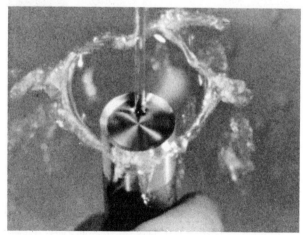

12. A water jet impinging on the end of a cylinder forms a liquid sheet. The edge of the liquid sheet continuously breaks up into droplets.

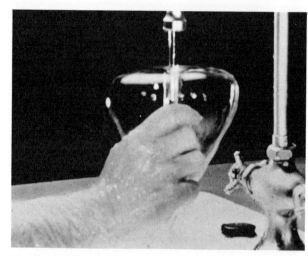

13. The liquid sheet of Fig. 12 can form a water bell, illustrating that surface tension can cause a moving liquid sheet to bend.

14. A moving liquid sheet bends back sharply upon itself, apparently violating conservation of momentum.

as in Fig. 12, the resulting liquid sheet breaks up into droplets. The rolling-up edge is unstable. The deflected sheet can have different forms, such as the water bell in Fig. 13. It can even bend back sharply upon itself, as in Fig. 14; the sheet cannot go out beyond the radius where the surface tension is equalled by the radial momentum flux of the liquid, and it is at that radius that the water changes direction sharply.

Impact of a Milkdrop

The interplay of surface forces and inertia forces is shown well in high-speed photographs of a drop of milk hitting a thin layer of water (Fig. 15). The depth of water is about half the radius of the drop. A sequence of events occurs. The impacting drop spreads sideways, meets the stationary sheet of water, and sends up a cylindrical sheet of liquid, the base of which increases in radius at a rate approximately half that of the original drop velocity. The sheet projected upwards has a leading edge, which is pulled back by the surface tension of the two sides, causing a rolled-up

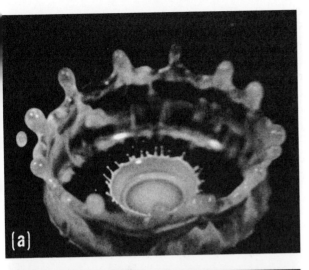

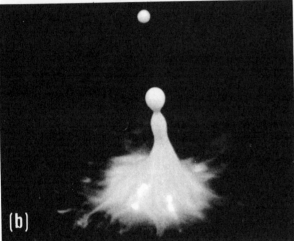

15. A drop of milk hitting a thin layer of water spreads radially, and sends up a cylindrical sheet that surface tension changes into a coronet, (a). Surface tension pulls the liquid down, some impacting at the center to send up a spike of liquid, (b), which in turn is pinched off by surface tension. All this happens in a fraction of a second.

edge. This edge is unstable, and its tendency to become drops results in a coronet-like shape of surprising regularity. This coronet is pulled down, by surface tension, and about half of its liquid moves inwards, focusing on the center where it becomes a liquid spike. This spike, like all liquid cylinders, is unstable, and the top pinches off to form a drop. Because of its upward momentum, the drop continues to move up, but the remainder of the spike is pulled back down by surface tension.

Motions Caused by Surface-Tension Gradients

Liquids must move when the surface tension changes along their free surfaces. A liquid cannot stay still with unbalanced forces acting on it. Consideration of the forces on a unit area of interface between a liquid, *A*, and another fluid such as gas, *B*, indicates that the

surface must move in the direction of higher surface tension pull at such a velocity as to produce viscous forces in the two fluids, just balancing the gradient of surface tension:

$$\nabla \sigma = \tau_A + \tau_B.$$

These shear forces produce movement in the fluids adjacent to the moving surface, and the moving surface in this way drags along its bounding fluids.

Motions Caused by Concentration Gradients along Surfaces

A simple example of surface-tension variation occurs with a limp loop of thread on water. It snaps into a circle when the water inside is touched by soap (Fig. 16).

Continuous, rapid motion is experienced by bits of camphor scraped onto water. Camphor molecules lower the surface tension. Each particle dissolves unevenly, and the higher pull on the side where it dissolves the least pulls the particle along the surface. A bit of wood with camphor at one end moves like a boat (Fig. 17).

16. Loop of thread on water snaps into a circle when surface tension inside is reduced by soap.

17. A bit of wood with camphor at one end moves like a boat. The greater surface tension on the bow pulls the boat ahead.

18. Wine tears. Surface tension differences continually pump alcohol-water mixtures up the sides of cocktail glasses.

Another continuous motion occurs in cocktail glasses. A water-alcohol mixture sloshed around in a glass leads to "wine tears," drops that move up and down the sides of the glass (Fig. 18). Evaporation of alcohol lowers the alcoholic content of the liquid film on the glass and increases its surface tension. The surface is therefore continuously pulled from the bulk liquid up the side of the glass, pumping up liquid which accumulates to form the wine tears.

This is also why one can blow bubbles with soapy water, but not with a pure liquid. The soap film in Fig. 1 holds itself up against gravity by continuous self-adjustment, to maintain a lower proportion of soap, and therefore a larger surface tension, in the upper region of the film.

Motions Caused by Electrical and Chemical Effects at Surfaces

Electrical charges also affect the magnitude of surface tension. Electrical charges usually concentrate at surfaces where their repulsive forces on each other subtract from the surface tension. For example, mercury and dilute sulphuric acid with a bit of potassium dichromate added will form a chemical battery if another metal, an iron nail, is brought in as a second

19. Self-excited oscillation of mercury pool. Two metals, mercury and iron, forming a battery in an electrolyte, can set up an oscillation that lasts for hours, driven by repeated surface-tension changes as oscillations repeatedly short-circuit the "electrodes."

electrode. When the two metals touch, the battery is short-circuited and the charge density on the mercury surface is reduced. The mercury drop pulls up because of increased surface tension. With just the right position of the nail, the mercury will set itself into oscillation, repeatedly short-circuiting the electrical charges that build up when the mercury is not touching the nail (Fig. 19).

Mercury, in nitric rather than sulphuric acid, experiences electrical and chemical effects at the surface that cause the mercury to swim toward potassium dichro-

20. A pool of mercury in nitric acid reaches out to touch a crystal of potassium dichromate.

mate crystals (Fig. 20), at times so rapidly that it fragments itself into several drops. Surface motions are so violent that the mercury drop acts as an effective agitator, forcing convection and diffusion of the potassium dichromate throughout the liquid.

Motions Caused by Temperature Gradients along Surfaces

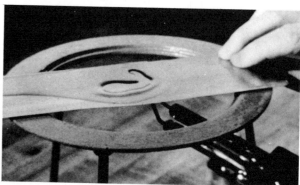

21. Liquid film on a thin sheet of metal is pulled away from hot spot above soldering iron by higher surface tension of surrounding, colder liquid.

Temperature gradients cause surface-tension gradients. If a layer of silicone oil is put on a thin metal plate, a hot soldering iron moved underneath the plate causes a bare spot in the liquid above. As the iron is moved around under the plate, the bare spot moves along with it (Fig. 21). Increasing the temperature always lowers the surface tension, because it becomes easier to pull molecules up into the surface. Above the soldering iron, the hot liquid has a lower surface tension than the cold liquid around it, so the hot sur-

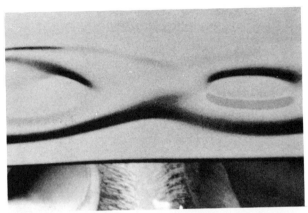

22. Liquid is pulled up over figure-eight path of ice cube underneath metal plate, by higher surface tension of cooler liquid.

face is pulled to the cold regions. In moving, the surface drags away the underlying liquid, leaving the bare spot. An ice cube instead of a soldering iron causes the liquid film above to hump up over the cold region (Fig. 22).

Air bubbles will "swim" in a liquid that is uneven in temperature. The cold side of the bubble will have a higher surface tension than the warm side. It will therefore pull surface from the warm side, where surface will be generated, around the bubble to the cold end, where the surface will disappear. This movement of surface, with its viscous drag upon the boundary fluid, will pick up a sheet of liquid and jet it off the cold back end. By jetting liquid one way, the bubble propels itself the other way. It will swim up the temperature gradient. Thermodynamically, such a self-propelling bubble is a heat engine. Wherever surface is created, heat is absorbed, and wherever surface

23. Bubbles in liquid-filled tube will swim toward tip of soldering iron, because bubble surfaces keep moving from hot to cold ends of the bubbles.

is destroyed heat is given off. Therefore a swimming bubble absorbs heat at its hot end and rejects heat at its cold end, and becomes a self-propelling heat engine. As an example of swimming bubbles, air bubbles in silicone oil in a horizontal glass tube move themselves toward the hot end of a soldering iron held against the glass (Fig. 23).

An electrically-heated horizontal wire in a mixture of acetone and about 2% water generates bubbles of acetone and water vapor at nucleation sites along the wire. At low heating rates, the bubbles, instead of

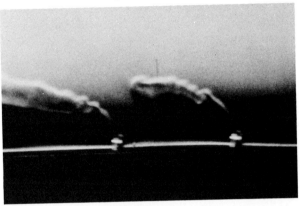

24. Vapor bubbles propel themselves along a heated wire in an acetone-water mixture.

rising off the wire at the nucleation sites, swim along the wire, jetting hot liquid away from their cold ends (Fig. 24). Once bubbles start moving, the wire is colder where they have been and hotter where they are going, so they just keep going to where it is hotter.

When Surface Tension Is Important

When dimensionless groups such as the Bond number, $\rho g L^2 / \sigma$, reflecting the ratio of gravitational to surface forces, and the Weber number, $\rho V^2 L / \sigma$, reflecting the ratio of inertial to surface forces, are small, surface effects can be expected to dominate. There are situations, such as in fuel tanks of orbiting space vehicles, where L, the characteristic length, can be measured in feet, yet because of a much reduced g, surface tension can be a controlling force. As a general rule, however, surface tension is controlling only in small systems, of which only a few representative cases have been described above.

References

1. Boys, C. V., *Soap Bubbles*, 1890. Reprinted by Doubleday & Company, Inc., 1959, and by Dover Publishing Co., 1959.
2. Adam, N. K., *Physics and Chemistry of Surfaces*, Oxford University Press, 1941.
3. Levich, V. G., *Physicochemical Hydrodynamics*, Prentice-Hall, 1962.
4. *Contact Angle, Wettability and Adhesion*, Advances in Chemistry No. 43, American Chemical Soc., 1964.
5. Prandtl, L., *Essentials of Fluid Dynamics*, Hafner, 1952.

Flow Visualization

S. J. Kline
STANFORD UNIVERSITY

I. Uses of Visualization

Moving fluids often form patterns so complicated that intuition fails when we try to imagine them. Some flows are so complicated that we cannot analyze all their details from the governing equations, even with the biggest computers now available. Visual images of the actual flows (Figs. 1, 2, 3, 4) can advise us of the real flow patterns.

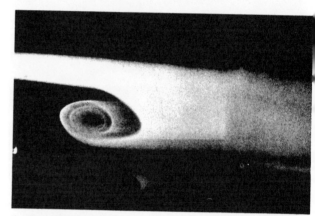

1. **Boundary-layer flow in a symmetric strut corner (visualization by hydrogen bubbles).**

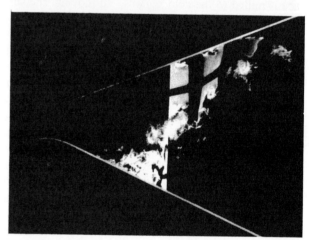

2. **Fully-developed stall in a diffuser (visualization by hydrogen bubbles).**

In time-dependent flows, different flow patterns succeed one another, making it very difficult to obtain understanding of the flow phenomena from the output of a few probes at fixed locations. Various visual techniques can then be used as experimental tools for determining the general nature of a complicated flow

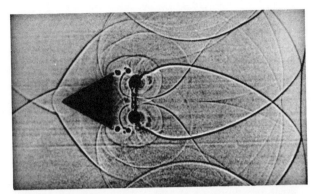

3. Transient shock pattern in flow over a wedge (shadowgraph visualization).

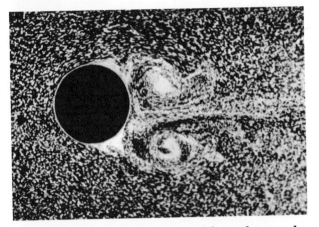

4. The wake of a cylinder visualized by surface powder.

be of great aid in perceiving the total flow structure, in finding instantaneous velocity profiles, and in determining the time sequences of events as a step in establishing causality. A good example is the sequence of Fig. 6, showing respectively laminar, transitional, and turbulent flow in a boundary layer on a smooth, flat surface.

Many flows are very sensitive to small changes in geometry or in other boundary conditions; a small change in conditions may cause a very large shift in

and also to provide quantitative measurements of such things as speed, frequency, density, time sequences, etc.

In a flow past an oscillating plate (Fig. 5), the motion is truly periodic; in principle, it could be constructed from point measurements if phase information were retained. Even so, construction from visual data, as in Figs. 5a and 5b, may be easier. In flows that are unsteady but not truly periodic, such as turbulent shear flows, the measurements are much more difficult. In such cases, motion pictures of the flow can

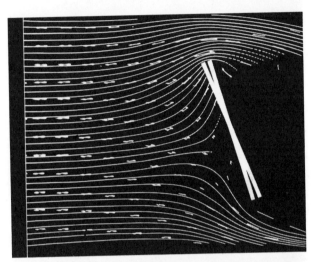

5b. Instantaneous streamlines* for oscillating plate flow, determined from the velocity-direction field of Fig. 5a.

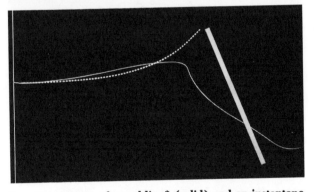

5c. Comparison of a pathline* (solid) and an instantaneous streamline (dashed) in oscillating plate flow.

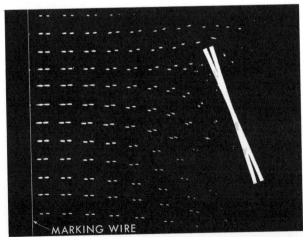

MARKING WIRE

5a. Combined-time-streak markers in flow past an oscillating plate. Two frames a short time apart are shown superimposed.

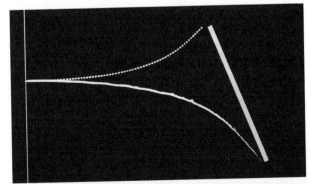

5d. Comparison of a streakline* (solid) and an instantaneous streamline (dashed) in oscillating plate flow.
*Defined on pages 3-4.

the flow pattern. For example, a slight increase in noise or roughness can cause a laminar flow to become turbulent. Such sensitivity to small changes

6a. Laminar flow in boundary layer shown by dye injection through a wall slit. The wall is in the plane of the paper.

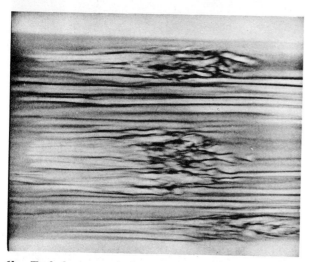

6b. Turbulent spots in boundary-layer laminar-turbulent transition.

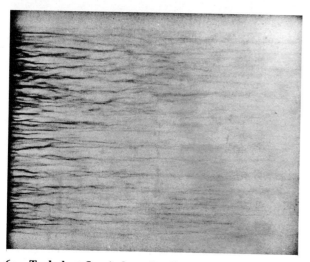

6c. Turbulent flow in boundary layer.

usually means that the theoretical equations cannot be solved with sufficient accuracy to predict the shifts in flow pattern, even if all the conditions affecting the flow were known with enough accuracy. It is then often expedient to determine the actual flow pattern for a given set of conditions using flow visualization.

Thus visualization can be of assistance for research, for direct solution of engineering problems, and for teaching.

Visualization has been used to provide significant information on such diverse flow phenomena as: waves in stratified fluids (FM-144, 145*), cells in rotating systems with thermal effects, flow in tee channels (FM-69), starting and stopping vortices (FM-10), supersonic and hypersonic wakes, start-up of compressible flow over sharp objects (FM-99), and flow models in turbulent boundary layers (FM-2).

A few examples where visualization has played a key part in solving scientific or engineering problems include subsonic diffuser flows (FM-49), boundary-layer interactions with shock waves (FM-27, 30), rotating stall in compressors (FM-7), wind-driven water waves (FM-148), vortex systems on wings of finite span (FM-24), flow models in transition to turbulence (FM-1), and magnus effect (FM-11). This list can be greatly extended by reading the titles of the loops in the NCFMF program.

II. Interpretation of Visual Images of Flow Patterns

To use any visualization method (e.g., smoke lines, wall tufts, schlieren, birefringence, etc.) effectively, we must always ask ourselves: "What physical property of the flow or fluid does this picture show, and how do we interpret the patterns in the picture?" For instance, in any method involving the marking of fluid particles (e.g., smoke, dye, neutrally buoyant particles, hydrogen bubbles, surface powder), it is important to understand the relationships between the observed pictures and four concepts: pathlines, timelines, streaklines, and streamlines.

A *pathline* is the locus of points traversed by a given fluid particle during some specified time interval; the pathline is the particle's actual path during this interval.

A *timeline* is a set of fluid particles that form a line at an instant in time. Thus we make a timeline visible by marking it with bubbles, by photo catalysis, or by other means, at a single instant. At later times both the shape and location of the timeline will generally have altered.

A *streakline* is the locus of particles which have

*FM numbers refer to film loops produced by the National Committee for Fluid Mechanics Films. They are available from Encyclopaedia Britannica Educational Corporation, 425 No. Michigan Avenue, Chicago, Illinois 60611.

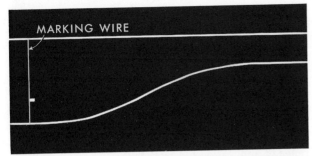

7a. **Water flow in a contracting channel. A small element is marked at time t_0 by hydrogen bubbles.**

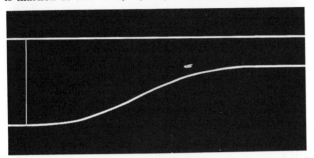

7b. **Element at time $t_0 + 4.7$ secs.**

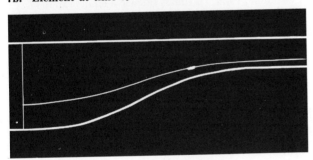

7c. **Element and superposed pathline.**

passed through a prescribed point during a specified time interval. Dye or smoke issuing slowly from a fixed injector shows the streakline passing through the injection point. In a steady flow new material points pass any given point continuously; the streakline for different time intervals will be composed of different particles, even in a steady flow.

A *streamline* is a line which at a given instant is everywhere tangent to the velocity vector. The streamline concept is essentially mathematical in nature. What we usually generate by visualization are streak-

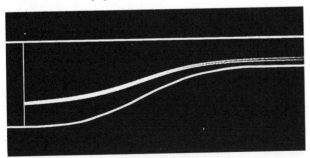

8. **Streakline made visible by steady DC current on marking wire.**

lines, pathlines, or timelines. A field of streamlines and individual streamlines can be found in various ways from streaklines, pathlines, and timelines.

One way to make pathlines, streaklines, and time-

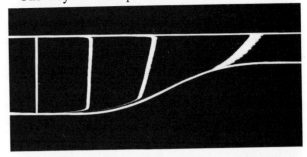

9. **Narrow timelines marked by hydrogen bubbles.**

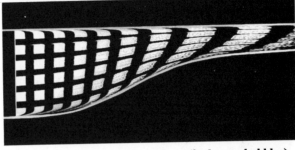

10. **Combined-time-streak markers (hydrogen bubbles).**

lines visible is the hydrogen bubble method, as illustrated in Figs. 7, 8, 9, and 10. (For more on the method see Section III.)

In a steady flow, a streakline, pathline, and the streamline which all pass through a single point are identical. This follows from the fact that the velocity field is unchanging in time; hence each particle traverses the same trajectory. Thus each particle passing a fixed point in space follows the same streakline; this is its pathline (by definition), and is also everywhere tangent to the velocity vector field. Thus the streamline pattern is conveniently shown by a set of dye or smoke streaklines in steady motions.

In unsteady flows, the pathline, streakline, and streamline generally differ from each other, as seen in Figs. 5c and 5d. In unsteady flows the instantaneous streamlines can be determined from a short time exposure or a multiple exposure, as in Figs. 5a and 5b. However, the procedure is a clumsy one.

Sometimes we want to determine the velocity vector field. In a steady, two-dimensional, incompressible flow, this can be done using either pathlines or streaklines alone since the flow is always parallel to them, and the velocity vector magnitude is inversely proportional to their separation. However, the numerical accuracy of such a method is not good, since we must measure small distances between lines that may not be very sharply defined.

If we use time and space markers simultaneously, we can extend the techniques to unsteady flows and also obtain more accuracy in steady flows. In the film this is called the "combined-time-streak marker," or "CTSM," method. An example of the CTSM method using moderate-sized squares as markers is shown in Fig. 10.

Using the CTSM method it is possible to obtain the velocity components in two directions, one of which must be the flow direction. One obtains a "section" through the flow. Measurements have been made in some cases where velocity components were obtained in two orthogonal planes at once. Similar results can be achieved in principle by photo-catalytic techniques. A great advantage of such techniques is that they provide measurements of instantaneous velocity profiles, although the accuracy is relatively low.

III. Techniques of Flow Visualization

Many kinds of flow visualization are known and have been employed in research, development, or teaching. Table 1 shows the more common methods grouped into five major types: (i) marker methods, (ii) optical methods, (iii) wall trace methods, (iv) bi-refringence, and (v) self-visible phenomena. Loops FM-21 and 22, Techniques of Visualization for Low-Speed Flows, illustrate some of the methods in Table 1.

The first column indicates whether the method is essentially a qualitative one or can be used to provide quantitative data on one or more variables of the flow field. In this regard, the opinions given are the author's. The second column suggests the normal range of application for each method.

References

1. Prandtl, L. and Tietjens, O. G., *Fundamentals of Hydro and Aero Mechanics*, McGraw-Hill, 1934, Ch. V.
2. Clayton, B. R. and Massey, B. S., "Flow Visualization in Water: A Review of Techniques," J. Sci. Instrs. *44*, 1967, pp. 2-11: (includes modern bibliography).
3. Schraub, F. A., Kline, S. J., Henry, J., Runstadler, P. W., and Littell, A., "Use of Hydrogen Bubbles for Quantitative Determination of Time-Dependent Velocity Fields in Low-Speed Water Flows," Trans. ASME, Jou. Basic Eng. *87*, 1965, p. 429ff.
4. Symposium on Flow Visualization, ASME Symposium Volume, 1960, (includes extensive bibliographies).

TABLE 1
Flow Visualization Techniques

Method	Measures	Fluid and Speed
MARKER METHODS		
Dye or smoke	Displacement; qualitative	Low-speed flows
Surface powder	Displacement; qualitative	Restricted to open surface liquid flow
Neutral density particles	Displacement; quantitative	Mainly liquids
Spark discharge	Displacement; qualitative	Limited to low density gases
Hydrogen bubble	Displacement; quantitative	Limited to electrolytic fluids
Aluminum flakes	Displacement, but orientation of flakes is shear-dependent; qualitative	Dense liquids
Photo-catalysis	Displacement; qualitative	
Electro-chemical luminescence	Velocity near surfaces; qualitative	Low speed, special solutions
OPTICAL METHODS		
Shadowgraph	$d^2\rho/dx^2$; quantitative	
Schlieren	$d\rho/dx$; quantitative	Hi-speed gases, or thermal or concentration gradients in liquids
Interferometer	ρ; quantitative	
WALL TRACE METHODS		
Tufts	Velocity direction; transition; separation; reattachment	No basic limit
Evaporative and chemical change at wall	Velocity direction; transition; separation; reattachment	No basic limit
BIREFRINGENCE	Shear stress; qualitative	Low-speed flows; special fluids only
SELF-VISIBLE		
Luminous	General motions; qualitative	Reacting or very high temperature
Phase interfaces	Displacements of interface	Two-phase fluids

Pressure Fields and Fluid Acceleration

Ascher H. Shapiro
MASSACHUSETTS INSTITUTE OF TECHNOLOGY

Fluid dynamics deals with the motions of gases and liquids and with how these motions are related to forces. For this film, we have designed *steady-flow* experiments in which all body forces (gravitational, electromagnetic, etc.) as well as viscous forces, are relatively unimportant, and in which the fluids are essentially *incompressible*. The main force accounting for the fluid acceleration is the pressure gradient.

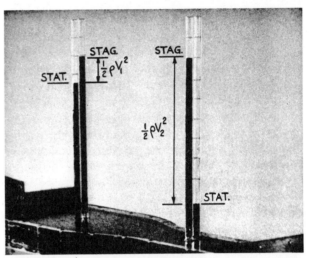

1. **Water flows from left to right. The manometer tubes marked "STAT"are connected to static pressure taps at the upstream and downstream cross-sections. Those marked "STAG" are connected to upstream-facing pitot tubes.**

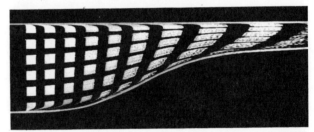

2. **Water flows from left to right. Tiny hydrogen bubbles are electrolyzed at a wire near the entrance. Segments of the wire are insulated, and the current is pulsed. Thus square patches of bubbles are released.**

Pressure Variation Along Streamlines

Our first experiments have to do with changes of pressure and velocity in the streamwise direction. Fig. 1 shows water flowing through a channel of decreasing cross-sectional area. The static-pressure manometers indicate that the pressure decreases in the direction of flow.

Since water is nearly incompressible, the volume flow entering the contraction must equal the volume flow leaving. But the volume flow Q is equal to AV, where V is the average velocity and A is the cross-sectional area; hence the area decrease should produce a velocity increase. The velocity field is shown in Fig. 2. Each fluid patch accelerates as it goes through the contraction. Because of the longitudinal velocity gradient, the front edge of each patch moves faster than the rear edge; hence each patch stretches out. Since water is virtually incompressible, and the flow is two-dimensional, the area of each patch cannot change; thus each becomes narrower as well as longer. The successive bubble patches mark out the streaklines of the flow and, (since the motion is steady) the streamlines as well.

The velocity increase inferred from Fig. 2 is related to the pressure decrease of Fig. 1 by Newton's law of motion. For some kinds of analysis, it is convenient to use distance along the streamlines, s, and distance along the normal trajectories to the streamlines, n, as curvilinear coordinates. Since viscous

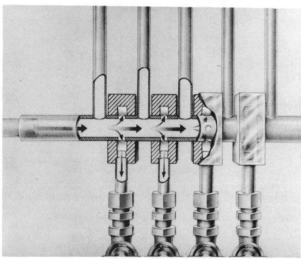

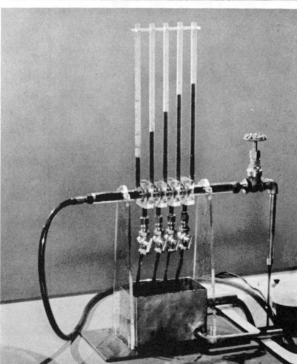

3. Water flows through a manifold from left to right. Five manometer tubes, open at the top, show the longitudinal pressure distribution when fluid is withdrawn through the open bleed valves.

stresses are negligible, the normal stress at a point is the same in all directions. It is the scalar hydrostatic pressure, p. The net force acting along s, arising from the pressure gradient $\delta p/\delta s$, is related to the fluid acceleration along s by Euler's equation of steady inviscid motion along the streamline:

$$\delta p/\delta s = -\rho V(\delta V/\delta s) \qquad \text{(Eq. 1)}$$

The minus sign tells us that a velocity *decrease* in the streamwise direction is accompanied by a pressure *increase*, and vice versa.

Fig. 3 shows flow through a manifold, with bleed valves installed to allow withdrawal of fluid at several stations between the manometer tubes. With the bleed valves closed, the average water velocity is the same at all cross-sections, and there is a barely perceptible pressure drop due to viscous forces. The volume flow is then reduced in each successive cross-section by bleeding water away through the valves. The pressure now *rises*. Since the area A is in this case constant, the velocity V is proportional to the volume flow Q; hence the velocity V decreases along s, and $\delta V/\delta s$ is negative. In agreement with Eq. 1, $\delta p/\delta s$ is positive.

In an incompressible, steady, inviscid flow, Euler's equation may be integrated with respect to distance along each streamline, to obtain the Bernoulli integral,

$$p + \rho V^2/2 = \text{constant} = p_{\text{stag}} \qquad \text{(Eq. 2)}$$

Along each streamline the sum of the static pressure p and the dynamic pressure $\rho V^2/2$ is a constant. This sum is called the *stagnation pressure,* or *total pressure.* On any one streamline, wherever the velocity is high, the pressure is low, and vice versa. The highest possible pressure, the stagnation pressure, occurs where the velocity is zero. A large reservoir supplying a fluid to a duct system is itself a stagnation region.

Fig. 4 shows a streaming flow past a blunt-nosed body, with a central stream tube marked by hydrogen bubbles. The widening of this stream tube shows that the flow is decelerating. Where the central streamline reaches the nose is a stagnation point. There the speed is zero. On both sides of the unique stagnation streamline the fluid decelerates, although not to zero speed, and then accelerates as it slides around the sides.

4.

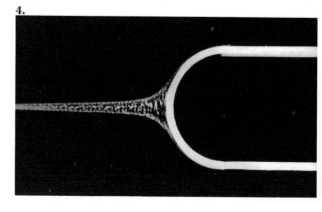

The experimental realization of a stagnation point is put to practical use in a pitot tube. The hole at the upstream-facing nose measures the stagnation pressure of the particular streamline reaching the hole, as defined by Eq. 2. In the contracting flow of Fig. 1, pitot tubes at the upstream and downstream cross-sections of the contraction show the respective stagnation pressures. By Eq. 2, the difference between the local stagnation pressure and the local static pressure is the local dynamic pressure. Thus (Fig. 1), we may determine the fluid speed through a measurement of this difference. Furthermore, Fig. 1 verifies that although both the dynamic pressure and the static pressure vary in the contraction, the sum of the two — the stagnation pressure — remains constant.

Bernoulli's integral also explains how suction can be produced by blowing, as in the aspirator experiment of Fig. 5. To illustrate in detail how a venturi can produce low pressure, Fig. 6 shows an experiment with flow through an unsymmetrical venturi in which one wall is straight. Since AV is constant, the average cross-sectional velocity increases to the throat; in accord with Eq. 1, the pressure decreases. Downstream of the throat, the area increase produces a velocity decrease, and the pressure rises. At the upstream and downstream ends, where the cross-sectional areas are

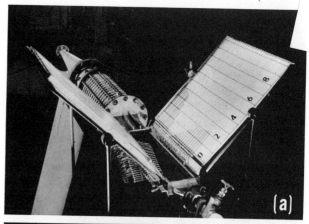

6. (a) A two-dimensional horizontal venturi-tube test section is installed at the exit of the large settling section. Pressure leads from each wall of the channel are connected to the two manometer boards inclined at 45°. (b) Overhead view of pressure distribution on the straight and curved walls of the venturi. Water flow is from top to bottom.

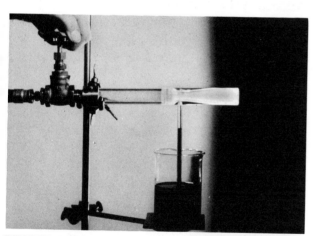

7. Pressure distributions for a venturi in which the diffuser diverges so rapidly that flow separation occurs and there is little pressure recovery. Compare with Fig. 6b.

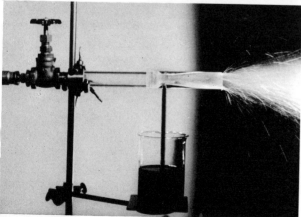

5. Air is blown from left to right through a venturi. The sub-atmospheric pressure at the throat sucks up water from a beaker.

equal, the average velocities are also equal. Bernoulli's integral predicts equal pressures, but the downstream pressure is substantially less than the upstream pressure. The static-pressure recovery in the diffuser is only about half of the static-pressure decrease to the throat. This difference is due to viscous boundary layers whose behavior we must always keep in mind

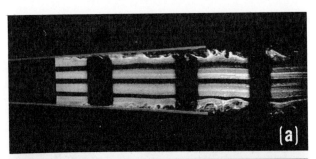

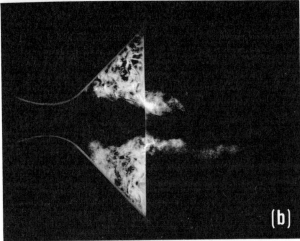

8. Hydrogen bubbles mark the flow in a diffuser. Flow is from left to right. (a) Small divergence angle. The flow remains attached. (b) Large divergence angle. The flow detaches.

even when we pretend temporarily that viscosity is absent. In the experiment of Fig. 7, the area divergence is too rapid, and the static pressure recovery in the diffuser is only about a quarter of the decrease in static pressure to the throat. Even a very small amount of viscosity, if it leads to boundary-layer separation (Fig. 8), can produce radical changes from a hypothetical non-viscous flow.

In the experiment of Fig. 9, water flows through a venturi that discharges to the atmosphere. Since the

9. Flow of water, from left to right, through a venturi which discharges to the atmosphere. The gauge shows the vacuum at the throat, in inches of Hg. below atmospheric. When the vacuum reaches nearly 30″ Hg., cavitation zones appear near the throat. These are seen in the photo as dark patches.

pressure at the exit is atmospheric, the throat pressure is sub-atmospheric. As we increase the flow, all pressure differences increase with the square of the velocity. When the absolute pressure at the throat goes below the vapor pressure of water (about one inch of mercury), boiling occurs. This is cavitation. Steam bubbles appear in the water. Their subsequent collapse creates enormous pressures which can be clearly heard as sound and which can produce very high stresses on nearby walls. Such mechanical stresses often do great damage to liquid-handling machinery, to marine hydrofoils and propellers, and to hydraulic structures.

10. The white plastic disk is free to move vertically. The thick clear disk at top is supported from the base on three posts. Air blown down through the vertical tube at the top escapes rapidly in the space between the white disk and the clear disk. The black streamers indicate the escaping air flow. (a) No air flow. The streamers are limp, and the white disk rests on stops. (b) The air flow is on, and the streamers are extended. The white disk is lifted upward by the downward air flow, until only a small gap separates it from the clear disk.

When air is blown from a hole in the middle of one disk against a parallel disk (Fig. 10), the resulting pressure distribution can be such as to force the disks together. To see how this happens, we have instrumented the lower plate (Fig. 11a) with pressure taps attached to manometer tubes below. Note that the water in each manometer tube is pushed down when the pressure goes above atmospheric. At the disk axis, the pressure is greater than atmospheric by an amount equal to the dynamic pressure of the air jet. Over most of the plate, though, the pressure is less than atmospheric, and rises to atmospheric pressure at the outer edge. The reason is that the through-flow cross-sectional area between the disks increases with radius. This area increase causes a fluid deceleration and, by Bernoulli's integral, a pressure rise to the atmospheric

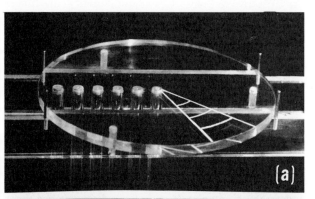

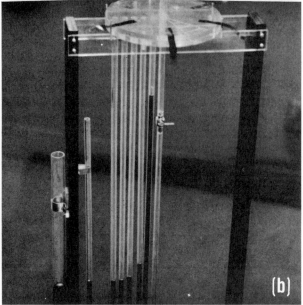

11. (a) A fixed lower disk for the experiment of Fig. 10, with six static pressure taps and manometer tubes attached below. Three small gap spacers are visible, on which the upper disk of Fig. 10 is placed. (b) The six manometer tubes are connected through a common header (out of frame at bottom) to the large reservoir at the far left. The single manometer tube at the left is open at the top and indicates atmospheric pressure. The manometers show the pressure in the gap is below atmospheric except near the axis of the air jet. There the pressure is so high that the liquid level is out of sight.

pressure at the outer edge. The sub-atmospheric pressure over most of the area accounts for the disks being forced together. Actually the viscous forces are by no means negligible. For the equilibrium position, however, the inertial forces are generally larger than the viscous forces and govern the shape of the pressure distribution.

Pressure Variation Normal to Streamlines

Thus far we have discussed the relationship between pressure changes and velocity changes *along* a streamline, or on the average along a channel. However, Fig. 7 shows how different the pressure distributions may be along the two walls of an unsymmetrical channel. To understand this, we must consider the particle dynamics in a direction normal to the streamline.

Let R be the local radius of streamline curvature. The pressure gradient $\delta p/\delta n$, acting on the fluid particle toward the center of curvature is related to the acceleration V^2/R by Euler's equation of inviscid motion normal to the streamline,

$$\delta p/\delta n = \rho V^2/R \qquad (\text{Eq. 3})$$

which shows that the pressure always increases outward from the center of curvature.

The channel bend of Fig. 12 produces curved streamlines. In the straight section approaching the bend the streamlines are nearly straight and the pressure gradient normal to the streamlines is virtually zero. The difference between the entering stagnation pressure and the static pressure at the upstream tubes is a dynamic pressure of about eleven inches of water. At the middle of the bend, where streamline curvature is pronounced, there is a pressure difference between the two sides of the channel, normal to the streamlines, of about three and a half inches of water. Downstream of the bend the pressure is almost uniform again (the slight variation is due to a complicated secondary flow induced by the bend).

Returning to Fig. 6b, we see that the throat pressure

12. Flow in a two-dimensional 80° bend, entering from large settling chamber at left and discharging toward viewer at right. Each set of three static pressure manometers shows the transverse pressure distribution in a plane normal to the flow. The first and last sets are in straight sections of the channel, while the middle set is in the mid-point of the bend. The single manometer tube at left shows the stagnation pressure.

on the curved wall is considerably less than at the straight wall, with the pressure increasing in the direction away from the center of streamline curvature, as demanded by Eq. 3. When we use venturis for flow measurement, the pressure we measure at the wall is thus not the average pressure at the throat.

When you place your finger near the side of a jet from a water faucet, the jet bends toward your finger. This is the Coanda effect. In Fig. 13 the jet attaches to and bends around a freely-suspended hollow cylinder which has in it a small static pressure hole connected to a manometer. Since the streamlines are curved, there is a normal pressure gradient, with the pressure at the cylinder surface less than the atmospheric pressure at the outside boundary of the jet. Because of this sub-atmospheric pressure, the jet is bent around the cylinder, and there is a net force on the cylinder acting toward the jet.

The Coanda effect is also part of the explanation for the fact that a billiard ball may be supported by a rather small air jet (Fig. 14).

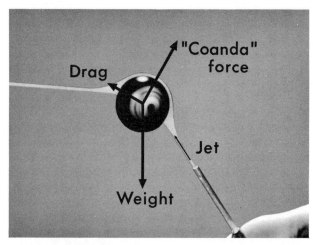

14. A billiard ball is supported by an air jet supplied by a ⅛″ diameter nozzle at 60 psig. The superposed drawing shows the forces on the ball accounting for its equilibrium.

Complete Pressure Field

We have now looked separately at the pressure gradients along and normal to the streamlines. To understand a complete flow pattern, however, we must consider both gradients simultaneously, and we must also consider the equation of continuity.

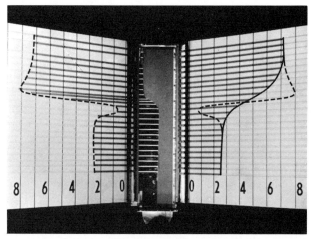

15. Wall pressure distributions for flow through a rapid unsymmetrical contraction. The dashed curve compares the distribution on the curved wall with that on the straight wall.

Fig. 15 compares the pressure distributions on the two walls of a rapid unsymmetrical contraction. Upstream, the pressures are equal on both walls. Downstream, they are also equal but lower. Because of the general area decrease, the average velocity increases. In agreement with Eq. 1, the average pressure falls.

On the straight wall the pressure falls continuously. But on the curved wall it first rises, then undershoots to a very low value, then rises once again to its final value.

These two pressure distributions may be interpreted using the two dynamical equations and the equation of continuity. The middle curve of pressure versus distance in Fig. 16 shows the streamwise variation of the

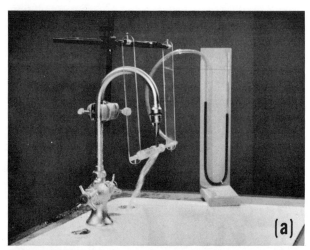

13. Jet of water from a slit nozzle. Behind and below it is a hollow closed cylinder hanging freely and vertically from pivots. When the cylinder is brought next to the jet, the latter attaches to and bends around the cylinder. The sub-atmospheric pressure on the cylinder surface is sensed at a pressure tap and transmitted to the manometer. The resulting pressure distribution on the cylinder produces on it a force directed toward the jet, as shown by the inclination of the pendulum support.

average pressure at each cross-section as determined (through Bernoulli's integral) by the average velocity at each cross-section. Far upstream and far downstream, where the streamlines have virtually no curvature, the pressure is uniform over the cross-sections.

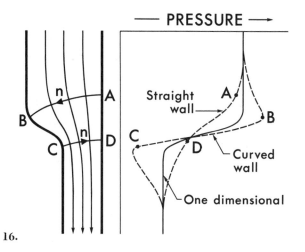

16.

From the shape of the channel, we may expect that the streamlines will be generally of the shape sketched in Fig. 16; concave to the right in the neighborhood of AB; concave to the left in the neighborhood of DC. The curves AB and DC are drawn normal to the streamlines in regions of different concavity. The arrows on these curves show the directions of increasing n. From one-dimensional considerations and a comparison of cross-sectional areas, the average pressure on AB is only slightly less than the upstream pressure. But, because of streamline curvature, the pressure increases from A to B. Hence the pressure at B is greater than the upstream pressure, and that at A is less. Similarly, the average pressure on CD is only slightly greater than the downstream pressure, but again, because of streamline curvature, the pressure increases from C to D. Hence the pressure at D is greater than the average, and that at C is less than the average. Putting all this together, we can understand why the pressure on the curved wall first rises to a maximum at B and then falls to a minimum at C, while on the straight wall it falls continuously from A to D.

According to Bernoulli's integral, which in this case

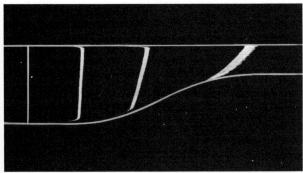

17. **Successive positions of a bubble line which is initially vertical.**

has the same stagnation pressure for each streamline, the velocity is a minimum where the pressure is highest and a maximum where the pressure is least. The hydrogen bubble lines of Fig. 17 verify the velocity distributions which may be inferred from the pressure distributions of Figs. 15 and 16 by using Bernoulli's integral. The tilting of these fluid lines shows that, near the beginning of the contraction, the velocity near the curved wall is less than that near the straight wall.

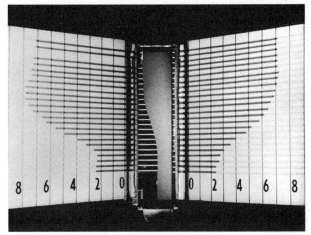

18. **Pressure distributions for half of a gradual symmetrical contraction. Compare with Fig. 15.**

To avoid separation of the flow from the walls due to adverse pressure gradients in wind tunnel nozzles, contractions are made gentle, as in Fig. 18. The streamline curvature here is much less than in Fig. 15, and the transverse pressure gradients which cause the distinctive peaks in the pressure distribution of Fig. 15 are reduced.

Bernoulli's Integral Is Not Always Valid

The statement "high velocity means low pressure" is only sometimes true. Viscosity or compressibility or unsteadiness can render Bernoulli's integral invalid, and the integral applies only to individual streamlines unless the motion is irrotational.

The straight duct of Fig. 19 has a partition. One side is free and clear; the other side has a flow resistance. The static pressures shown by the manometers are equal. However, to conclude from Bernoulli's integral that the velocities of the two streams are also equal would be wrong. The stagnation pressures of the two adjoining streams are actually quite different. From the two dynamic pressures we see that the velocity in the obstructed passage is less than that in the clear passage. The reason Bernoulli's integral cannot be used here to interpret the observed static pressures is that we are dealing with different streamlines having different stagnation pressures. We must instead use Eq. 3: because of the confinement of the channel walls, the streamlines have virtually no curvature; thus the normal pressure gradient is zero, and the

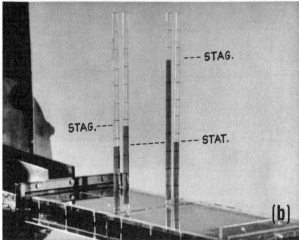

19. (a) A straight duct with a divider at the left. Both halves of the flow come from a common reservoir, but the lower half passes through a bank of small tubes which act as a flow resistance. Downstream of the partition the two halves of the flow are again in contact. (b) The flow is on. The two outer manometers show the static pressures, while the two inner manometers are connected to upstream-facing pitot tubes and show stagnation pressures.

static pressures in the two streams are the same even though the velocities differ. For the same reason, the static pressure across a viscous boundary layer is virtually constant.

The vertical tubes of Fig. 20 show the pressure dis-

tribution in a horizontal tank of water when we rotate it on a turntable. After viscosity forces the water into solid body rotation, the velocity increases linearly with radius, that is $V = \Omega r$. If we used Bernoulli's integral (which would be improper because we would be crossing streamlines), we might expect the pressure

20. A closed cylindrical tank of colored water is mounted on a horizontal turntable. The manometer tubes mounted in the cover are open at the top to atmosphere, and show how the pressure varies with radius. With the tank initially at rest, the level is the same in all the tubes. After the turntable has rotated at uniform speed for a long time, the water is in solid-body rotation, locked by viscosity to the tank. This picture was taken at an instant when the rotating bank of manometer tubes was normal to the line of view.

to *decrease* with radius. Actually, the pressure *increases* with radius. This means that the stagnation pressure also increases with radius, and that each circular streamline has a different Bernoulli constant. The right way to look at this is with Equation 3:

$$\delta p / \delta n = \rho V^2 / R = \rho \Omega^2 r^2 / R$$

But, for circular motions, $\delta p / \delta n = dp/dr$, and $R = r$. Thus, $dp = \rho \Omega^2 r dr$, which integrates to $p = \rho \Omega^2 r^2 / 2 +$ constant. This is the parabolic pressure distribution of Fig. 20.

Reference

Prandtl, L.: *Essentials of Fluid Dynamics,* Chapter 11. Hafner Publishing Co., New York, 1952.

Low-Reynolds-Number Flows

Sir Geoffrey Taylor

Introduction

Low-Reynolds-number flows are those in which inertia plays only a very small part in the conditions which determine the motion. The Reynolds number of the flow of a fluid which is characterized only by viscosity and density is defined as $R = LV\rho/\mu$. Here L is chosen as a length connected with the solid boundaries of the flow which may be expected to determine the scale of the fluid motion, V is a characteristic velocity, ρ and μ are the density and viscosity of the fluid. In low-Reynolds-number flows, the numerical value of R provides a rough estimate of the relative importance of inertia and viscosity. When R is small, the importance of inertia is small compared with that of viscosity. As examples, the movement of miscroscopic organisms for which L is very small and the movement of glaciers for which V is very small and μ very·large are given. In the latter case the flow was made apparent by a line of red flags which was initially straight and was bowed out by the slow motion of the glacier after two years.

Pulling a knife vertically out of a pot of honey for

1. **Honey flowing from a knife drawn out of a jar.**

which μ is large demonstrates two properties of a viscous liquid. It can resist both tangential and tensile stresses. The honey can be lifted by a tangential force exerted by a knife's surface, and the stretching of the stream as it falls gives rise to a tensile stress over a horizontal section such as AB in Fig. 1.

47   LOW-REYNOLDS-NUMBER FLOWS

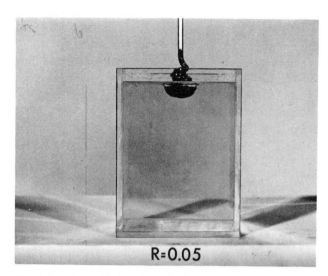

R=0.05

2. A low-Reynolds-number "jet" penetrates only slightly into a mass of the same fluid.

Jet Flows

Though the honey stream demonstrates some properties of viscosity, it does not provide a model showing the relative importance of viscosity and inertia. For this purpose the honey experiment is repeated, using a piston driven at a known speed to produce a controlled jet of colored fluid at a known velocity. This jet falls vertically into a transparent box containing the same fluid, but uncolored. Four fluids whose viscosities cover a very wide range are used, and in order to ensure that the differences between the flows observed in the four cases are, and can be seen to be, due only to differences in Reynolds number, the diameters D and velocities V are the same in all four cases. Figs. 2 and 3 show the flows, and the corresponding values of $R = DV\rho/\mu$ are printed below. In Fig. 2 ($R = 0.05$), the fluid is syrup which is so viscous that the vertical velocity is destroyed by the vertical stress in the jet before it reaches the free surface of the fluid in the box, so that it forms a pile which flattens slowly under the influence of gravity. In Fig. 3a ($R = 10$) the fluid is glycerine. The depth to which the jet penetrates before losing its velocity and spreading out into a mushroomlike head is only a few jet diameters. The angle of the cone which forms the stalk of the mushroom is an indication of the rate at which the momentum of the jet is being retarded by viscosity. In Fig. 3b ($R = 200$) the fluid is a mixture of glycerine and water, and the jet penetrates many diameters before being stopped. In fact, it is not stopped until it reaches the bottom of the box. Comparing Figs. 3a and 3b, it appears that the angle which the roughly conical jet assumes when it is retarded by viscosity is of the order of magnitude $1/R$. Fig. 3c shows what happens when the Reynolds number is sufficiently high. At $R = 3000$ the jet has become turbulent.

Flow Through Long Tubes

This film is concerned with flow like that shown in Fig. 2, for which R is small compared with 1.0 so that inertia plays no appreciable part in the situation, flow being determined only by the balance of viscous and pressure stresses brought into play by gravity or forces applied at the boundaries. The geometry associated with the flow shown in Fig. 2, however, is too complicated for complete mathematical analysis, and a simpler case is discussed in greater detail — namely, the flow through long tubes of uniform bore. Here, though the Reynolds number is not necessarily small, there is no change in the inertia of the flow as it passes through the tube, so that the results of calculation of the kind used in discussing low-Reynolds-number

3. Comparison of the penetrations of the first two jets with Fig. 2 demonstrates the effect of increasing Reynolds number. The third jet is turbulent.

(a)   **(b)**   **(c)**

R=10

R=200

R=3000

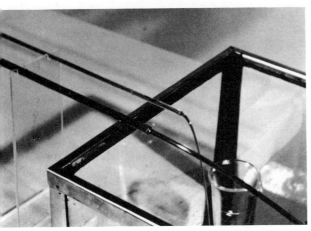

4. Laminar flow through two tubes with inner diameters in the ratio 2:1.

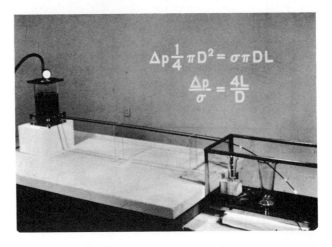

$$\Delta p \tfrac{1}{4}\pi D^2 = \sigma \pi D L$$

$$\frac{\Delta p}{\sigma} = \frac{4L}{D}$$

flows are applicable. The two tubes shown in Fig. 4 (left) have the same length, but the bore d of one of them is twice that of the other. Compressed air at the top of the reservoir drives the fluid at constant pressure through the two tubes into two receptacles with marks at one ounce and 16 ounces respectively (Fig. 4, right). Calculation shows that the discharge from the large tube should be 16 times that of the other, so that if the flows into both receptacles are started at the same time, they reach the marks simultaneously. In this experiment the driving pressure $\triangle P$ acting over the area of cross section $\tfrac{1}{4}\pi d^2$ balances the tangential stress σ acting over an area $\pi d \times L$ so that $\triangle P\tfrac{1}{4}\pi d^2 = \sigma \pi d L$ or $\triangle P/\sigma = 4L/d$. When the length L is large compared with the diameter d, as it is in the experiment shown, a small tangential stress can produce a large change in pressure. This is the principle on which hydrodynamic lubrication is based.

Hydrodynamic Lubrication

A very simple experiment can demonstrate that the coefficient of friction, which is the ratio of the tangential to the normal force between two solid bodies, can be much reduced by hydrodynamic lubrication. If a sheet of note paper is held horizontally above a smooth polished table and dropped onto it with a small horizontal velocity, it can be made to slide smoothly many times as far as if the table were not smooth. This is because, as the paper settles, the air under it must flow outwards and this outflow produces a tangential stress which enables enough pressure to be built up in a very narrow space between paper and table to support the paper out of contact with the table. This tangential stress is much smaller than solid contact friction stress. When the layer gets so thin that lack of flatness or smoothness of paper or table

permits actual contact, the paper stops gliding. In this experiment the layer of air continually decreases in thickness, but if the sheet could be held at a very small angle to the table and moved horizontally with the wide gap in front, air would be swept in there and would be forced to flow out through the narrower parts of the gap. In this way a layer of air would separate paper and table so long as the horizontal motion was maintained. This principle is illustrated by means of the toy shown in Fig. 5, which can be called a teetotum. This consists of three equal flat laminae made of mica, which can be very flat. These are stuck to the lower sides of three light arms which are rigidly set to form a 120° triad from the center of which rises a light vertical rod. The laminae are mounted so that they are inclined upwards at an angle of half a degree or less with a plane normal to the axis rod. The teetotum is spun between finger and thumb in a counterclockwise direction and dropped on a horizontal table. It will spin for a very large number of revolutions, but if it is spun in a clockwise direction it stops instantly. In the film the direction of rotation

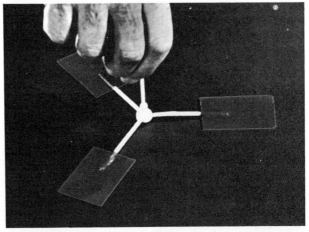

5. The three laminae of the spinning toy are inclined slightly upwards. When spinning counterclockwise the front edges are higher.

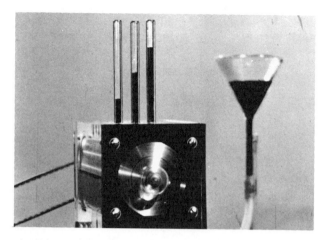

6. **Pressure distribution in a journal bearing with minimum gap at the top. Shaft is rotating in a counterclockwise direction.**

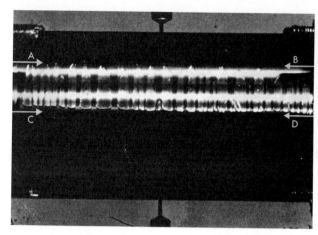

7. **Cavitation bubbles in the low-pressure region of a journal bearing.**

is partly obscured by a stroboscopic effect when the teetotum is spinning.

The principle illustrated by the teetotum is employed not only in slipper bearings, but also in the lubrication of a journal bearing in which a cylindrical shaft rotates inside a cylinder of slightly greater diameter. The shaft assumes a slightly eccentric position so that there is a narrowing gap into which the oil is dragged, so producing a high pressure. Beyond the point where the clearance is a minimum, the gap is expanding and the pressure is reduced.

This is illustrated in Fig. 6, which shows a journal bearing with a fixed eccentricity in which the top of the inner cylinder is the position of minimum clearance. Three manometer tubes are connected with holes in the outer cylinder, one just before the minimum clearance, one at this position, and the third as the same distance beyond it. In Fig. 6 the shaft is seen rotating in the counterclockwise direction. The level of the lubricant in the right-hand manometer is just as much above that of the central one as that in the left-hand manometer is below it. The situation is reversed when the direction of rotation is reversed.

Very large pressures which can support large loads can be produced if the fluid is very viscous or the gap very small. The fluid, however, cannot support the large negative pressures which the last experiment might lead us to expect, and cavitation bubbles may occur. These can be seen in Fig. 7, which was taken through a transparent journal and shows the fluid downstream of the minimum clearance at the top of the picture. The bubbles form at the position of minimum pressure marked AB in Fig. 7, and extend into the widening gap to the position CD where the pressure begins to rise again.

Kinematic Reversibility

The reversibility of the pressure when the direction of rotation is changed in the experiment of Fig. 6 implies reversibility of flow. Some surprising results of this reversibility are shown with the apparatus of Fig. 8, in which the space between two concentric cylinders is filled with glycerine. Dye is introduced

8. **Kinematic reversibility in an annulus, (a) initial dyed element, (b) inner cylinder turned 4 turns forward, (c) inner cylinder turned back 4 turns.**

(a)

(b)

(c)

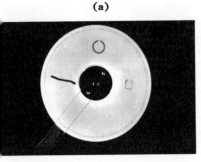

(a)

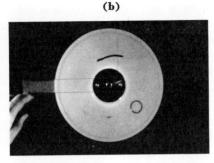

(b)

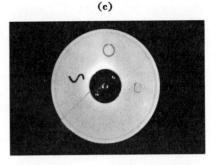

(c)

9. Kinematic reversibility is achieved for the dyed fluid square and the rigid ring, but is not achieved for the bit of yarn.

into the annulus which forms a compact colored volume (Fig. 8a). The inner cylinder is turned through, say, *N* revolutions. When observed from the side, the colored area seems to mix with the uncolored glycerine (Fig. 8b), just as milk mixes with tea when stirred in a cup, but on reversal of the motion the dye suddenly collects into a compact mass when the cylinder has been turned exactly *N* turns in reverse (Fig. 8c). To understand the reason for this peculiar behavior, one can observe what happens when looking through the fluid in a direction parallel to the axis of rotation. One sees the dyed areas being drawn out into long, thin streaks. On reversal of the motion of the boundary, every particle retraces exactly the same path on its return journey as on the outward journey, and at every point its speed is the same fraction of the boundary speed as it was at the same point on its outward journey, so that when the boundary has returned to its original position every particle in the fluid has also done so and the original pattern of dye is reproduced. Of course, molecular diffusion, which is irreversible, is negligible during the time of this experiment.

The motion of a rigid body suspended in a fluid is also reversible, but that of a flexible body is not, because when the stresses in a flexible body are reversed it changes its shape. This is illustrated in Fig. 9, where

the rigid body is a small plastic ring with a gap to mark its orientation. In Fig. 9a the gap is in the 12 o'clock position. The flexible body, consisting of a piece of wool, is on the left and a rectangular pattern of dyed fluid on the right. The inner cylinder was then rotated in a clockwise direction (Fig. 9b), the dye, ring and wool moved round in a clockwise direction, and the ring rotated in a counterclockwise direction about its center. The wool remained nearly straight, because it was in fluid which was moving in such a way as to stretch it. The dye almost disappeared. After the motion was reversed until the inner cylinder was in its original position (Fig. 9c), the rigid ring returned to its original position and orientation. The rectangle of dye reconstituted itself, but the wool curled up because on the return path the viscous stresses gave rise to a compressive stress along its length which naturally made it collapse.

Falling Bodies and Sedimentation

The resistance of similar solid bodies moving at low Reynolds numbers through a fluid are proportional to their linear dimensions. The weight of a sphere is proportional to the cube of its diameter. When it is falling through the fluid, it weight is supported by the fluid resistance, so it will fall at a rate proportional to the square of its diameter. Fig. 10

10. Two balls with ¾-in. and ⅜-in. diameters fall in syrup.

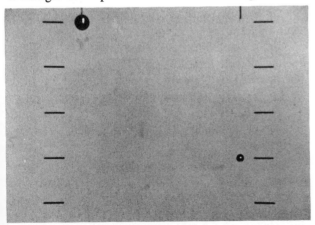

(a)

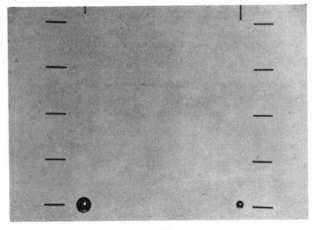

(b)

shows two brass balls of diameters ¾ inch and ⅜ inch falling in syrup. They can be released mechanically. Background marks at equal intervals make it possible to demonstrate their relative falling speeds. The bigger ball is released when the smaller one has traversed ¾ of the distance to the lowest mark (Fig. 10a). The two balls reach this mark simultaneously (Fig. 10b).

At low Reynolds numbers the disturbance produced by a moving ball extends many diameters. Beads suspended in a fluid at considerable distance from it are moved, conversely, a solid body such as a wall

11. The sphere near the wall falls more slowly than an identical one far from the wall.

can affect its rate of fall. Fig. 11 shows two identical balls which were released simultaneously from the same height. The one nearer the vertical wall falls more slowly, but owing to the reversibility of low-Reynolds-number flow it remains at a constant distance from it.

The retarding effect of neighbors makes a dispersed suspension of particles fall more slowly than a single one. Thus, when a suspension of particles falls in a fluid it develops a sharply defined top. This happens

12. Two boxes of assorted beads left to settle at the same time. The box at left has a larger particle density.

even for a suspension of particles of assorted sizes, as in this experiment. A particle which has a terminal velocity rather lower than its neighbors does not get left behind, because if it did it would find itself isolated and would fall faster and catch up the rest. Fig. 12 shows two boxes, each containing fluid and sediment. They were both shaken and left to settle at the same moment. The left-hand box has many particles and settles much more slowly than the right-hand box, which has few.

Resistance of Long Thin Rods

When a body is not spherical, its resistance at low Reynolds numbers is not the same for all directions of motion. A long, thin body of revolution, for instance, has twice the resistance to lateral motion that it has to motion parallel to the axis. This was first proved for the special case of a long ellipsoid, but is true generally. Fig. 13 shows two identical rods which

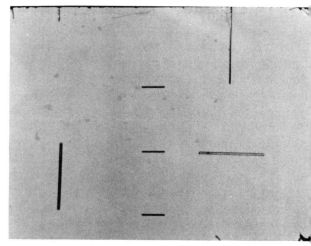

13. Identical cylinders falling in syrup. When released, the bottom of the vertical cylinder was at the same height as the horizontal cylinder.

were released simultaneously in syrup with the bottom of the vertical rod at the same level as the horizontal rod. This level was that of the uppermost of three equally spaced marks. The photograph was taken when the bottom of the vertical rod had just reached the lowest mark and the horizonal rod was level with the intermediate mark.

When a rod of uniform section and density is released obliquely, it does not change its orientation but drifts sideways. At the terminal speed the net weight is just balanced by the drag, which therefore acts vertically. This drag is the resultant of two forces, one parallel to the long axis and one perpendicular to it. In the triangle of forces *ABC* (Fig 14a) these are represented by *CA* and *BC*, while the weight is represented by *AB*. Since at low Reynolds number the velocity of a body is proportional to the applied force, and since for long cylinders a force moves the ody only half as fast when applied laterally as when applied longitudinally, the triangle of velocity is ACD where D is the midpoint of BC. AD therefore is the direction of motion when the force on the cylinder

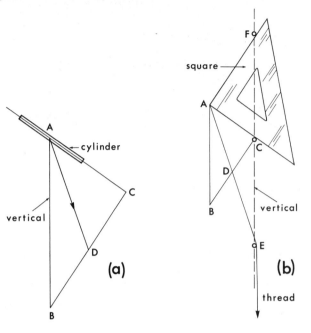

14. **Geometric construction of the "flight path" of an obliquely oriented cylinder.**

acts in the direction of AB. This geometrical construction can be mechanized, using three drawing pins F, C and E (Fig. 14b) on a vertical line FE of which C is the midpoint. A draftsman's square whose rectangular corner is A can slide around making contact on its two perpendicular sides with pins F and C. The line joining A and E (which in a model could be a thread attached to the corner A of the square) is the direction of motion when the axis of the cylinder lies parallel to AC. The lines AB and CB in Fig. 14b are drawn parallel to AB and CB in Fig. 14a to reveal the geometry of the model. By observing the slope of AE to the line AB as the square is moved, it can be seen that the maximum angle of inclination of the flight path to the vertical is about 19 degrees.

Self-Propelling Bodies

All the familiar types of self-propelling bodies such as airplanes, boats, or fish derive their thrust from the inertial reactions of air or water to their propulsive mechanisms. Even a swimming snake derives its propulsion by sending waves down its body so that each section of it contributes a forward force component by inertial reaction. The relevant Reynolds numbers for these cases are often as high as many millions. Even tadpoles, for which the Reynolds number is of order 10^2 to 10^3, derive their propulsion almost en-

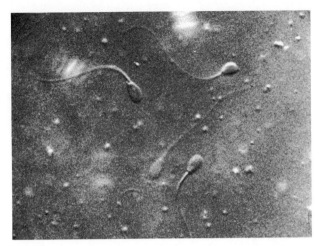

15. **Bull spermatozoa. Each sperm is about .005 cm long.**

tirely from inertial reaction. On the other hand, microscopic organisms such as the bull sperms shown in Fig. 15, though they make motions like those of tadpoles, have such low Reynolds numbers (of the order 10^{-3} if the over-all size is used in defining Reynolds number or 10^{-5} if the diameter of the tail is chosen) that they cannot derive any appreciable thrust from inertia. They derive their forward thrust from viscous reaction due to oblique motion of thin tails, just in the way that has been demonstrated with an obliquely moving cylinder.

To illustrate the difference between inertial and viscous propulsion, the two models shown in Fig. 16 were constructed. The "engines" of both consist of twisted rubber bands. In the lower one a tail organ

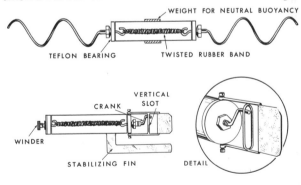

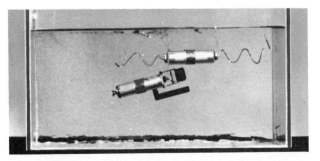

16. **Mechanical swimming models in a vat of syrup.**

is made to oscillate by means of a crank, and in the upper one two similar spiral wires, one right-handed and the other left-handed, are driven in opposite directions of rotation by the twisted band to the ends of which they are attached. When the oscillating-tail model is put into the water it swims well, because, as is well known in the case of a boat rudder, such a motion at a large Reynolds number gives rise to a backward flow, the reaction to which propels the boat. The spiral model propels itself much more slowly, because the area of the wires is much smaller than that of the oscillating blades so that it gets only a feeble grip of the water and produces a much smaller backward stream.

When the oscillating-tail model is wound up and put into viscous syrup, the tail waves backwards and forwards but produces no resultant motion because, owing to the reversibility of low-Reynolds-number flow, the forward motion of the blade is exactly neutralized by the backward motion when it returns through the same position.

The spiral model swims when put into syrup, because every element of each spiral is behaving like the obliquely moving rod. Since lateral resistance is greater than longitudinal resistance, motion of every element at right angles to the axis of the spiral contributes a resultant longitudinal component.

Hele-Shaw Cell

This apparatus consists of two parallel rectangular glass plates fixed 0.020 inch apart. A viscous fluid is driven through it under pressure applied at one side, the two neighboring sides being sealed. Thus all particles move parallel to the sealed sides when there is no obstruction, though at speeds which depend on the distance from the plates. Colored fluid is injected at points along the injection side so that an observer sees a set of parallel straight lines when the flow is unobstructed. When an obstruction is placed in the cell, the streamlines spread out around it and join together again downstream. The obstruction shown in Fig. 17 is a circular disc of the same thickness as the distance of separation

17. Hele-Shaw flow past a circular disk.

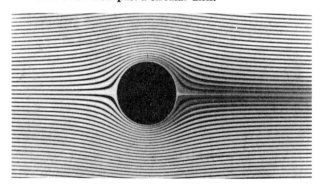

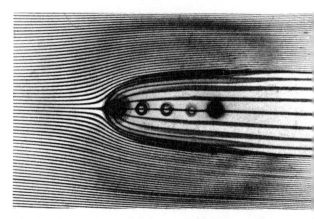

18. Blurring of streamlines of a Hele-Shaw source and uniform flow when the source strength is varied.

of the plates. That the visible streamlines are sharply defined is due to two causes. The first is that motion is steady, so that dye particles follow one another along a fixed streamline. The second is that all particles on a line perpendicular to the sheet move in the same direction, though at different speeds, so that to an eye observing the pattern along a line perpendicular to the glass sheets the streamlines at all depths are superposed when the motion is steady. If the obstacle were to move or change its shape during the experiment, the visible lines of colored particles would no longer be superposed and would appear blurred. This effect is shown in Fig. 18 where the obstacle has been replaced by a source flow. Internal streamlines are marked with red dye emitted from eight small holes at a small radial distance from the source. As long as the rate of delivery at the source is constant the boundary of the fluid originating there is like a fixed obstacle and the streamlines around it are sharply defined. If the rate of delivery is changing, as it was when Fig. 18 was recorded, the colored streams are no longer sharply defined, but they re-establish their definiteness in a new position when the rate of delivery at the source becomes constant at a new value. The interest of the Hele-Shaw cell is that the streamlines observed in this low-Reynolds-number flow have exactly the same shape as those predicted theoretically for two-dimensional flow of a fluid with no viscosity and therefore infinite Reynolds number.

References

Fundamental equations of viscous flows:
  Lamb, H., *Hydrodynamics*, Dover Publications, New York, 1945
  Goldstein, S., *Modern Developments in Fluid Dynamics*, Dover Publications, New York, 1965

More advanced techniques of solving particular problems:
  Happel, T. and Brenner, H., *Low Reynolds Number Hydrodynamics*, Prentice-Hall, Inc., New Jersey, 1965

Hydrodynamic lubrication:
  Tipei, N., *Theory of Lubrication*, Stanford University Press, 1962

Channel Flow of a Compressible Fluid

Donald Coles

CALIFORNIA INSTITUTE OF TECHNOLOGY

Introduction

The purpose of this film is to demonstrate several effects of compressibility which are important in any attempt to produce or control a supersonic internal flow. The heart of the subject is the behavior of a compressible gas flowing at high speed through a converging-diverging nozzle. The film therefore begins with the phenomenon of *choking,* or sonic flow at a throat. Some examples of shock waves and expansion waves are shown in passing. Finally, the film takes up the problem of bringing a supersonic stream efficiently to rest.

So that the whole flow can be viewed at one time, the demonstrations are performed in a channel small enough to fit within the 30-inch diameter beam of a conventional schlieren optical system. The apparatus is shown in Fig. 1. Room air enters the channel at the left and flows from left to right through the channel, through a control valve, and into a vacuum pump downstream. Each column on the manometer board shows the static pressure at the point directly above on the flat lower wall of the channel. The manometer is so connected that the mercury level rises when the pressure in the channel rises. The first column always

1. **Converging-diverging channel and manometer board at Aerophysics Laboratory, M.I.T.**

shows the upstream stagnation pressure (the pressure in the room), and the last column always shows the downstream stagnation pressure, just ahead of the valve.

The manometer display is readily interpreted with the aid of the one-dimensional momentum equation for steady flow. If viscous forces are neglected, the fluid acceleration depends on pressure forces alone:*

$$\rho V dV + dp = 0 \qquad \text{(Eq. 1)}$$

*Strictly speaking, Eq. (1) holds (except at discontinuities) along each streamline in steady inviscid flow. Provided that the streamline radius of curvature is large compared to the lateral extent of the flow region of interest, lateral changes in p, ρ, V, etc. can be neglected compared to changes in the direction of the flow. Some local effects of streamline curvature for the case of an incompressible fluid are demonstrated in the film "Pressure Fields and Fluid Acceleration" (Ref. 1).

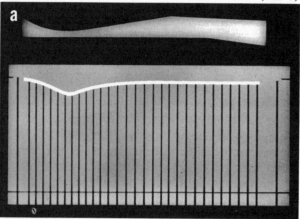

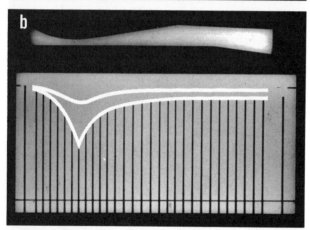

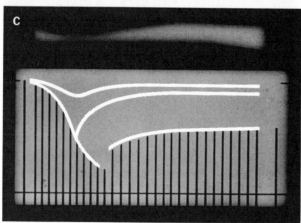

2. (a) Subsonic flow in entire channel; note symmetry of pressure distribution about the throat. (b) Choking at throat; note corner in pressure distribution. (c) Supersonic flow downstream of throat; pressure distribution no longer symmetrical. Farther downstream, the pressure rises abruptly. Note large loss of stagnation pressure.

Consequently, changes in pressure and velocity always have *opposite* sign, no matter what happens to the density. A falling pressure always means a rising velocity and vice versa.

Flow Near a Throat

At low rates of flow, the highest velocity in the channel is at the point of minimum area, called the throat (Fig. 2a). As this throat velocity is increased by opening the valve, a critical flow rate is reached, and the pressure distribution develops a sharp corner at the throat (Fig. 2b). As the valve is opened further, the symmetry disappears (Fig. 2c), and the pressure and velocity upstream of the throat no longer change. Just downstream of the throat, the falling pressure means that the velocity is increasing in the diverging part of the channel. Then the pressure rises almost discontinuously, but does not return to its original upstream value.

To show what is happening in the channel, the schlieren optical system* can be used to make density variations visible. As the valve is opened, a *shock wave* is seen to appear at the throat and to move into the diverging part of the channel. The position of the shock wave matches the discontinuity in pressure (Fig. 3).

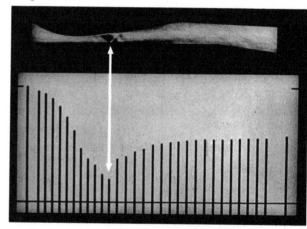

3. Schlieren visualization shows that position of shock wave matches discontinuity in pressure distribution. Subsonic flow exists downstream of shock wave and supersonic flow exists between throat and shock wave.

The appearance of a shock wave in the channel means that the flow is locally *supersonic*. The difference between the symmetrical flow in Fig. 2a and the unsymmetrical flow in Fig. 2c is that the velocity in the latter case is *sonic* at the throat. This transition

*Schlieren and other optical methods for flow visualization are described in many places; e.g., Ref. 2 & 3. For all the illustrations in these notes, the knife edge is parallel to the general flow direction. The boundary layers usually show up clearly, but the shock patterns in some cases do not. Other knife-edge positions could not be used because the thickness of the glass side walls, which were also in the schlieren beam, was highly non-uniform in the flow direction.

o sonic flow at the throat, with all of its consequences, is called *choking*.

Choking

To see why choking plays an important part in compressible channel flow, consider the one-dimensional momentum equation, Eq. (1), for steady flow without viscosity. For isentropic flow, this relation becomes

$$\frac{d\rho}{\rho} + M^2 \frac{dV}{V} = 0 \qquad \text{(Eq. 2)}$$

where M is the Mach number (the ratio of the velocity of the fluid to the local speed of sound).

For nearly parallel flow, changes in density and velocity can be related to changes in channel area by the one-dimensional continuity equation $\rho AV = $ constant, or, in differential form,

$$\frac{d\rho}{\rho} + \frac{dV}{V} + \frac{dA}{A} = 0 \qquad \text{(Eq. 3)}$$

At the throat, where the area is locally constant, the last equation becomes

$$\frac{d\rho}{\rho} + \frac{dV}{V} = 0 \quad \text{(at throat)} \quad \text{(Eq. 4)}$$

Eqs. (2) and (4) are a pair of homogeneous algebraic equations for $d\rho$ and dV at a throat. If the Mach number at the throat is *not* equal to one, the two equations can only be satisfied if $d\rho$ and dV are zero. That is, the velocity, density, and pressure are locally constant at the throat, and must be going through a maximum or a minimum there. The flow has a kind of local symmetry about the throat (Fig. 2a).

On the other hand, if the Mach number at the throat is *equal* to one, then Eqs. (2) and (4) become identical. *This is the special case called choking.* Since both $d\rho$ and dV cannot be determined from a single equation, they need no longer be zero at the throat. The velocity, density, and pressure can increase or decrease continuously through a sonic throat, and the flow need not be symmetrical. In particular, the highest velocity need not occur at the throat (Fig. 2c).

Subsonic and Supersonic Flow

To illustrate these ideas, and to test the accuracy of the one-dimensional approximation, a theoretical pressure distribution can be computed* for the choked

*The details of this calculation (for one dimensional isentropic flow in a channel of specified area) are described in any standard text on compressible fluid flow; e.g., Ref. 3 or Ref. 4. The nozzle used to illustrate choking in the early part of the film, incidentally, has one flat wall and is best described as a half nozzle. This geometry allows pressure changes along the nozzle to be spread out sufficiently so that details of the flow behavior near the throat can be easily seen.

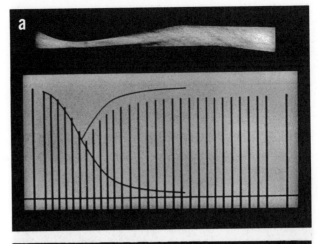

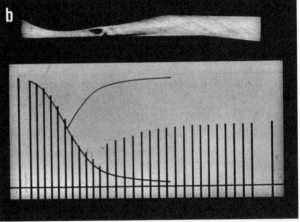

4. (a) Comparison of theory and experiment for subsonic flow just at choking. (b) Comparison of theory and experiment for supersonic flow after choking.

channel (Figs. 4a, 4b). The intersection of the curves marks the conditions for which air, accelerating from rest, should reach sonic speed ($p/p_o = 0.528$).

Downstream of the throat, there are two theoretical curves. The upper branch is *subsonic,* with the Mach number decreasing from one toward zero as the channel area increases. Such a subsonic flow behaves qualitatively as if the density were constant. The continuity equation, Eq. (3), is then satisfied (to a first approximation) by a balance between *area* and *velocity*. In a subsonic flow, the velocity and area change in opposite directions.

The small channel can also be used to demonstrate some properties of supersonic flow. As an aid in flow visualization, the flat wall of the channel is roughened with strips of plastic tape. There is no visible effect until the shock wave appears. Then characteristic disturbances, called Mach waves, appear in the region between the throat and the shock wave (Figs. 5a,

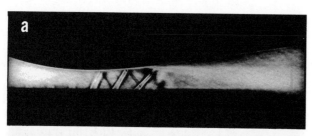

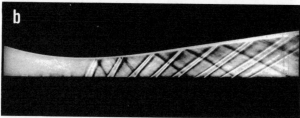

5. (a) Wave patterns confirming existence of supersonic flow downstream of throat. (b) The same flow as in (a) but with shock wave farther downstream.

5b).* These Mach waves, which originate at the edges of the plastic tape, are evidence of supersonic flow in the diverging part of the channel. So is the pressure distribution (Fig. 4b), because at first the pressure in the diverging part of the channel is decreasing, and hence the velocity is increasing [cf. Eq. (1)]. Downstream of the shock wave, the pressure is increasing and the velocity is decreasing in the same diverging channel, showing subsonic flow. The supersonic region ends at the shock wave.

The Mach waves and the pressure distribution thus confirm that the lower theoretical branch downstream of the throat (Fig. 4b) is *supersonic,* with the Mach number increasing from one as the channel area increases. To help in understanding this experimental fact of acceleration in a diverging channel, think of the extreme case where the Mach number is so large that the velocity is close to its maximum value for the energy available. But if the velocity is almost constant, the continuity equation, Eq. (3), is satisfied (to a first approximation) by a balance between *area* and *density*. The density goes down as the area goes up; the gas *expands* to fill the channel.

These results can be summarized in a single equation, obtained by eliminating $d\rho$ between Eqs. (2) and (3),

$$\frac{dV}{V} = \frac{1}{M^2 - 1} \frac{dA}{A} \qquad \text{(Eq. 5)}$$

*Strictly speaking, Mach waves are infinitesimal disturbances, and therefore cannot be seen. The waves in Figs. 5a and 5b obviously have finite strength; in fact, they remain visible after reflection from the upper wall. Note also that the waves appear to become weaker in the downstream part of the channel. This is partly because the density (and hence the schlieren sensitivity) is decreasing rapidly in the downstream direction in the supersonic region; but it is also partly because the initial disturbances (produced at the base of the boundary layer on the lower wall) become weaker as the boundary layer increases in thickness.

Eq. 5 is equivalent to the statement that *for steady acceleration from subsonic to supersonic velocity the area of the channel — or of any stream tube — must first decrease and then increase. There must be a throat. Since the fluid is accelerating at this throat, the Mach number there must be equal to one, and the channel is necessarily choked.*

Upstream Influence

The absence of upstream influence in a supersonic region is illustrated by the fact that the Mach waves in Fig. 5a are steady, although the shock wave itself is not steady.* When the downstream pressure in the choked channel is changed by the control valve, the shock wave moves, but there is no other influence on the flow upstream of the shock wave.** Because the area, velocity, and density are all fixed at the throat, so is the mass flow. A choked channel is therefore a convenient device for maintaining a specified mass flow from a reservoir which is held at constant pressure and temperature.

Flow at a Nozzle Exit

Even if the wall is cut away at some point downstream of the throat (Fig. 6), there is no change in the mass flow in a choked channel. As long as the back pressure, or chamber pressure, is lower than the nozzle exit pressure, downstream conditions do not affect the flow in the closed part of the channel. In

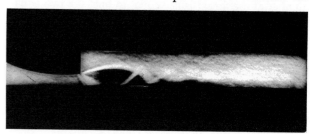

6. Flow in underexpanded nozzle, with wall cut away in supersonic region.

Fig. 6, the Mach number is about 1.2 at the nozzle exit, and the flow is adjusting itself to low back pressure through an expansion wave. A nozzle operating under these conditions is said to be *underexpanded.*

*The shock wave is not a single *normal shock,* but a complicated system of non-steady oblique shocks. This occurs because the boundary layers in the channel are relatively thick at the low Reynolds numbers which are typical of the small channel, and are separating from the walls. For more information about boundary layers and separation, see the films "Fundamentals of Boundary Layers," "Flow Visualization," and "Boundary-Layer Control" (Refs. 5-7).

**The sound track of the film includes, as background, the aerodynamic noise picked up by a microphone placed at the channel entrance. Noise from the valve and piping can only propagate upstream inside the channel if the entire flow is subsonic. The presence of supersonic flow is therefore accompanied by a decreased sound level.

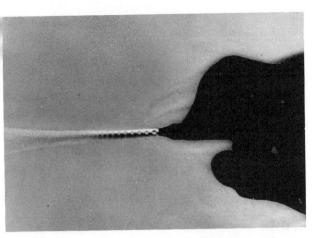

7. Wave pattern in jet from shop air hose and nozzle.

Fig. 7 shows an ordinary air hose and nozzle from a machine shop. When the jet is viewed with a schlieren optical system, a typical pattern of expansion and compression waves appears. These waves are the mechanism of adjustment to the ambient pressure.

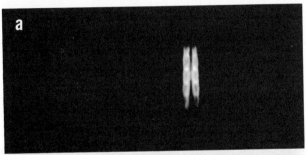

8. (a) Rocket exhaust just after liftoff. (b) Rocket at 30 miles altitude. (Courtesy of NASA, Marshal S.F.C.)

The same pattern can be seen in a rocket exhaust (Fig. 8a) when the outside pressure is not equal to the exit pressure. At very high altitudes, a poor match in pressure is unavoidable (Fig. 8b), and the expansion wave at the exit of an underexpanded nozzle may turn the flow through a spectacular angle.

Shock Waves

Converging-diverging nozzles are a means of *generating* supersonic flows. They can also be used to *slow down* supersonic flows. For example, large quan-

tities of air may have to be brought almost to rest in the engine inlet passages of a supersonic air-breathing vehicle. Especially at high Mach numbers, the engine power is likely to depend strongly on the efficiency of this deceleration and compression process.

Without careful design, supersonic flow in a channel tends to slow down abruptly, through a strong shock wave, with a large loss in stagnation pressure. The deceleration in a normal shock wave (Fig. 9) is

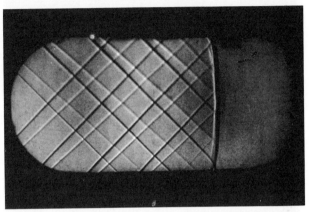

9. Normal shock wave at a Mach number of 1.5.* (Courtesy of von Karman Gas Dynamics Facility, ARO, Inc., Tullahoma, Tennessee.)

typically thousands of millions of gravities, and the word "shock" is certainly appropriate. There are very large gradients of velocity and temperature inside such a shock wave, and large losses in stagnation pressure due to viscous dissipation and heat conduction.

Supersonic Diffusion

Consider the problem of *controlling* a supersonic flow, in the sense of bringing it to rest as efficiently as possible. Just as a supersonic stream can be speeded up by increasing the channel area, it can be slowed down — at least in principle — by decreasing the area. A converging channel used for this purpose is called a *supersonic diffuser*. A suitable geometry for demonstrating this process in the small channel is shown in Fig. 10. Downstream of the inlet there is a short symmetrical nozzle, designed to give uniform flow at a Mach number of about 2.3 in the *test section* which follows. Then there is a supersonic diffuser, consisting of an adjustable *second throat* which can be opened or closed to vary the amount of supersonic compression. As a result, there are now *two* means of control over the flow in the channel; these are (a) the geometry of the second throat, and (b) the original control valve.

If the second throat is smaller than the first one, the

---

*This shadowgraph shows the central portion of a normal shock wave in a channel much larger than that of Fig. 1. The flow is therefore much less subject to boundary-layer separation and other adverse effects of viscosity. The Mach waves remove any uncertainty about the direction of flow.

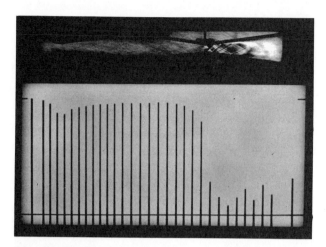

10. Choking at a second throat.

channel chokes at the *second* throat as the valve is opened (Fig. 10). The upstream flow is subsonic and independent of downstream conditions. All that the control valve can do is move the shock wave in the region downstream of the second throat without affecting the flow upstream of the shock.

If the second throat is slightly larger than the first one, the channel chokes at the *first* throat as the flow rate increases, and a shock wave appears in the nozzle

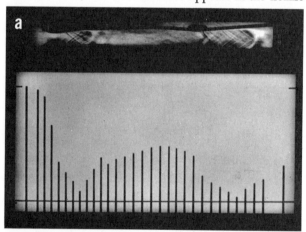

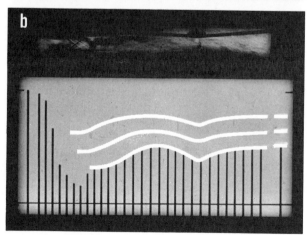

11. (a) Blocking in a channel with two throats. (b) Evolution of pressure distribution during blocking as control valve is opened.

and moves downstream. Then the second throat chokes (Fig. 11a), and it is no longer possible to affect the first shock wave with the control valve. Under these conditions the channel is said to be *blocked,* with the first shock wave *trapped* in the supersonic nozzle.

Blocking

To understand blocking, consider the original shock wave after it has formed just downstream of the first throat. This shock wave moves downstream as the control valve is gradually opened. The Mach number of the supersonic flow entering the shock wave there-

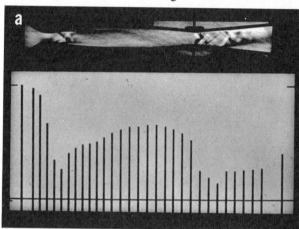

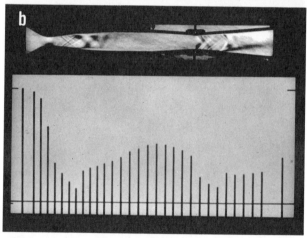

12. (a) Blocked channel with reduced second throat area; first shock has moved upstream (compare Fig. 11a). (b) Same flow with larger second throat area; first shock moves downstream.

fore increases, and the dissipation of energy in the shock wave also increases. As a result, the downstream stagnation pressure gradually decreases. Several stages in the evolution of the pressure distribution are shown in Fig. 11b. Since the *mass flow* is fixed — because the first throat is choked — and since the *area* of the second throat is constant, the fact that the *density* is going down means that the *velocity* at the second throat must be going up. Finally the second throat goes sonic, or chokes, and the channel is blocked (Fig. 11a).

The alternative means of control over the flow can

be demonstrated by changing the area of the second throat while the channel is blocked (Figs. 12a, 12b). The shock wave in the nozzle moves, even though the motion of the walls is hardly visible. As long as the flow is sonic at the second throat, the density there has to change when the area does, because of the fixed mass flow. The shock wave in the nozzle must therefore move in such a way that the proper stagnation pressure and density through the second throat are maintained.*

Starting

To avoid blocking the channel, the second throat must be opened a little more, so that it will not choke even for the strongest possible shock wave (in this case at the nominal test-section Mach number of 2.3). As the control valve is opened, the first throat chokes, but

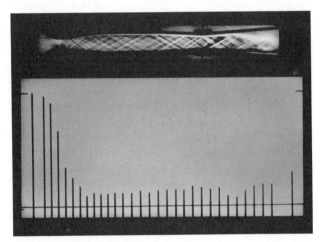

13. Started or running channel.

the second throat does not. As the back pressure continues to decrease, the shock wave moves downstream in the channel . . . and suddenly jumps through the second throat to the subsonic diffuser (Fig. 13). The channel is now said to be *started,* or running; the original shock wave has been *swallowed* by the second throat. The wave pattern ahead of the second throat, and the pressure distribution, both show that the flow is now supersonic all the way from the first throat to the terminal shock wave.

If the control valve is now slowly closed, the terminal shock pattern is moved upstream toward the second throat, to a position where the Mach number is lower than the test section value because of the supersonic compression. Too large an increase in back pressure will unstart the channel, or disgorge the shock wave, but the channel can readily be restarted by again opening the valve.

*Notice that the control valve, which is a sonic orifice of variable area, has been operating in this manner all along, acting as a second or third throat in the flow channel. The different function of such a control valve in a choked channel **and** in an unchoked one should be carefully noted.

Changes of Second Throat Area

When the channel is running, the argument about blocking (in terms of the relative area of the second throat) does not apply, since there is no strong shock wave in the nozzle to lower the stagnation pressure of the flow. However, if the second throat is gradually closed, the flow becomes blocked almost immediately. A close-up view (Fig. 14a) shows part of the reason why the second throat cannot be closed to the same area as the first one. Room must be allowed for the

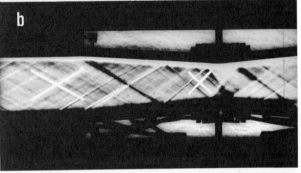

14. (a) Overlaid dashed lines show thickness of boundary layers in supersonic diffuser. Boundary-layer region is dark on the lower wall and light on the upper wall because the schlieren knife-edge is parallel to the walls. (b) Overlaid crossed lines call attention to change in wave angle and therefore to decrease in Mach number in supersonic diffuser.

wall boundary layers, which are very thick in the supersonic region and which are not carrying their share of the mass flow. With the second throat closed almost to the point of blocking, it can be deduced from the wave angles (Fig. 14b) that the Mach number in the test section is about 2.1, while the lowest supersonic Mach number obtainable at the second throat is about 1.4.

In normal operation of a supersonic channel, the channel is first started, and then the second throat is closed almost to the point of blocking the channel (Fig. 15a). Advantage is then taken of the supersonic compression by closing the control valve a little, to move the terminal shock pattern upstream to a lower Mach number. The result is an optimum running condition, with a relatively small dissipation of energy (Fig. 15b). The advantage of using an adjustable second throat to

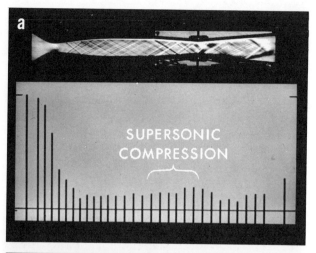

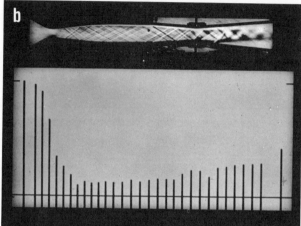

15. (a) **Started channel with second throat closed almost to point of choking.** (b) **Same flow but with optimum pressure ratio; note increase in downstream stagnation pressure compared to 15a.**

control the air flow is that the stagnation pressure losses, and hence the power required to run the channel, can be reduced below the requirement for starting.

Summary

This film has illustrated three important phenomena which occur in adiabatic channel flow. *Choking* implies sonic flow at a throat, with an upstream flow which is independent of downstream conditions (e.g., Figs. 5a, 10). *Blocking,* or choking at two throats in succession, occurs if the second throat area is too small (e.g., Fig. 11a). *Starting* implies swallowing of a shock wave (e.g., Fig. 13), and is a necessary step in the achievement of supersonic compression or diffusion in channel flow of a compressible fluid.

References

1. Shapiro, A. H. "Pressure Fields and Fluid Acceleration" (a NCFMF film)

2. Beams, J. W. "Shadow and Schlieren Methods," in Vol. IX of *High Speed Aerodynamics and Jet Propulsion,* Princeton U. Press, 1954

3. Shapiro, A. H. *The Dynamics and Thermodynamics of Compressible Fluid Flow,* Ronald Press, New York, 1953

4. Liepmann, H. and Roshko, A., *Elements of Gasdynamics,* Wiley, New York, 1957

5. Abernathy, F. H. "Fundamentals of Boundary Layers" (a NCFMF film)

6. Kline, S. J. "Flow Visualization" (a NCFMF film)

7. Hazen, D. C. "Boundary-Layer Control" (a NCFMF film)

Vorticity

Ascher H. Shapiro
MASSACHUSETTS INSTITUTE OF TECHNOLOGY

Introduction

The experiments in this film illustrate the concepts of vorticity and circulation, and show how these concepts can be useful in understanding fluid flows.

The vorticity is defined as the curl of the velocity vector: $\omega = \nabla \times \mathbf{V}$. Thus each point in the fluid has an associated vector vorticity, and the whole fluid space may be thought of as being threaded by vortex lines which are everywhere tangent to the local vorticity vector. These vortex lines represent the local axis of spin of the fluid particle at each point. In two dimensions, *the vorticity is the sum of the angular velocities* of any pair of mutually-perpendicular, infinitesimal fluid lines passing through the point in question. For rigid body rotation, every line perpendicular to the axis of rotation has the same angular velocity: therefore the vorticity is the same at every point, and is twice the angular velocity.

Vorticity is related to the moment of momentum of a small spherical fluid particle about its own center of mass. Given some very complicated motion of a liquid, suppose that it were possible — by magic — suddenly to freeze a small sphere of the liquid into a solid, while

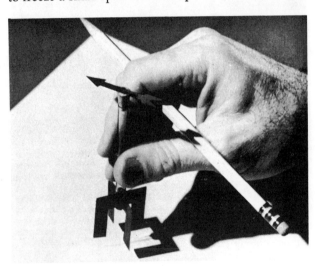

1. **Vorticity meter. The four vanes at the bottom are rigidly attached at right angles to the vertical glass tube. The arrow is fixed to the tube and rotates with approximately the average angular speed of the pair of mutually-perpendicular fluid lines which coincide with the vanes. Thus the rate of rotation of the arrow is approximately half the vertical component of vorticity of the lump of water in which the vanes are immersed. Note that since the vanes are rigidly connected, the float does not respond to shear deformation of the two fluid lines, but only to their average angular velocity.**

conserving the moment of momentum. The angular velocity of the solid sphere at the moment of its birth would be exactly half the vorticity of the fluid before freezing. Several dynamical theorems in effect relate the changes in vorticity of a fluid particle — and thus of its moment of momentum — to the moments of the forces acting on that fluid particle.

Fig. 1 shows a "vorticity meter" which floats in water with its axis vertical. Placed in a liquid which is in solid-body rotation (Fig. 2), the vorticity float

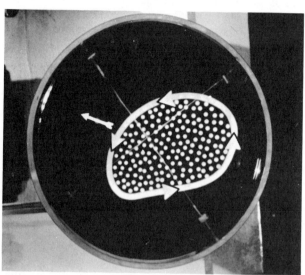

2. **View from above of an open cylindrical tank containing water. It is mounted on a turntable which rotates about a vertical axis. The cross hairs are scribed on the bottom of the tank. After a long time the water is brought into solid-body rotation by viscous forces, and the vorticity float (arrow) moves as though it were rigidly affixed to the cross hairs. The drawing superimposed shows a closed curve C on which the circulation Γ is reckoned, and the associated vorticity flux through the area bounded by C.**

moves as though it were rigidly attached to the rotating tank. It has the angular velocity of the tank; the vertical component of vorticity at every point in the fluid is twice this angular velocity.

Sometimes the word "rotation" is used as a synonym for vorticity, but this does not mean that a flow has to be curved for vorticity to be present. For instance, Fig. 3 shows water flowing in a straight channel. The

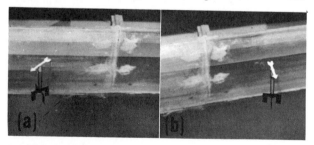

3. **Water flowing from left to right in a channel with straight vertical walls. The vorticity meter is placed in the viscous boundary layer near one wall. As it moves downstream, the arrow turns counterclockwise. Picture (b) was taken a short time after picture (a).**

streamlines are essentially straight and parallel to the side wall. But the rotation of the arrow shows that vertical vorticity is present. Near the wall is a viscous boundary layer in which the velocity increases with

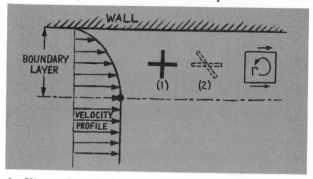

4. **Viscous boundary layer near plane wall. Fluid cross at position 1 (solid lines) changes in form as it moves to position 2 (dashed lines). Square at right shows the closed circuit on which circulation is reckoned.**

distance from the wall (Fig. 4). Examine the fluid cross. One leg moves downstream parallel to the wall while the other leg rotates counterclockwise owing to the non-uniform velocity distribution. Thus there is a net vorticity, and the vorticity meter of Fig. 3 turns counterclockwise.

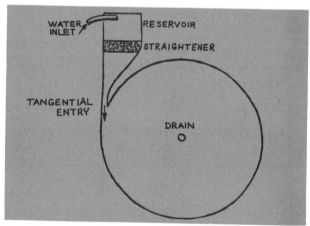

5. **Plan view of sink-vortex tank. Water from the reservoir passes through the flow straightener, and thence into the tank through the tangential entry. After spiralling round and round in a very tight vortex, the water leaves vertically through the drain at the center.**

On the other hand, the flow may be without rotation even though the streamlines are curved. Fig. 5 shows in plan view a tank for producing a sink vortex in which the streamlines are tight spirals and nearly circular. As shown in Fig. 6, the vorticity meter moves in a circular path but does not rotate. It moves in pure translation — as would a compass needle on a phonograph turntable. Consider a fluid cross at a point on a circular streamline (Fig. 7). Leg A follows the streamline, hence it rotates counterclockwise. Since the angular momentum of the fluid is conserved as it flows toward the drain, the tangential velocity varies

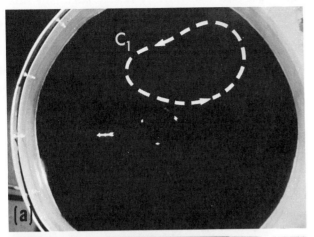

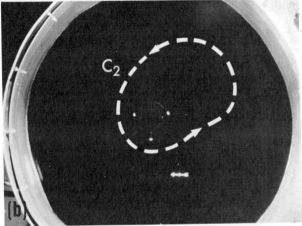

6. Plan view of vorticity meter in the sink-vortex tank. Picture (b) was taken a short time after picture (a); the arrow direction has not changed. The drawings superposed show closed curves, C_1 and C_2, on which the circulation is reckoned for circuits which respectively do not enclose, and do enclose, the center of the vortex.

inversely with the radius. Thus the velocity of the inner part of leg B is greater than the velocity of the outer part, and leg B turns clockwise. The clockwise turning rate of B is just equal and opposite to the counterclockwise turning rate of A. Hence the vorticity is zero. The vorticity meter, in averaging the rotations of legs A and B, translates, without rotation, on a circular trajectory.

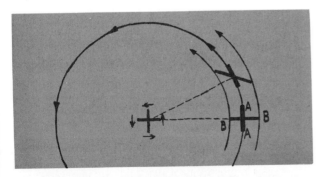

7. Fluid crosses in a two-dimensional vortex. One is on a circular streamline, the other at the center.

Crocco's Theorem

For the special case of steady motion of an incompressible, inviscid fluid acted on by conservative body forces, *Crocco's theorem* has the form

$$\mathbf{V} \times \omega = \frac{1}{\rho} \nabla p_0 \, ; p_0 \equiv p + \tfrac{1}{2}\rho V^2 + \rho U \quad (1)$$

where $\mathbf{V}$ is the vector velocity, ω the vector vorticity, and ρ the density. The stagnation pressure p_0 is the sum of the static pressure p, the dynamic pressure $\rho V^2/2$, and the potential energy per unit volume $\rho\, U$ associated with the conservative body-force field.

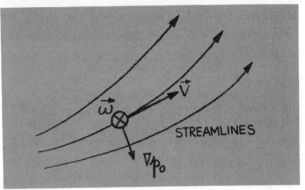

8. In a two-dimensional flow, $\mathbf{V} \times \omega$ is in the plane of the flow and perpendicular to the streamlines.

When a flow is two-dimensional in the plane of the paper, the vorticity vector is normal to the paper while the velocity vector lies in the paper and along the streamline (Fig. 8). By Crocco's theorem, the gradient of stagnation pressure is normal to both the velocity vector and the vorticity vector; thus it lies in the plane of the paper and normal to $\mathbf{V}$. Consequently the stagnation pressure, p_0, is constant along each streamline and varies between streamlines only if vorticity is present.

To illustrate, consider again the straight boundary layer of Figs. 3 and 4. The static pressure is uniform across the boundary layer but the velocity is variable. Thus the stagnation pressure is variable, and, by Eq. 1, vorticity is present. The velocity gradient is strongest near the wall and so is the gradient of stagnation pressure. When the vorticity meter is near the wall, the rate of spin is relatively large. With the vorticity meter farther out in the boundary layer, the rate of spin is smaller.

When the vorticity is zero, as in the sink-vortex tank (Figs. 5 and 6), Crocco's theorem says that the stagnation pressure must be everywhere the same. The spiral of the vortex is so tight (Fig. 9) that it is not much of a liberty to think of the streamlines as being concentric circles. One may verify that the uniformity of angular momentum, $Vr = \text{const.}$, is the condition for, (1) the stagnation pressure to be constant throughout, and (2) the free surface to be a hyperboloid

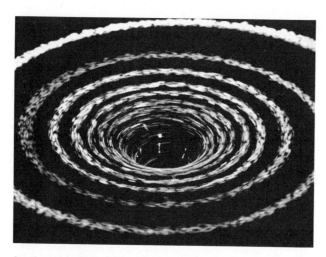

9. A streamline of the sink vortex. Note hyperboloidal depression in center, over the drain.

of revolution. For an inviscid fluid, the hole in the core would extend downward indefinitely. However, the high-velocity gradients and strain rates near the axis produce large viscous forces which reduce the depression to a deep dimple having a bottom.

A flow which is otherwise without rotation may contain small regions where the vorticity is very large. In the sink-vortex flow, for instance, the vorticity is generally zero (Fig. 6), except for a highly-concentrated core of vortical fluid right at the center. When the vorticity float finally drifts into the center, its motion, which hitherto was purely translational, is immediately changed to a pure and rapid rotation. Only at the singular point in the center of Fig. 7 do both arms of a fluid cross rotate in the same direction and thus produce a net vorticity.

Fluid Circulation

The fluid *circulation* Γ is defined as *the line integral of the velocity* $\mathbf{V}$ *around any closed curve* C. The circulation theorem — which is purely geometrical — equates the circulation Γ around C to the flux of the vorticity vector ω, through any surface area bounded by C.

$$\Gamma = \oint \mathbf{V} \cdot d\mathbf{r} = \iint \nabla \times \mathbf{V} \cdot d\mathbf{A} = \iint \omega \cdot d\mathbf{A} \quad (2)$$

If there is a definite circulation around C, then the fluid lying in any surface bounded by C must have vorticity. When the circulation is zero for *every* curve in a certain region, the fluid in that region must be entirely free of vorticity: the motion is then called *irrotational*.

Returning to the boundary-layer flow of Figs. 3 and 4, consider the circulation for the small square circuit in Fig. 4. Because of the non-uniform distribution of speed, there is a net circulation which, by Eq. 2, is related to the vertical vorticity of the enclosed fluid.

Vorticity may be distributed throughout the entire fluid. But often the vorticity is very large only in a thin thread of fluid while the remaining fluid is virtually without vorticity. Then we can simplify our thinking by lumping all the vorticity into a concentrated vortex line around which the fluid spins (Fig. 10), and by pretending that the remaining fluid is entirely free of vorticity. The finite amount of circulation around the core requires that the vorticity be infinite in the vortex line, which has zero cross-sectional area. In cross section, a straight vortex line with non-vortical fluid outside would appear as a point around which the fluid moves in concentric circles, the circumferential velocity varying inversely with radius.

In the solid-body rotation tank (Fig. 2), the vertical vorticity is everywhere equal to twice the angular velocity of the tank. Every horizontal circuit therefore has a circulation equal to twice the product of the angular velocity and the area bounded by the circuit.

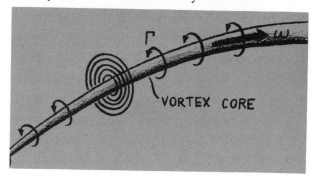

10. Schematic of a vortex core of strength Γ imbedded in otherwise irrotational fluid.

In the sink-vortex tank the flow is non-vortical except for the concentrated vortex core which accounts for the whole circulatory motion. All fluid circuits not surrounding the core (Fig. 6a) have zero circulation because they contain no vorticity flux. All fluid circuits surrounding the core (Fig. 6b) have the same circulation because they contain the entire vorticity flux.

A wing generates lift because of the higher pressure below and the lower pressure above. According to Bernoulli's integral, the velocity on the upper surface must be greater than the velocity on the lower surface. This means that there is a net circulation around a lifting wing. Often we model this circulation as being produced by a fictitious vortex which is "bound" in the wing and which accounts for the circulatory movement (Fig. 19). The vorticity is really present, but it is distributed throughout the viscous boundary layer rather than concentrated in a single vortex line.

Kelvin's Theorem

The concept of circulation is important mainly because of a powerful theorem evolved by Lord Kelvin from the dynamical laws of motion. It shows how the time rate of change of circulation Γ_c associated with closed curve C always made up of the same fluid particles is governed by the torques produced by all the forces acting in the fluid:

$$\frac{D\,\Gamma_c}{D\,t} = -\oint \frac{dp}{\rho} + \oint \mathbf{G}\cdot d\mathbf{r} + \oint \frac{\mu}{\rho}\nabla^2\mathbf{V}\cdot d\mathbf{r} \quad (3)$$

The three terms on the right represent torques due to pressure forces, body forces, and viscous forces, respectively.

Viscous Torques

Let us consider first the torques produced by viscous forces acting on a fluid particle. A force diagram for a fluid particle (Fig. 11) shows that viscous forces are

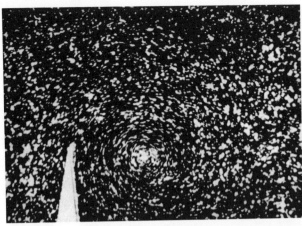

12. Fluid flows from left to right past a sharp edge. The photo shows conditions soon after the flow has started impulsively. The flow separates behind the edge, and fluid in the surface of discontinuity, in which strong viscous forces act, forms a starting vortex which moves downstream. (After Prandtl.)

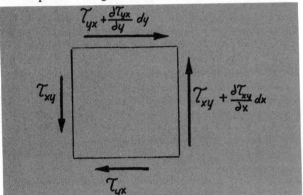

11. Viscous stresses on a fluid particle.

indeed capable of producing torques about the center of mass. Such viscous torques change the vorticity of the fluid particle and thus the circulation on a bounding circuit. For instance, in the straight channel of Fig. 3, a vorticity float inserted just outside the boundary layer moves downstream for a while without turning. But vorticity is diffusing outward from the wall, and eventually the fluid in which the float travels reaches a position where a viscous torque produces vorticity. Then the float begins to spin.

When the water in the cylindrical tank illustrated in Fig. 2 is at rest, $\Gamma = 0$ for all circuits. When the turntable is started, the water in the middle does not at first move, because none of the forces which create circulation come into play there for a while. Next to the wall, the fluid moves promptly because of viscous stresses. These viscous forces (aided by outward flow in the boundary layer on the bottom of the tank) gradually accelerate fluid farther out from the wall and more and more of the fluid moves. In the end, viscosity brings all the fluid into a perfect solid-body rotation. At this limit, paradoxically, viscous forces have vanished altogether and there is nothing to force further changes in circulation. If the turntable is then stopped, the fluid continues to rotate as a solid body except near the walls of the tank where the viscous forces are large. Viscosity again causes the vorticity change to diffuse inward from the wall, decreasing the circulation more

and more until finally the whole fluid is brought to rest.

When a fluid flows around a sharp edge (Figs. 12 and 23b), viscous and pressure forces in the boundary layer lead to a separated flow. The fluid which has been affected by viscous forces forms a concentrated vortex.

In Fig. 23b, two vortices are made by pulling a plate with sharp edges through the water. We can make the vortex visible in this experiment by placing a ping-pong ball in the dimple of the vortex. It remains there stably. In a channel of constant width and depth, the spin of the ball decreases with time because viscosity diffuses the vorticity of the core into the surrounding fluid.

Body-Force Torques

If the body force $\mathbf{G}$ is *irrotational* (i.e. curl $\mathbf{G} = 0$), that is, *conservative,* the body-force term in Eq. 3 is zero. But, for *rotational* body forces, that is, *nonconservative* forces, this term is, in general, not zero.

Whenever the net body force $\mathbf{G}$ passes through the center of mass of a small sphere of fluid, it produces no torque to change the circulation. Centrally-directed forces like gravity are of this type. They are irrotational.

There are two important rotational body forces in fluid mechanics which can change circulation: (1) *Coriolis forces,* $(-2\,\Omega\times\mathbf{V})$, in rotating reference frames, and (2) *Lorentz forces,* $(\mathbf{J}\times\mathbf{B})$, due to the flow of an electric current at an angle to a magnetic field. In both these cases the line of action of the resultant body force need not go through the center of mass of a spherical particle. Because of such forces the oceans and the atmosphere are full of vorticity, as are magnetohydrodynamic flows.

Does the vortex in the bathtub always turn in the same direction? Does it depend on which hemisphere you are in? You can't really tell in the bathtub, because the Coriolis force due to the earth's rotation, for

13. A tank six feet in diameter and six inches high, with a drain hole ⅜ inch in diameter at the center, is filled with water swirling *clockwise*. It is then covered to minimize motions induced by air currents, by buoyancy, and by impurities on the surface causing non-uniform surface tension, and it is allowed to stand for 24 hours. The flow is started by pulling a plug from the end of a hose, several feet long, attached to the drain. The experiment is carried out at latitude 42° N near Boston, Mass.

a speed toward the drain of about 0.2 inches per minute, is only about a billionth the force of gravity! Other effects all too easily mask that of the earth's rotation. However, with care, one can do an experiment dominated by the earth's Coriolis force (Fig. 13). Immediately after starting the flow, a small vorticity float with its vanes entirely submerged is placed over the drain. For the first 10 or 15 minutes there is no perceptible rotation of the float. But at about 15 minutes a distinct counterclockwise motion begins. At 24 minutes, with the tank nearly empty, the float is turning at about 0.3 rev./sec. This represents a 30,000-fold amplification of the earth's rotation at Boston. The reader can verify that this agrees, in order of magnitude, with the angular momentum being conserved.

The Coriolis force acting on a fluid particle in the northern hemisphere as it moves radially inward toward

the drain is circumferential and counterclockwise. This force integrated around a circle contributes a counter clockwise torque in Kelvin's theorem. This tends to make the circulation increase counterclockwise with time. In the reference frame of the earth a fluid circle which starts at one radius with zero circulation therefore acquires counterclockwise circulation as time proceeds.

Although the earth's Coriolis forces are small compared with gravity, they are extremely important to our everyday weather. They can also generate hurricanes. If the conditions of temperature and humidity are such that there is a strong local up-draft in some region, air must rush in from the sides to the "sink" forming the up-draft. This is like the water-tank experiment, but upside down, and a strong vortex is formed (Fig. 14).

Pressure Torques

When the fluid is effectively incompressible, or more generally when the density depends upon pressure alone, the term $\oint dp/\rho$ is zero, and thus does not change the circulation.

To see the physical significance of this term, consider the fluid particle in the circular region of Fig. 15. The lines of constant pressure (isobars) are shown by the solid lines. The pressure forces acting on the particle parallel to these lines exactly cancel each other, hence the net pressure force on the particle is perpendicular to the isobars and passes through the geometric center. The dashed lines represent the contours of constant density (isochors) in the fluid particle. If the isochors are parallel to the isobars — a situation described as *barotropic*, which means that ρ is a function of p alone — the line of action of the net pressure force goes through the mass center M, and produces no torque about M. But, if the isochors are *not* parallel to the isobars (Fig. 15), the net pressure force pro-

14. Funnel of a tornado.

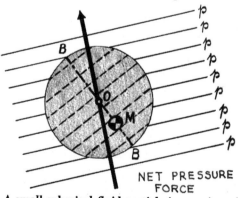

NET PRESSURE
FORCE

15. A small spherical fluid particle in a region where the isobars are the solid lines marked *p*. The net pressure force on the particle is perpendicular to the isobars and passes through the geometric center *O*. The dashed lines represent the isochors in the fluid. The center of mass *M* lies on the line *BB*, which is perpendicular to the isochors and passes through *O*.

luces a torque about M, and acts to change the circulation.

If the fluid is at rest in the gravity field of the earth, the isobars are horizontal. Since the circulation in a stationary fluid is forever zero, the surfaces of constant density must coincide with the surfaces of constant pressure. This is why the free surface of water in a pail is horizontal. If the pail contains oil floating on water, the interface is also horizontal. When we tilt the free surface and the interface by tipping the pail, the surfaces do not remain tilted: the isobars and isochors are now misaligned and the term $\oint dp/\rho$ in Eq. 3 creates a circulation which tends to make the surfaces horizontal again.

The circulations arising in natural convection systems are driven by the pressure-density term in Kelvin's theorem. One example is the hot-water or hot-air heating system in a house. Another is the principal circulation of the earth's atmosphere between the cold regions at the poles and the hot regions at the equator.

Origin of Irrotational Flow

Kelvin's theorem shows how irrotational flows may arise. Consider a motion which began from a state of rest. With no motion, there is no vorticity, and with no vorticity, there is no circulation. Suppose that the fluid is barotropic, that body forces if present are conservative, and that viscous forces are negligible. Then the circulation must forever remain zero on every fluid circuit, and the vorticity must also everywhere and forever remain zero.

Let an airfoil begin to move suddenly through a fluid initially at rest. In the absence of viscosity, the circulation around any arbitrary fluid element is zero to begin with and therefore remains zero; thus the flow remains everywhere irrotational. There can be no circulation around the airfoil, and hence no lift. Fortunately, viscous friction, no matter how small, together with the no-slip condition at the solid surface, make lift generation possible. When the airfoil begins to move, viscous effects near the trailing edge result in the shedding of a so-called "starting vortex" (Fig. 16), and thus to a circulation on the curve $ABCDA$. But the fluid along the larger curve $ABCDEFA$ is not subject to viscous forces, being outside the viscous boundary layer and wake. By Kelvin's theorem, the circulation on this curve must be zero. For this to be true there must be along the curve $ADEFA$ surrounding the airfoil a circulation equal and opposite to that on the curve $ABCDA$ surrounding the vortex. This circulation around the airfoil may be ascribed to a fictitious vortex bound in the airfoil, and is necessary for the production of lift.

When the airfoil stops, the bound vortex is shed as

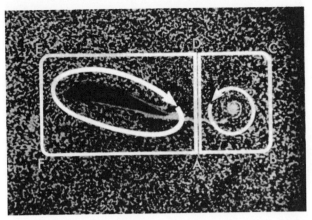

16. An airfoil impulsively started from right to left sheds a "starting vortex" at the sharp trailing edge. (After Prandtl.)

a stopping vortex (Fig. 17), again because of viscous forces at the sharp trailing edge. On a circuit around either vortex, viscosity has acted and circulation is present. But on a circuit enclosing both vortices, and passing through fluid on which friction has never acted, the circulation remains zero. The equal and opposite vortices produce zero net flux of vorticity through any area containing both vortices.

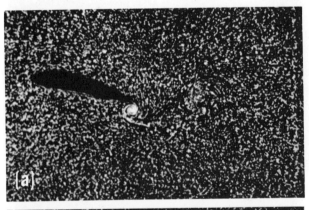

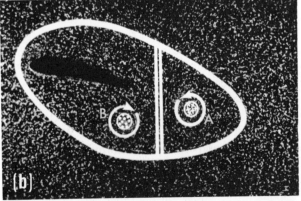

17. Shortly after the airfoil of Fig. 16 is impulsively started, leaving a *starting vortex* "A", it is impulsively stopped, and sheds the *stopping vortex* "B" of opposite sign to "A" but of equal strength (after Prandtl). (a) Immediately after stopping. (b) A short time later.

Helmholtz's Vortex Laws

When all the torque-producing factors in Kelvin's theorem are absent, fluid dynamics can be given a beautiful geometrical interpretation in terms of Helmholtz's laws:

(1) Vortex lines never end in the fluid. They either form closed loops or end at a fluid boundary, and the circulation is the same for every contour enclosing the vortex line.

(2) A fluid line which at any instant of time coincides with a vortex line will coincide with a vortex line forever. (The vortex lines are, as it were, frozen to the fluid.)

(3) On a vortex line of fixed identity, the ratio of the vorticity to the product of the fluid density with the length of the line remains constant as time proceeds ($\omega/\rho l =$ const.). Thus, if the vortex line is stretched, the vorticity increases.

The vortices A and B of Fig. 17b are of equal and opposite strength. They move downward together, because the velocity field of B displaces the fluid at A, and vice versa, and because each vortex core is convected with the fluid to which it is frozen.

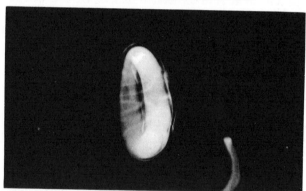

18. Smoke ring.

One can never see a smoke ring which is broken somewhere, because vortex lines can never end (Fig. 18). The fact that the smoke ring propels itself shows that the vorticity is frozen in the fluid: each element of fluid is propelled forward by the induced velocity fields of all the other elements of the ring vortex, and the whole vortex is thus convected by itself. The smoke, which marks the fluid, is carried with the vortex core, showing that the vorticity is locked to the fluid. When a smoke ring approaches a wall normal to the axis of the ring, it spreads out and slows down. This may be explained in terms of the induced velocity field of the fictitious image vortex on the other side of the wall, which, in effect, takes the place of the wall itself.

Fig. 19 shows the vortex system for a wing of finite span. The circulation required to produce lift may be considered as originating from vorticity bound in the wing. But the bound vortex lines cannot end at the wing tips; they form vortex loops, closed by trailing

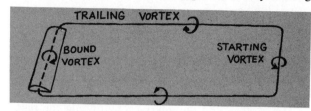

19. The vortices associated with a wing of finite span.

vortices and the starting vortex left at the airport. The trailing vortices from the tips of the lifting wing are made visible in Fig. 20. As the angle of attack of the wing is increased, the tip vortices grow in strength as the lift, the circulation, and the strength of the bound vortex also increase.

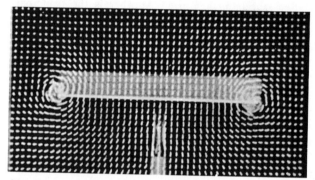

20. A view looking upstream along the axis of a wind tunnel. Between the camera and the trailing edge of the wing is a rectangular grid of fine wires, in a vertical plane, with wool tufts attached at the net points. The wool tufts align themselves with the flow, which is more or less perpendicular to the paper. The relatively heavy, horizontal white line is the trailing edge of the wing, and the dim white region above is the upper surface at incidence. One sees the projections of the wool tufts in a plane transverse to the direction of free flow and downstream of the trailing edge, and thus obtains an impression of the transverse velocity field. (Courtesy NASA.)

The little vorticity meter of Fig. 21 shows streamwise vorticity. Put right behind the wing tip, it spins very fast; if we move it slightly inboard or outboard of the tip, it hardly turns. The trailing vorticity is strongly concentrated at the wing tip.

The concentration of trailing vorticity in strong tip vortices results in very low pressures in the center of these vortices. Behind marine propellers, the water in the vortex cores may boil (cavitate), and this makes the trailing vortices from the blades easily visible (Fig. 22). They form a helical pattern.

The induced velocity field of the vortex loop of Fig. 19 produces downwash velocities within the enclosed region. These downwash velocities, which are observable in the tuft pattern of Fig. 20, act at the wing itself. They make the wing appear to be flying through air which is itself descending, and this results in what

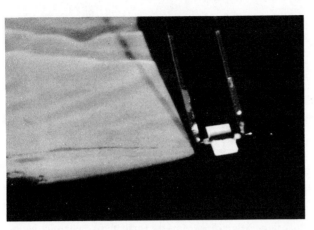

21. A vorticity meter whose axis is aligned with the flow shows the streamwise component of vorticity. It is here located behind the trailing edge of a wing, near the wing tip. Flow is from left to right.

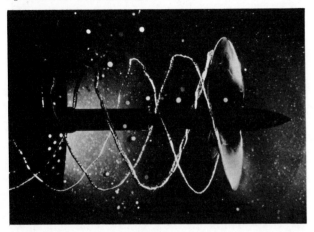

22. Trailing vortex system from a marine propeller in a water tunnel, made visible by cavitation in the vortex cores. (Courtesy M.I.T. Propeller Tunnel.)

is called induced drag. The work of the forward-moving wing as it moves against this induced drag force accounts for the kinetic energy being fed into the constantly-lengthening system of trailing vortices.

A plausible explanation for the V formation of migrating birds is that each bird takes advantage of the upwash velocities in the trailing vortex systems of the ones forward of it. Each bird behind the leader flies on an ascending induced air current, while the leader has not only his own induced drag, but additional induced drag due to the downwash of all the birds behind him.

The vortex laws of Helmholtz come from the dynamical equations of motion. Therefore anything we deduce from the vortex laws can also be deduced, although perhaps not as conveniently, from the pressure field. With the lifting wing, for example, air can leak around the tip from the high-pressure region below to the low-pressure region above. This leakage produces a transverse flow which accounts for the tip vortex.

To show the effect of stretching vortex lines, water

is caused to flow over a hump (Fig. 23), and a vortex with vertical axis is observed. As it approaches the crest of the hump, the spin of the ping-pong ball decreases. This agrees with Helmholtz's third statement,

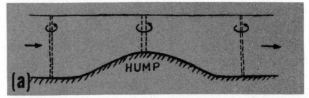

23a. Water flows sub-critically from left to right over a hump in an open channel. The water depth first decreases to the top of the hump, then increases.

23b. Two vertical vortices are made by a plate with sharp edges, which is then withdrawn, and a ping-pong ball is placed in the dimple of one of the vortices. The rate of spin of the ball gives an indication of the vorticity in the vortex core.

inasmuch as the lengths of the vertical lines of fluid to which the vorticity is attached are also decreasing. When the vortex goes down the hump, the lengths of the vortex lines increase, and the rate of spin of the ball is seen to increase. The change in spin rate as the ball goes down the hump, however, is not as strong as the change going up the hump. That is because viscosity is always acting to *decrease* the rate of spin of the ping-pong ball. There is a simple mechanical explanation of this experiment. When a vertical thread of fluid moves up the hump, its length decreases, but the volume of the thread remains the same; hence its diameter increases. For moment of momentum to be conserved, the spin rate must decrease. When the vertical thread of fluid goes down the hump, it stretches and becomes thinner. Accordingly, the rate of spin increases. A figure skater or ballet dancer knows this mechanical trick instinctively. She speeds up in a pirouette by moving her arms and legs inward to decrease her moment of inertia.

Turbulent flows are full of vorticity. The vortex lines are like tangled spaghetti. The mutually-induced velocities of these vortex lines cause some of them to lengthen, and this lengthening produces a finer-grained turbulence with higher velocity gradients. This makes for added viscous dissipation. As L. F. G. Richardson has put it, "Big whirls make little whirls,/which feed on their velocity./Little whirls make lesser whirls,/and so on, to viscosity."

Secondary Flow

The generation of secondary flows is illustrated in the curved channel flow of Fig. 24. Upstream, the flow is parallel to the walls of the channel. Because of

24. Water flows around a bend, from upper right to lower left, in an open channel of trapezoidal cross-section. (a) A floating ping-pong ball entering the bend, near the inside of the bend. (b) The same ball a few seconds later. It is approaching the outer wall.

the boundary layer on the bottom of the channel the upstream flow has a horizontal component of vorticity in a direction transverse to the flow, but the vorticity components in the vertical and streamwise directions are both zero. When the streamwise vorticity meter shown in Fig. 21 is aligned with the flow at the exit of the bend, it spins, showing that somehow a streamwise component of vorticity has been generated in the bend. The same result is shown by the drift of the ping-pong ball to the outer side of the bend, while heavy oil droplets rolling along the floor drift to the inside of the bend. To see why, consider three perpendicular fluid lines A, B and C, in the upstream position (Fig. 25). If viscous forces are neglected compared with inertial forces in the bend, the vortex lines may be

thought of as frozen to these three fluid lines. Notice, though, that upstream there is no vertical component

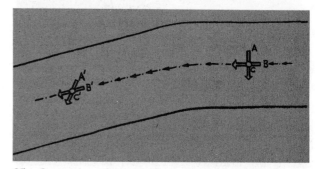

25. Generation of streamwise vorticity in a channel bend.

of vorticity C, nor any streamwise component of vorticity B. There is only a transverse component of vorticity A associated with the viscous boundary layer on the floor of the channel. Now, to first order, C moves along the center streamline to C', so there is no vertical component of vorticity at C'. This means that the *average* turning of the lines A and B must be zero. Line B follows its streamline, rotating counterclockwise, hence line A must rotate clockwise by an equal amount into position A'. One consequence is that the velocity at the inside of the bend is greater than at the outside. But, more interestingly, the vortex line A' now has *two* components: one transverse to the local flow, which existed upstream, and a new component *along* the flow. This streamwise component — the secondary vorticity — swirls the flow clockwise as one looks downstream, and explains the observed motions of the vorticity meter, the ping-pong ball, and the oil droplets.

Such secondary flows often occur in curved channels. When a river goes around a bend (Fig. 26), the secondary flow erodes the outer bank and deposits sand and pebbles on the inner bank. This tends to accentuate the curve of the bend. This may be one of the mechanisms by which exaggerated cases of river meandering occur (Fig. 27). When a pipe is curved, there

26. Secondary flow currents in a river bend.

27. Meandering of a river.

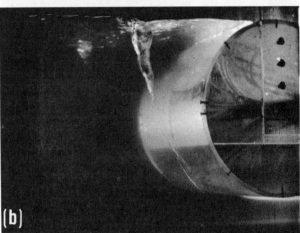

are two cells of secondary flow (Fig. 28). These secondary flows carry high-energy fluid from the middle of the pipe to the walls, thereby increasing the frictional losses.

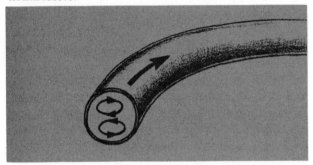

28. Secondary flow in a curved pipe.

The vortex laws are also illustrated by flow in an open channel when it is forced to pass under a transverse circular cylinder (Fig. 29). Strong vortices are developed from the vertical vorticity present in the viscous boundary layers on the two side walls. When a vertical vortex line generated by viscosity on a side wall is convected downstream with the fluid to which it is frozen, it is not only bent around the cylinder but also greatly stretched. The stretching of the vortex line intensifies the vorticity and makes a vertical vortex core visible upstream of the cylinder. The bending of the line produces a streamwise component of vorticity downstream of the cylinder.

Vertical vorticity was introduced by stirring in Fig. 30. Then the plug was pulled from the bottom. A strong vortex formed very quickly. A vertical column of fluid initially on the axis is also a vortex line. When the flow begins, this column is enormously lengthened and the vorticity increases proportionately, according to Helmholtz's third law. A similar mechanism underlies the formation of tornadoes as air with angular momentum flows inwards to the ascending vortex core. The centrifugal pressure field of the vortex creates such

29. Water flowing from left to right in a horizontal channel passes under a transverse circular cylinder which acts as a sluice gate. Vertical vorticity is present in the boundary layers on the vertical side walls, having been generated by viscosity over a considerable distance upstream. The stretching of vortex lines is evidenced by the vertical vortex core upstream of the cylinder (b). The bending of the vortex lines is evidenced by the spinning of the streamwise vorticity meter held in the downstream flow (a).

30. A beaker of water with a small hole in the bottom, initially plugged. The water is stirred with a rod, and the plug is then removed.

a very low pressure at its center that houses over which the eye of the tornado passes literally explode.

In the bathtub vortex experiment of Fig. 13, the fluid initially had a very small vorticity in inertial space due to the turning of the earth. When the plug was pulled the vertical fluid thread on the axis was enormously lengthened and the vorticity in inertial space was strengthened proportionately, finally to the point where we could see it in the reference frame of the earth. Photographs of the earth's cloud cover taken from orbiting satellites show that similar events occur in the earth's atmosphere on a grand scale.

References

1. Prandtl, L., *Essentials of Fluid Dynamics*, Chapter II, Hafner Publishing Co., N. Y., 1952.

2. Prandtl, L., and Tietjens, O. G. (Translated by Rosenhead, L.): *Fundamentals of Hydro- and Aeromechanics*, Chapter XII, McGraw-Hill Book Co., N. Y., 1934.

3. Lamb, H., *Hydrodynamics*, Chapters III and VII, Dover Publications, N. Y., 6th edition, 1945.

4. Eskinazi, S., *Vector Mechanics of Fluids and Magneto Fluids*, Academic Press, N. Y., 1967.

5. Batchelor, G. K., *An Introduction to Fluid Dynamics*, Cambridge University Press, 1967.

Fundamentals of Boundary Layers

Frederick H. Abernathy

HARVARD UNIVERSITY

Introduction

In potential flows, which assume an ideal fluid without viscosity, only pressure and inertia forces determine the flow dynamics (Fig. 1). Real fluids do have viscosity, and the flow field can be very different (Fig. 2). *Boundary layers,* thin layers of fluid in which vis-

1. Potential flow streamlines about a thin plate attached to a cylinder.

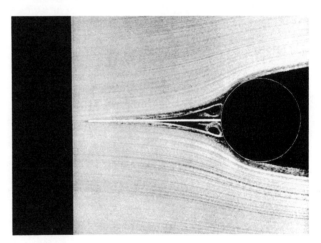

2. Hydrogen-bubble visualization of water flowing past the object in Fig. 1.

cosity effects are significant, are formed along solid boundaries. In some cases these boundary layers, under the influence of pressure gradients, significantly affect the entire flow field.

The streamline pattern of a real fluid, air, flowing past an airfoil at a small angle of attack (Fig. 3) is very nearly what one would predict from inviscid flow

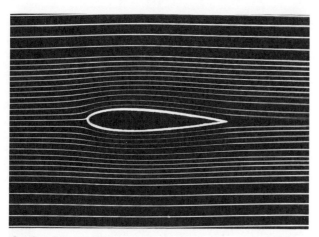

3. Smoke visualization of air flow past an airfoil at a small angle of attack.

theory. Because the Reynolds number is large, the influences of viscosity are confined to a narrow region close to the surface of the wing. The primary effect of viscosity is to create a drag force on the wing through the integrated effect of surface shear stresses. When the angle of attack of the airfoil is increased, viscous effects become very pronounced and change the flow field in a qualitative manner. Pressure gradients imposed on the boundary layers become so large

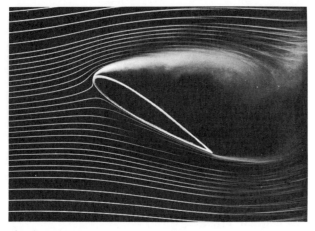

4. Same airfoil, at a large angle of attack.

that separation of the boundary layer occurs on the upper surface (Fig. 4). A region of recirculating flow is formed over most of the upper surface of the wing, which is then said to be stalled. An understanding of how viscous forces can influence the entire flow field, as shown in Figs. 2 and 4, is intimately related to an understanding of the behavior of boundary layers. The film shows the causes of boundary layers, how they grow, how they respond to pressure gradients, and the differences in the behavior of laminar and turbulent boundary layers.

Flow Along a Flat Plate

We first examine a boundary-layer flow where pressure gradients are negligible — two-dimensional uni-

form flow over a long flat plate (Figs. 5 and 6). The flow is visualized by hydrogen bubbles generated by electrolysis along wires oriented perpendicular to the plate. Upstream of the plate the front and back edges of hydrogen-bubble patches remain perpendicular to the streamlines, showing that the flow is uniform and free of vorticity (Fig. 5). Downstream of the leading

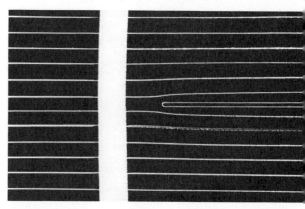

5. Flow approaching a flat plate in a water channel.

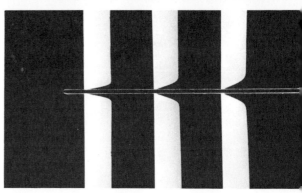

6. Timelines produced at wires perpendicular to the plate correspond closely to velocity profiles.

edge of the plate the flow is still uniform and free of vorticity except in a narrow region adjacent to the plate (Fig. 6). This narrow region containing vorticity is the viscous boundary layer. In this layer, both viscous forces and inertial forces are important. Outside of this boundary layer viscous forces can be neglected.

The experimental fact that there is no slip between the plate and the layer of fluid immediately adjacent to it is shown in Fig. 6. The velocity of the liquid at the surface of the plate is zero. This is called the *no slip boundary condition* of viscous flow.

The thickness of the boundary layer increases along the length of the plate. Physically, fluid deceleration is transferred successively from one fluid layer to the next by viscous shear stresses acting in the layers. The boundary-layer thickness is sometimes defined as the distance, δ, from the surface to where the velocity, U, reaches some fixed percentage (say 95%) of the free-stream value (Fig. 7a). The local shear stress, τ, is related to the velocity gradient normal to the surface

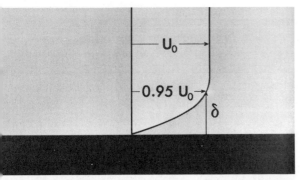

a. Definition of boundary-layer thickness δ.

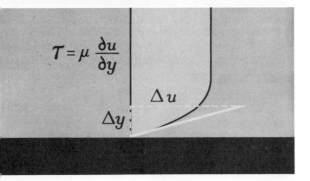

b. Relationship of shear stress τ to velocity gradient
t wall.

. Boundary-layer velocity profiles near the leading edge
left) and far downstream (right).

$y \, \tau = \mu \, \dfrac{\partial u}{\partial y}$, where μ is the fluid viscosity (Fig. 7b).
"he composite photograph of Fig. 8 compares the
elocity profiles at upstream and downstream stations
along the plate. The velocity gradient at the wall is
:ss downstream than upstream, indicating that the
~all shear stress decreases along the plate.

One way to understand the mechanism of boundary-
ayer growth is to consider the time history of the
orticity within the boundary layer. Stokes' theorem
:ates that the area integral of the vorticity vector, ω,
ounded by a closed contour, is equal to the line inte-
ral of the velocity vector around the bounding con-
our, which is called the circulation, Γ. (See Fig. 9.)

$$\oint \omega \cdot d\mathbf{a} = \oint \mathbf{V} \cdot d\mathbf{s} = \Gamma$$

n other words, the circulation around a closed contour

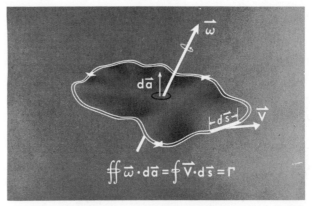

9. The area integral of vorticity ω equals the line integral
of the velocity around the bounding contour.

is the sum of the vorticity enclosed within it. The
contour shown in Fig. 10 is at the upstream station;
it is of unit length along the plate, and more than a
boundary-layer thickness high. The free-stream velocity is parallel to the top of the contour but is directed
in the opposite sense. This contributes $-U_o$ times a
unit length to the value of the circulation. The components of vertical velocity along the right and left
parts of the contour are virtually zero and, because
there is no slip, the velocity contribution to the circulation at the surface is exactly zero. Therefore, the
total circulation is $-U_o$ times a unit length. At any
downstream station the circulation is also equal to
$-U_o$ times a unit length. Therefore the total amount
of vorticity within each contour is the same. Because
there is no vorticity upstream of the plate and because
the circulation per unit length along it is constant, we
conclude that all of the vorticity in the boundary layer
is introduced at the leading edge as a consequence of
the no-slip boundary condition.

Even though the total amount of vorticity contained
in the boundary layer per unit length of the plate is
the same, the distribution of vorticity normal to the
plate does change along its length. Viscosity acts,
through the mechanism of molecular diffusion, to
spread the vorticity transversely as it is convected
downstream. The local boundary-layer thickness can
be thought of as a measure of the distance vorticity has

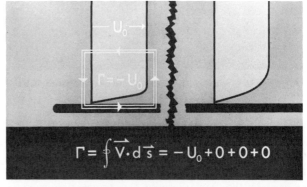

10. Evaluation of the circulation per unit length about
a contour at the upstream station.

diffused away from the plate. We can relate the factors controlling this growth process in the following approximate way by considering that the transverse diffusion length, δ, is of the order of $\sqrt{vt}$, where v is the kinematic viscosity and t is the time of diffusion. At

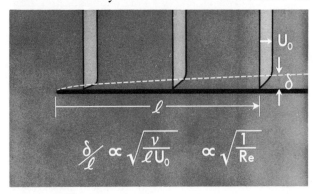

11. Boundary-layer growth along a flat plate.

a distance l from the leading edge, the time during which vorticity has diffused is approximately $t \simeq l/U_o$ (Fig. 11). Thus $\delta/l \propto \sqrt{\dfrac{v}{lU_o}} = \sqrt{\dfrac{1}{\mathrm{Re}}}$. This relationship is valid only at high Reynolds numbers, where $\delta/l << 1$. As an example of the Reynolds-number dependence, note that increasing the flow velocity decreases the boundary-layer thickness at a given station along the plate. With a higher main-stream velocity, at any position along the plate the boundary layer thickness is less because it has had less time to grow.

Favorable Pressure Gradients

The pressure gradients in the flow direction along the flat plate in Figs. 5 and 6 were negligibly small. In most other flow situations there are regions of decreasing pressure and regions of increasing pressure in

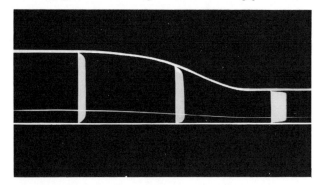

12. Flow in a converging channel (two-to-one contraction ratio).

the flow direction. By using the two-to-one contracting flow channel of Fig. 12 and observing the behavior of the boundary layer along the flat side, we can examine the effects of a pressure distribution which decreases in the flow direction (a *favorable* pressure gradient).

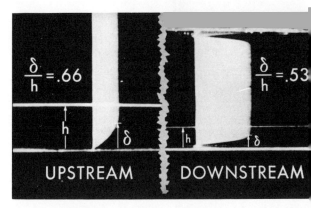

$\dfrac{\delta}{h} = .66$ $\quad h \quad \delta \qquad\qquad h \quad \delta$ $\dfrac{\delta}{h} = .53$

UPSTREAM DOWNSTREAM

13. Composite blowup of upstream and downstream boundary-layer profiles from Fig. 12.

The boundary layer upstream of the contracting portion of the flow channel is much thicker than the boundary layer emerging from it (Figs. 12 and 13). Most of this decrease in boundary-layer thickness through the flow contraction is attributable to the two-to-one decrease in flow area. However, using the local distance h from the lower wall to the nearby stream line as a reference dimension (Fig. 13), we see that the boundary-layer thickness relative to this dimension has also decreased. This decrease in relative thickness of the boundary layer can be explained using the vorticity arguments just developed.

The amount of vorticity contained in a contour of a unit length along the plate and of height h is twice as large downstream as it is upstream because the free-stream velocity has doubled through the contraction. This new vorticity is of course added to the boundary-layer fluid at the wall. It is as though a new boundary layer were being created within the older one at each increment along the way. The combined profile at the exit is relatively thinner because there has been little time for lateral diffusion of the new vorticity in the boundary layer. Downstream therefore, a larger percentage of the total vorticity is near the wall than upstream. This results in a relatively thinner boundary layer.

Instead of discussing vorticity concentration and diffusion, the same conclusions can be reached using force arguments. At each differential increment in distance along the contracting portion of the channel the pressure gradient causes a corresponding incremental increase to the main flow velocity, according to Bernoulli's equation. In the outer portions of the boundary layer, where changes in shear stress are small, the velocity increases by almost the same increment. It is only very near the wall that the incremental increase in velocity is substantially different from the free-stream value. The fluid velocity at the wall remains zero because of the no-slip condition. It is as though a new and therefore thin boundary layer were being added to the existing one at each step along the contraction.

The integrated effect is to enhance the already high shear stress near the wall and decrease the lateral distance required for the velocity to attain 95 per cent of the free-stream value.

Unfavorable Pressure Gradients

In the slightly divergent channel (a *diffuser*) of Fig. 14, the free-stream static pressure *increases* in the flow direction, thereby subjecting the wall boundary layers to a positive (or *unfavorable*) pressure gradient. If the unfavorable gradient is small enough (as it is in

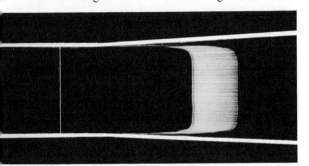

14. Flow in a small-angle diffuser.

the flow of Fig. 14), then the increasing pressure in the free stream causes a corresponding decrease in the free-stream velocity, increases the boundary-layer thickness, and decreases the wall shear stress, *without causing flow separation*. These effects can be deduced from either of the two arguments used in the previous section. Using pressure and velocity arguments it follows that the positive pressure gradient decreases the free-stream velocity and decreases the boundary-layer velocity by almost the same increment except very near the wall. The size of the increment decreases rapidly near the wall and must be zero at the wall. A major consequence of this incremental decrease in velocity is to decrease the velocity gradient

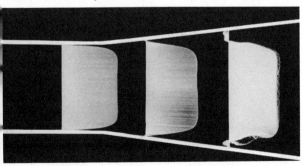

15. A diverging channel with a larger diffuser angle.

or shear stress at the wall. This change in boundary-layer profile can be seen by comparing the profiles at the first two stations of Fig. 15.

The deceleration of the flow imposed by a positive pressure gradient cannot be very large or sustained too long by the boundary-layer fluid without the wall shear stress going to zero, followed downstream by local

flow reversal. For the small-angle diffuser of Fig. 14, the positive pressure gradient is very small, and no flow separation occurs. The large-angle diffuser of Fig. 15 imposes a larger positive pressure gradient which the boundary layer cannot sustain without separating from the wall between the second and third stations in Fig. 15. At the second station the flow near the wall is to the right, while at the third station the flow near the wall is to the left. The point on the wall where the fluid in the upstream boundary layer meets the fluid from the region of flow reversal is called the separation point. The wall shear stress is zero there. Downstream of this point the fluid which was in the upstream boundary layer is no longer in contact with the wall, and is separated from it by the region of reversed or

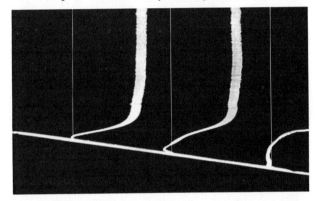

16. Bubbles generated at three wires in the downstream section of the diffuser of Fig. 15 show reversed flow near the wall.

recirculating flow. The boundary layer is said to have *separated*. The point of separation of the laminar boundary layer in Fig. 16 is just upstream of the first bubble wire. A comparison of the flow fields of Figs. 3 and 4 illustrates the enormous changes that boundary layer separation can cause.

Laminar to Turbulent Transition

In most practical situations the Reynolds number is large and the boundary layers are turbulent rather than laminar. Stages in the transition from a laminar to a turbulent boundary layer are shown in Fig. 17. In Fig. 17 a slight adverse pressure gradient causes transition to occur within the field of view. The steps in the transition are complicated and interdependent. First, there is the growth of nearly two-dimensional waves, Tollmien-Schlichting waves, followed by the appearance and growth of three-dimensional disturbances, which contain streamwise vorticity. Further downstream turbulent spots can be seen. Finally, fully turbulent flow appears.

The transition process is influenced by many factors: free-stream disturbances, plate roughness, pressure gradients, vibration, sound, etc. Therefore, the position where the transition process starts varies with time in a random way.

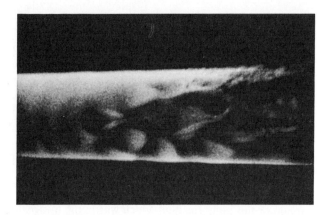

17. Side view of a long cylinder* with its axis aligned with an air flow. The faired nose of the cylinder is out of view to the left (upstream). A sheath of smoke generated upstream develops patterns which show stages of boundary-layer transition. (Courtesy F. N. M. Brown, University of Notre Dame.)

*The cylinder appears to be tapered because the camera is looking at a slight angle upstream.

Placing an obstruction in a boundary-layer flow stimulates the naturally occurring processes and hastens the onset of transition. In Fig. 18, the boundary layer on the lower wall of the diffuser has been made turbulent by inserting a trip rod upstream. The turbulent boundary layer is able to withstand the adverse pressure gradient in the diffuser and does not separate, while the laminar boundary along the top wall is separated, with reverse flow along the wall (Fig. 18a).

18a. Flow in channel of Fig. 15 with lower boundary layer made turbulent.

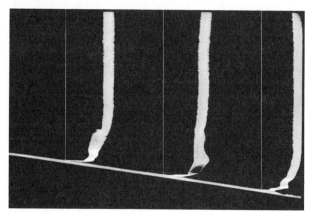

18b. Three bubble wires show unseparated, turbulent flow along the bottom wall.

In the turbulent boundary layer the flow is downstream (compare Figs. 18b and 16), and no flow reversal is evident.

A Turbulent Boundary Layer Along a Flat Plate

Laminar and turbulent boundary layers are different, and the differences explain why a turbulent boundary layer is able to withstand without separating a larger unfavorable pressure gradient than a laminar boundary layer. Consider again the flow along a long

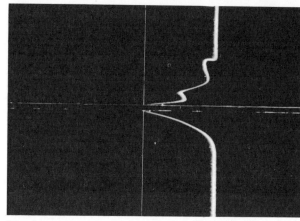

19. Instantaneous displacement profiles for flow along a thin plate. The boundary layer on the upper surface has been made turbulent, while the flow along the lower surface is laminar.

flat plate. In Fig. 19, the boundary layer on the lower side is laminar and two-dimensional; the boundary layer on the upper side has been tripped by a wire upstream and is turbulent. The motions in the turbulent boundary layer are unsteady and three-dimensional. Some motions are perpendicular to the plane of view. Because the displacement of a bubble line

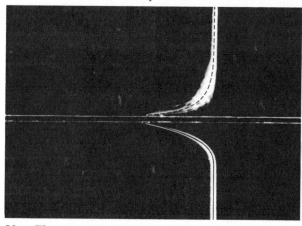

20a. The upper boundary layer is turbulent; the lower, laminar. Superposition of many instantaneous velocity profiles suggests mean velocity profiles.

corresponds closely to an instantaneous velocity profile, superimposing a number of individual displacement lines provides a method of obtaining a mean velocity profile for the turbulent layer. The superposi-

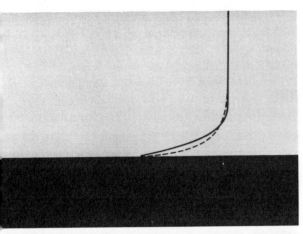

20b. The mean laminar (solid) and turbulent (dashed) profiles are compared.

tion also gives an experimental notion as to where the turbulent fluctuations occur and how large they are in the plane of mean motion. Figure 20a was constructed by such a superposition — Fig. 20b compares the mean laminar and turbulent profiles.

The velocity gradient perpendicular to the plate is larger for the turbulent layer than for the laminar layer (Fig. 21), and therefore the turbulent layer has

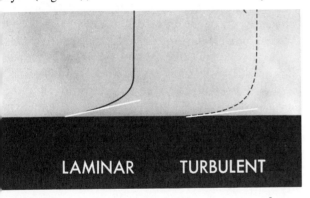

LAMINAR   TURBULENT

21. Velocity gradients for the profiles are compared.

the larger wall shear stress or drag. The circulation is the same for both layers, since the free-stream velocity is the same. Both boundary layers therefore contain the same total amount of vorticity per unit length of the plate. However, the distributions of vorticity in the two layers are very different. In the turbulent layer more vorticity is concentrated near the plate, even though some vorticity has also spread farther from the plate (Fig. 20b).

The distribution of momentum in the two boundary layers is also different. In the turbulent layer high-momentum fluid is transported toward the plate, and low-momentum fluid is transported away from the plate, by unsteady random rotary motions associated with vorticity aligned in the flow direction. There is more momentum near the wall in the turbulent boundary layer, even though the turbulent boundary layer is thicker. In the diffuser experiment (Figs. 18a and

b), the extra momentum near the wall in the turbulent boundary layer along the bottom wall enabled it to withstand the unfavorable pressure gradient without separating.

Similarly, turbulent boundary-layer flow on the upper surface of an airfoil delays large-scale separation, or stall, until higher angles of attack are reached. Vortex generators, small blades set perpendicular to the surface of airplane wings, are often used to delay the onset of separation. They are so named because they introduce additional axial vorticity which enhances the naturally occurring rotary momentum interchange in already turbulent boundary layers, and thereby increase the momentum of the fluid near the surface.

Summary

At large Reynolds numbers, boundary layers, thin layers of fluid in which viscosity effects are significant, are formed along solid boundaries, because viscous fluids cannot slip at solid boundaries. In the absence of pressure gradients the boundary layer along a flat surface increases in thickness as $l\sqrt{\dfrac{1}{\mathrm{Re}}}$. Negative (or favorable) pressure gradients in the flow direction, which accelerate the flow, decrease the boundary-layer thickness and increase the velocity gradient at the wall. Positive or unfavorable pressure gradients tend to decelerate the flow, to increase boundary-layer thickness, and to decrease the velocity gradient at the wall. Unfavorable pressure gradients can cause boundary-layer separation, which often results in drastically altered flow patterns and losses in performance of such devices as airplane wings and diffusers.

At relatively low values of Reynolds number, boundary layers tend to be laminar. At higher Reynolds numbers, a boundary layer is unstable to small disturbances. The disturbances grow, resulting in transition to a turbulent boundary layer. Most practical flow situations involve high Reynolds numbers and turbulent boundary layers. Because of three-dimensional interchanges of momentum, a turbulent boundary layer is thicker and has a larger wall velocity gradient than a laminar layer at the same Reynolds number. The increased momentum near the wall allows a turbulent boundary layer to withstand a larger unfavorable pressure gradient than a laminar layer without separating, but results in higher wall shear stress and drag.

References

1. Schlichting, H., *Boundary Layer Theory*, McGraw-Hill, 1960.
2. Hazen, D. C., "Boundary-Layer Control" (An NCFMF Film).
3. Goldstein, S., *Modern Developments in Fluid Mechanics*, Dover, 1965.

Turbulence

R. W. Stewart
UNIVERSITY OF BRITISH COLUMBIA

Introduction

"Turbulence" is not easy to define, but it is nearly ubiquitous. Tobacco smoke, industrial smoke, milk mixed into tea, all reveal turbulent motion.

Turbulent flows have common characteristics, one of the clearest of which is *disorder*. Fig. 1 shows a sheet of tiny bubbles advected by a channel flow. The

1. Channel flow visualized by hydrogen bubbles released uniformly from a wire stretched across the flow.

disorder is so central that no matter how carefully one reproduces the boundary conditions, the flow is never reproduced in detail. On the other hand, *averages,* such as the mean speed of flow or correlation functions, are very well defined and "stable."

There are disordered fluid motions — for example some fields of water waves or of acoustic waves — which we prefer to exclude from the definition of turbulence, since they do very little *mixing* and mixing is an essential feature of turbulence. Thus disorder is necessary but not sufficient for description. A further characteristic of turbulence is the presence of vorticity, distributed continuously but irregularly in all three dimensions.

We can borrow a word from pathology and give a defining *syndrome,* or set of symptoms, for turbulence. It has disorder, irreproducible in detail, performs efficient mixing and transport, and has vorticity irregularly distributed in three dimensions. This distinguishes turbulence from various kinds of wave motion and excludes two-dimensional flows. Something like turbulent motion *can* occur in two dimensions; large-scale weather systems have some of this character. However, in strictly two-dimensional flows vorticity behaves as a scalar, and there is no vorticity production by vortex line stretching.* Thus the characteristics of two-

*The constant density vorticity equation is

$$\frac{D\omega}{Dt} = \omega \cdot \nabla \mathbf{V} + \nu \nabla^2 \omega.$$

In two dimensions the first term on the right must vanish, since ω is everywhere perpendicular to the plane of the flow. The equation then becomes exactly analogous to that for a conservative diffusible scalar like heat.

imensional flows are quite different from those of hree-dimensional turbulent flows.

Reynolds Number

Some flows are clearly turbulent. Others, with similar boundary conditions, are equally clearly not. What determines whether a flow is turbulent?

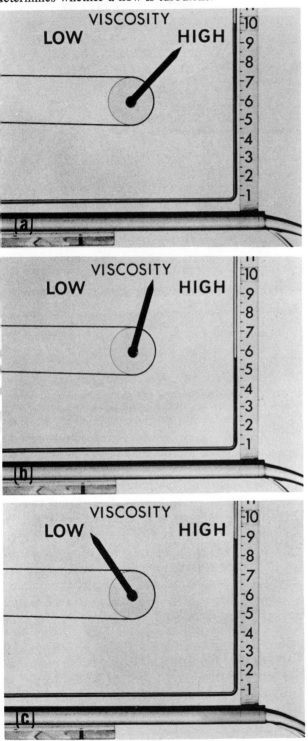

2. Pressure drops across a tube at constant flow rate and variable viscosity. In (a) and (b) the flow is laminar. In (c) it is turbulent.

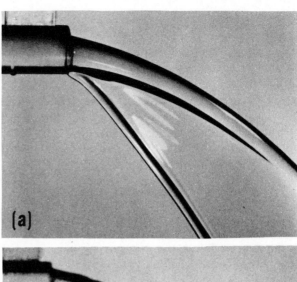

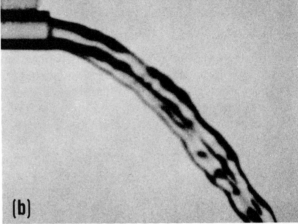

3. Flow issuing from the end of the pipe shown in Fig. 2. In (a) the flow corresponds to Fig. 2(a) and (b). In (b) the flow corresponds to Fig. 2(c).

In the film this question is discussed with the aid of an apparatus like that used by Hagen in the middle of the nineteenth century for a study of flow through pipes. A mixture of glycerin and water is pumped at a constant rate through a tube some 4 m long and 3 cm in diameter. At the downstream end the liquid issues into the free atmosphere. The pressure drop in the tube is shown by a manometer (Fig. 2a) which is tapped at an upstream position, and carried down to the open end. By varying the ratio of glycerin and water, the viscosity of the fluid can be controlled. According to the Hagen-Poiseuille law for laminar flow of a Newtonian fluid flowing through a circular pipe of length L and radius r, $(L \gg r)$, the pressure drop is given by $\Delta P = 8\mu LQ/\pi r^4$, where μ is the viscosity of the fluid and Q the volume flow rate.

As is seen in Fig. 2 (a) and (b), when the viscosity is reduced somewhat the pressure drop decreases, consistent with this formula. However, when the viscosity is reduced still further, as shown in Fig. 2 (c), we find that the pressure drop *increases*.

At the higher viscosities the flow issuing from the end of the pipe (Fig. 3a) is smooth and steady. At the lowest viscosity (Fig. 3b) high-speed photography reveals a time-dependent irregularity of the edges of the stream. Thus when we pass to the lowest viscosity we find that the Poiseuille law is not obeyed; instead of decreasing, the pressure drop increases. The flow in the pipe has become turbulent, revealed both by the irregular motion of the outcoming stream and by the greatly increased pressure drop down the tube.

In the early 1880's Osborne Reynolds did a series of experiments on flow through tubes and came to the conclusion that the criterion for the onset of turbulence depended upon a dimensionless function of the flow parameters which has since been called Reynolds number.

There is usually some arbitrariness in the choice of parameters for the definition of Reynolds number. For pipe flow we may take it as: $\mathrm{Re} = \dfrac{VD}{\nu}$, where D is the tube diameter, V is the average speed of the flow and ν is the kinematic viscosity. Although the question is still under investigation, it seems that if the Reynolds number so defined is appreciably less than 2000, the flow is not turbulent and perturbations are damped out by viscosity. At higher Reynolds numbers the flow may or may not be turbulent. Poiseuille's relation corresponds to a solution of the dynamical equations which is valid at all Reynolds numbers. At sufficiently large Reynolds number, however, this flow is unstable to certain perturbations. Whether or not a particular pipe flow is turbulent depends upon the length of the pipe and upon the nature and amplitude of perturbations, as well as upon the Reynolds number. If great care is taken to reduce such perturbations it is possible to push the Reynolds up to the neighborhood of 100,000 without turbulence.

Mixing

If a thin streamer of dye is introduced into the flow, as in Fig. 4 (a), mixing can be examined. At low Reynolds number the dye filament maintains its identity with very little change right to the end of the tube, as shown in Fig. 4 (b). The only mixing is molecular, so the process is very slow. If the Reynolds number is increased, perturbations can be seen in the dye flow, and at the onset of turbulence it seems to explode; the dye is rapidly mixed across the tube, as in Fig. 4 (c).

We can regard the increase in pressure drop with the onset of turbulence, shown in Fig. 2, as a manifestation of mixing too — mixing of momentum. When the flow is laminar, slow-moving fluid from close to the wall produces the steeply dropping portion of the stream shown in Fig. 3 (a). Flow in the center of the tube is much more rapid and produces the flat trajectory which forms the upper right boundary of the

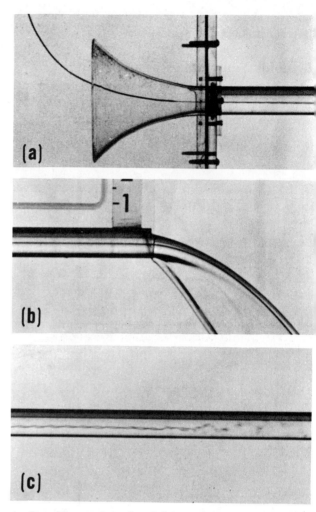

4. Dye filament introduced (a) at the entrance to a tube retains its identity (b) in laminar flow but "explodes" and mixes rapidly across the flow (c) when the flow becomes turbulent.

stream. When the flow becomes turbulent, the mixing of momentum causes the flow speed in the pipe to be much more uniform. The fastest fluid is not quite so fast, and there is so little slow fluid that it can be dragged along with the rest, producing the trajectory shown in Fig. 3 (b). The fluid motion vanishes at the wall, so we can regard it as the sink for momentum. The turbulence increases the rate at which momentum is transferred toward the wall. Thus, with turbulence we need a larger pressure gradient to replace the momentum lost to the wall.

Turbulent Transport and Reynolds Stress

Although the principal motion of the fluid in the channel of Fig. 5 is downstream, because of the turbulence there is appreciable cross-stream motion. Fluid moving across the stream tends to carry its properties with it. Thus the darker dye which marks fluid originally in the center of the stream has moved, in some

5. Dye injected near the center and near the wall of a turbulent channel flow. The walls of the channel have been deliberately roughened to increase the ratio of the turbulent to the mean flow speeds.

places, quite close to the wall. The lighter-colored dye, marking fluid originally close to the wall, has moved toward the center of the channel. This ability of turbulence to carry fluid properties is referred to as *turbulent transport,* and occurs whenever there is some gradient of a mean property, be it momentum, dye concentration or whatever, within the turbulent fluid. For example, in the flow shown in Fig. 5 the region near the wall continuously gains momentum at the expense of the region near the center of the flow.

Analytically, in tensor form, we may write the Navier-Stokes equation for the velocity component V_1 as

$$\frac{\delta \rho V_i}{\delta t} + \frac{\delta}{\delta x_j}\rho V_i V_j + \frac{\delta P}{\delta x_i} - \frac{\delta}{\delta x_j}\,\mu\!\left(\frac{\delta V_i}{\delta x_j} + \frac{\delta V_j}{\delta x_i}\right)$$
= body force

Now if we define some suitable average* velocity U_i (space, time, or ensemble, depending upon the situation) we can put

$$V_i = U_i + u_i,$$
$$<V_i> = U_i \qquad <u_i> = 0$$

and write the equation as

$$\frac{\delta \rho U_i}{\delta t} + \frac{\delta \rho U_i U_j}{\delta x_j} + \frac{\delta}{\delta x_j}<\rho u_i u_j> + \frac{\delta <P>}{\delta x_i}$$
$$- \frac{\delta}{\delta x_j}\,\mu\!\left(\frac{\delta U_i}{\delta x_j} + \frac{\delta U_j}{\delta x_i}\right) = <\text{body force}>$$

the expression $<-\rho u_i u_j>$ appears in the equation in the same way as does the viscous stress

$\mu\left(\frac{\delta U_i}{\delta x_j} + \frac{\delta U_j}{\delta x_i}\right)$. Thus it acts like a stress. It is called the *Reynolds stress.*

Consider now the mixing of a scalar. If two miscible liquids are carefully placed in a vessel, one floating on top of the other, after a week or two molecular diffusion does a fair job of mixing. However, much more

*Symbolized by the bracket $< - >$.

thorough mixing can be accomplished in less than a minute if we make the fluid turbulent. In this case too, the end result is intimate mingling on a molecular scale — although the turbulent motions themselves are not much smaller than a millimeter. The role of the turbulence is to make inhomogeneities more vulnerable to the effects of molecular diffusion. This is illustrated schematically in Fig. 6.

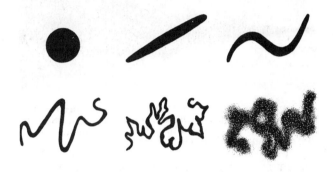

6. Schematic representation of the turbulent mixing of a scalar. The turbulent motions stretch and distort a blob of inhomogeneous fluid, until both the increase in surface area and the increase in property gradients enable molecular effects to occur rapidly.

Analytically, the transport of a scalar can be described as follows: the Eulerian equation for the concentration C of a conservative scalar property is

$$\frac{\delta C}{\delta t} + \nabla \cdot \mathbf{V}C = K\nabla^2 C$$

where K is the diffusivity appropriate to C.
If we again break the fluid velocity $\mathbf{V}$ into mean and turbulent parts $\mathbf{U}$ and $\mathbf{u}$, we get

$$\frac{\delta <C>}{\delta t} + \nabla \cdot \mathbf{U}<C> + \nabla \cdot <\mathbf{u}C> = K\,\nabla^2 <C>.$$

The vector $<\mathbf{u}C>$ represents the turbulent transport of the property C.

Notice that in both this example and the previous one, which showed the origin of the Reynolds stress, the turbulent effects were analytically derived from the "advection" terms $\nabla \cdot \mathbf{V}C$ and $\frac{\delta}{\delta x_j}\,\rho V_i V_j$. In the Navier-Stokes equation this term is non-linear. It is this essential non-linearity that leads both to the complexity of turbulence and to the great difficulty of treating it analytically. Typically, in turbulent situations, the non-linear term is as important as any other in the equation and so cannot be treated adequately by the usual perturbation methods.

The Influence of Reynolds Number on Fully Developed Turbulent Flows

One of the curious properties of turbulence is the fact that, although the Reynolds number is very im-

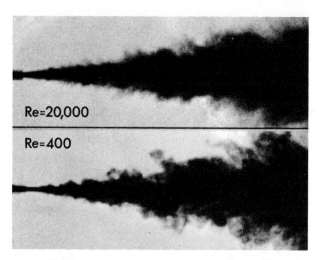

7. **Turbulent jets showing that the Reynolds number does not much affect the appearance, so long as it is sufficiently large that the jet is indeed turbulent. The upper jet has a Reynolds number 50 times that of the lower.**

portant in determining whether or not a particular flow will be turbulent, once it has become turbulent the Reynolds number is of very little importance so far as the *large-scale motion* is concerned. This is illustrated in Fig. 7, which shows two jets, identical in every way except for the viscosity of the fluids (and therefore the Reynolds number), which differs by a factor of fifty. Evidently the large-scale features of the flow are comparatively insensitive to Reynolds number.

However, the small-scale motion, as revealed in shadowgraphs such as those in Fig. 8, is markedly affected. The higher Reynolds number jet has a much finer scale structure than the other. This can be understood if we consider the energy dissipated. These two jets differ only in viscosity; all other conditions are the same, including the rate of energy input into the jet. Therefore they dissipate energy at the same rate.

Dimensionally the dissipation rate must be given by $\nu V^2/\lambda^2$, where λ is a characteristic scale important to

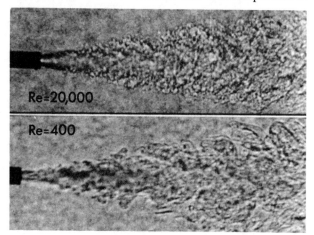

8. **Shadowgraphs of the jets shown in Fig. 7. Note how much finer grained is the structure in the high-Reynolds-number jet than that in the low-Reynolds-number jet.**

the dissipation process, and V a characteristic speed. Clearly, the larger ν is, the larger λ must be.

Turbulent Energy Cascade and Small-scale Similarity

This leads to one of the most important concepts in the study of turbulence: the idea of the *energy cascade*. As we have seen, under certain circumstances a large-scale motion can become turbulent. Some of the energy in the large-scale motion is converted into turbulent energy. The largest scales of the turbulence are usually smaller than, although comparable with, the scale of the basic flow, as can be seen in Fig. 1 and 7. However, usually these large-scale motions are themselves unstable and break into smaller-scale motions which take energy from them. Finally the energy passes down to scales like those revealed in the shadowgraphs of Fig. 8, which are so small that their Reynolds number is too low for instability. Their energy is then dissipated by the action of viscosity.

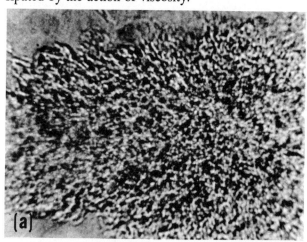

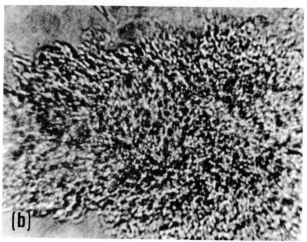

9. **Shadowgraph of a high-Reynolds-number jet. The only difference between these two photographs is the fact that a circle from the center of Fig. 9(a) has been rotated through some 80 degrees to produce Fig. 9(b). The fact that this circle is hard to locate in Fig. 9(b) indicates that the small-scale turbulence is approximately both isotropic and homogeneous.**

In this turbulent energy cascade, at the smaller scales of motion it is only the rate of energy dissipation which is of any consequence. Other information associated with the large-scale motion is lost in the transfer process. Thus at high-enough Reynolds number the small-scale turbulence loses all directional orientation. It becomes locally isotropic, as is illustrated in Fig. 9.

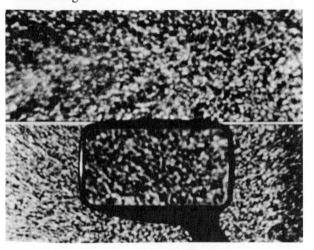

10. Shadowgraphs show similarity of small-scale structure. The upper half of the frame is a shadowgraph from a jet, like that of Fig. 9. The lower half is from a channel flow, in part magnified so that the scale will be comparable to that of the jet.

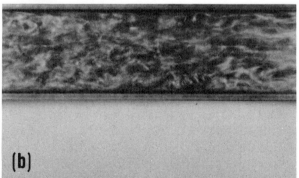

11. Flow visualization of decaying turbulence behind a grid. Photograph (b) is taken several seconds later than photograph (a). Although the energy-transfer mechanism passes energy largely from large scale to small, the decay of the small-scale motion is comparatively rapid, so that it is the large-scale motions that are last to die.

Moreover, at high Reynolds number the small-scale turbulent structure ceases to depend upon the nature of the large-scale flow. Macroscopically the difference between a jet and a channel flow is marked. However on the very small scale revealed by shadowgraphs, the difference in structure disappears, as is shown in Fig. 10. Because of the size difference, the similarity between the small-scale structures may not be obvious unless suitable magnification is used, as in Fig. 10. This is a kind of "similarity": similar structure despite differences in scale. (The velocity scales may differ, as well as the length scales.) We have already seen that the large-scale motion does not depend much upon the Reynolds number. We now find that the structure of the small-scale motion is similar for all kinds of turbulence. What the Reynolds number does is to determine the ratio of the largest scales to the smallest scales.

In decaying turbulence, energy seems, paradoxically, to pass from small scales to large. In fact, however, the energy transfer is still mostly from large scale to small. The large-scale motions are the last to die, because the small scales dissipate more rapidly. Fig. 11 illustrates the effect.

Effect of Buoyancy on Turbulence

The Reynolds number is not the only important parameter in determining the likelihood of turbulence. In some cases the Reynolds number may be enormous, many millions, and no turbulence will exist, because of the presence of some other influence like rotation, density stratification or, for conducting fluids, magnetic fields.

Of these, buoyancy effects are easiest to understand. If the fluid at the bottom is less dense than that at the top, convective activity sets in and can greatly increase the turbulence present — or even produce turbulence when none would otherwise exist. On the other hand, if the fluid on top is less dense, turbulence is inhibited, because the buoyancy effects operate in the other direction and take energy out of the turbulence.

In the atmosphere both stable and unstable buoyancy effects occur frequently. In Fig. 12 we see a smoke layer in an atmosphere which is stable because the air close to the ground is colder and heavier than the air above it. This situation is called an *inversion* by meteorologists. Vertical turbulent motions are strongly inhibited and any motion which occurs tends to be almost horizontal. Smog can accumulate when an inversion at some height above the city prevents pollution from mixing upwards.

On the other hand, air close to the ground is often heated. This can produce vigorous convection, as shown in Fig. 13. Buoyant convection occurs nearly

12. Smoke layer in a stable atmosphere. Because the air above is warmer and lighter than the air below, turbulence is greatly inhibited. All motion tends to be horizontal.

13. When the air is heated from below, convection is likely to result. In the case of cumulus clouds of this type, the convective activity is greatly enhanced by the release of latent heat when water vapor condenses into droplets to form a cloud.

always when a fluid is heated from below, whether in a porridge pot or in the surface layers of the sun.

Small-scale Intermittency

In the defining syndrome of turbulence we did not employ the word "random," although it would seem to be apropos, because, to some at least, it carries with it the connotation of a Gaussian process. Turbulent distributions are more complicated than that. In Fig. 14 we see the output of a hot-wire anemometer operated in an atmospheric boundary layer. The large-scale motion, as shown by the horizontal velocity component u, is closely Gaussian. However, if we differentiate the signal, or examine any other property that is strongly dependent upon the small-scale mo-

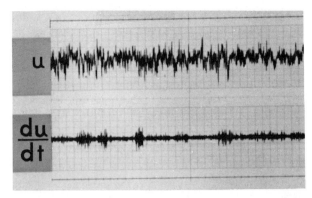

14. Chart recordings of hot-wire anemometer measurements of the downwind turbulent velocity component in an atmospheric boundary layer. The upper trace shows the measured velocity. The distribution is very nearly Gaussian. The lower trace is the time derivative of the signal. (It is best to interpret this signal as a spatial derivative, since the time rate of change is mostly produced by the turbulent structure blowing past the probe, rather than by changes in the structure itself. Thus this signal is related to vorticity.) The distribution here is very intermittent, and clearly non-Gaussian. This effect increases with increasing Reynolds number.

tions, we find that activity seems to be distributed in concentrated bursts separated by regions which are comparatively quiescent. The effect is illustrated in Fig. 14.

The non-Gaussian, intermittent character of the small-scale structure becomes more marked as the Reynolds number increases. It seems to be fundamental to the nature of the turbulent cascade, but as with many other aspects of turbulence we do not have a fully satisfactory theoretical explanation. It is another manifestation of its baffling but fascinating complexity.

References

1. Batchelor, G. K., *The Theory of Homogeneous Turbulence*, Cambridge University Press, 1959
2. Townsend, A. A., *The Structure of Turbulent Shear Flow*, Cambridge University Press, 1956
3. Hinze, J. O., *Turbulence*, 568 pp. McGraw-Hill, 1959
4. Monin, A. S. and Yaglom, A. M., *Statistical Hydromechanics*, Vol. 1, 1965; Vol. 2, 1967. In Russian, izdatelctvo "Nauka." (English translation, M.I.T. Press, 1968.)

Boundary-Layer Control

David C. Hazen
PRINCETON UNIVERSITY

Introduction

Although potential theory can be used to explain many aerodynamic phenomena, there are cases in which the boundary layer — the thin layer of fluid next to a solid surface in which effects of viscosity may be considered concentrated — significantly alters theoretical predictions. A simple example is the flow past an airfoil. The airfoil section in Fig. 1 is in a narrow wind tunnel at zero angle of attack. Kerosene smoke provides the flow visualization. At low angles of attack, the streamline pattern about such a shape is very close to the predictions of inviscid theory. However, a drag force not accounted for by such a theory exists. This drag is largely due to viscous shear forces and is called *skin-friction drag*.

In regions over the surface in which the boundary-layer flow is laminar, the fluid mixing and viscous skin friction are low. However, such laminar flows are often unstable and develop into turbulent flows. Turbulent flows involve more rapid mixing, which produces higher skin-friction drag. On occasion, the combined action of viscous forces and an adverse pressure gradient produces a reversal of the flow next to the surface which, in turn, causes separation of the

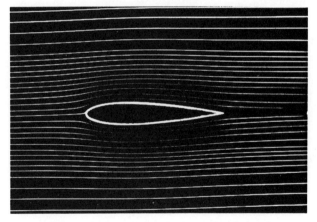

1. **Smoke flow past airfoil at zero angle of attack.**

adjacent flow from the surface. This situation is exemplified in Fig. 2, where the flow on the top surface is separated and the airfoil is said to be *stalled*.

The presence of the boundary layer has produced many design problems in all areas of fluid mechanics. However, the most intensive investigations have been directed towards its effect upon the lift and drag of wings. The techniques that have been developed to manipulate the boundary layer, either to increase the

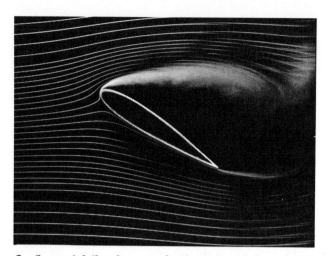

2. Same airfoil at large angle of attack with flow separation.

lift or decrease the drag, are classified under the general heading of *boundary-layer control.*

Two boundary-layer phenomena for which controls have been sought are the *transition* of a laminar layer to a turbulent flow and the *separation* of the entire flow from the surface. By maintaining as much of the boundary layer in the laminar state as possible, one can reduce the skin friction. By preventing separation, it is possible to increase the lifting effectiveness and reduce the pressure drag. Sometimes the same control can serve both functions.

Controlling Transition by Shaping the Airfoil

Transition to turbulence is associated with instability of the laminar boundary layer. When studied with the aid of high-speed photography (Fig. 3), disturbances in the laminar flow are seen to amplify to the point of forming large eddies. These in turn produce the highly disorderly motion of turbulent flow. The

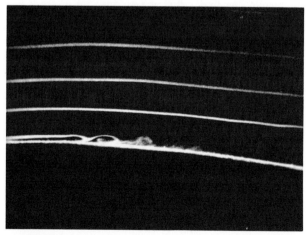

3. High-speed photograph of boundary layer on airfoil undergoing transition. Large eddies are formed prior to breakdown into turbulence.

location on the surface at which transition occurs depends both upon the stability of the laminar boundary layer and upon the nature of the disturbances. Factors producing disturbances, such as surface roughness, noise, vibration, heat, or airstream turbulence, can sometimes be avoided or isolated. The stability of the laminar boundary layer may also be influenced by manipulating the pressure gradient produced by the flow over the surface.

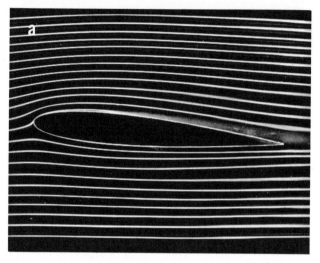

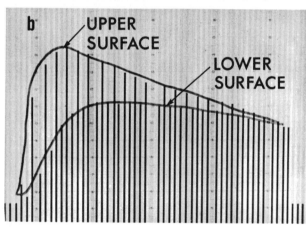

4. Conventional airfoil profile at moderate angle of attack. (a) Smoke visualization shows laminar flow extending to about 25% of the chord length. (b) Manometer bank shows pressure distribution on upper and lower surfaces.

Combinations of favorable and unfavorable pressure gradients occur over the surfaces of aerodynamic shapes. An airfoil with its point of maximum thickness located about 25% of its chord length aft of the leading edge, placed at a moderate angle of attack (such as might correspond to cruising flight of an airplane), has a minimum-pressure point near the leading edge (Fig. 4).* Thus, the upper surface aft of this

---

*The pressure taps on the airfoil surface are connected to the manometer tubes to show pressure distribution. As with a soda straw, a pressure *decrease* causes a *rise* in the fluid level in the tube. Thus a high level indicates a low pressure.

point is subjected to an *adverse* gradient. Increasing the angle of attack increases the adverse gradient. Such gradients are destabilizing, so that the amount of laminar flow over a structure decreases with increasing adverse pressure gradient.

A different airfoil with its maximum thickness at 50% chord (Fig. 5), when placed at an angle of attack

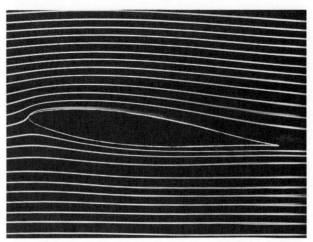

6. **High-speed profile at larger angle of attack than in Fig. 5. Transition to turbulence occurs near leading edge.**

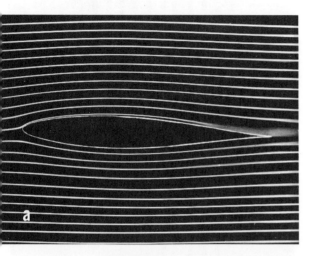

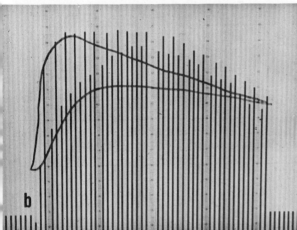

5. **High-speed profile at angle of attack producing same lift as airfoil in Fig. 4. (a) Note greater extent of laminar flow on upper surface. (b) Comparison of pressure distribution with that of conventional airfoil (marked outline).**

producing the same amount of lift as the airfoil of Fig. 4, has a minimum pressure point farther aft, and the extent of adverse gradient is therefore less. Flow studies show that the extent of the laminar boundary layer is correspondingly increased.

As the angle of attack of the airfoil of Fig. 5 is increased, the situation changes radically. The leading-edge radius of such "laminar-flow" profiles is necessarily small. Therefore, as the forward stagnation point moves downward, the flow passing over the leading edge must accelerate rapidly, producing there a sharp pressure minimum followed by a strong adverse pressure gradient. Under these circumstances the transi-

tion point abruptly moves to the leading edge, covering the entire surface with turbulent flow (Fig. 6).

Turbulent boundary layers thicken more rapidly and produce greater skin friction than laminar layers. Thus, reducing the extent of turbulent flow reduces the drag of the profile. This can be verified experimentally by measuring the flux of momentum deficiency in the wake of the airfoil (by means of an array of total-head tubes located far enough behind the profile that the static pressure in the wake is nearly the same as that in the main flow). The loss in total head reflects a change in streamwise momentum flux, which is a measure of the *profile drag*. Seen on a bank of manometer tubes (Fig. 7), the momentum defect produced by the profile with the greater extent of laminar flow is markedly less than that created by the profile with its maximum thickness at 25% chord, at the same lift coefficient.

Changes in the geometry of an aerodynamic shape provide a method of adjusting pressure gradients so as

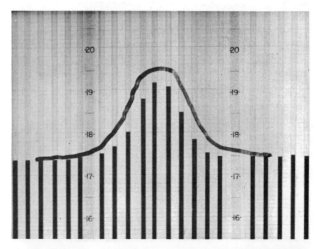

7. **Manometers connected to total-head tubes illustrate difference in wake momentum loss between conventional profile (marked outline) and high-speed profile at same lift coefficient.**

to favor the existence of laminar boundary layers. However, regions of adverse gradient can never be completely avoided. On occasion, changes in the geometry, while helpful at some angles of attack, worsen the situation at other angles.

Controlling Transition by Suction

A rather different means of stabilizing the boundary layer is the use of *suction*. Suction may be applied either through porous surfaces or through a series of finite slots, as in Fig. 8. When applied in this manner, suction reduces the thickness of the boundary layer by removing the low-momentum fluid next to the surface. A more stable layer results, and transition to turbulence is delayed.

To achieve stabilization of the boundary layer at various angles of attack, compromises must be made as to the number of slots, their location, and the amount of suction flow through each slot.

A wake survey (Fig. 9) shows that, even with such compromises, suction can be effective. Without suc-

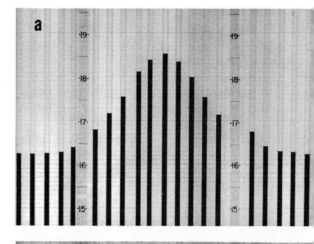

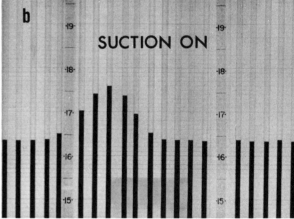

9. **Manometers connected to total-head tubes illustrate wake momentum loss of multi-slotted profile. (a) No suction. (b) Suction on.**

tion the wake is broad, indicating high drag. The application of suction greatly reduces the streamwise momentum loss in the wake. If the suction is applied only on one surface, as in the case demonstrated, the wake reduction will not be symmetrical (cf. Fig. 7).

Power is needed to achieve this drag reduction. The optimum condition occurs when the total drag — the aerodynamic drag plus the suction power converted to an equivalent drag — is a minimum.

Controlling Separation by Suction

There are many cases in which control of boundary-layer separation is important. Suction can be used for this purpose, too. If a profile equipped with suction slots is placed at a high enough angle of attack (Fig. 10a), suction will not be able to maintain the entire boundary layer in the laminar state. It can, however, exert a profound effect upon the turbulent layer, frequently keeping the flow attached well beyond the angle at which stalling occurs without suction (Fig. 10b). In general, more suction power is required to attach a flow that is already stalled than to maintain attached flow at the same angle of attack.

Separation control by suction is accomplished by

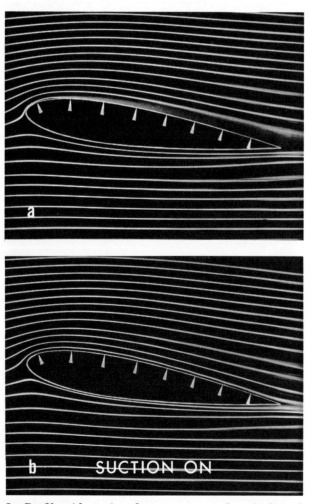

8. **Profile with suction slots on upper surface (indicated by arrowheads). (a) Suction off. Boundary layer is turbulent over most of upper surface. (b) Suction on. Laminar flow restored to upper surface.**

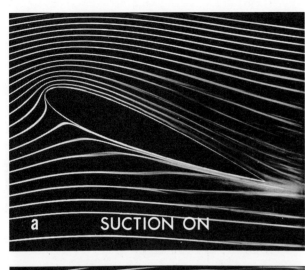

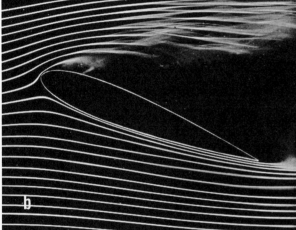

10. Multi-slotted profile of Fig. 8 at high angle of attack. (a) Suction on. Separation is prevented. (b) No suction. Flow totally separates from upper surfaces.*

drawing the low-momentum layers from the bottom of the boundary layer into the suction slots. This draws the higher-energy air from the outer layers closer to the surface.

Controlling Separation by Variable Geometry and by Blowing

Separation control can also be accomplished by other techniques. Laminar separations such as those that occur at the sharp leading edge of a thin profile can frequently be avoided by a change in geometry that alters the pressure field, such as the deflection of a *nose flap* (Fig. 11).

Vortex generators (Fig. 12) can help to delay separation by mixing high-momentum fluid from the outer flow with low-momentum fluid next to the airfoil surface.

Blowing jets directed into critical areas are also

*The blurring of the smoke lines well above the airfoil is a result of disturbances from the side walls of the tunnel (where suction is *not* applied).

useful. Frequently these can be created by utilizing the pressure differences that exist on the aerodynamic bodies themselves. A *leading-edge slot* is an example (Fig. 13). When open it leads air from the region close to the stagnation point through a converging channel and ejects it at high speed at a point of low

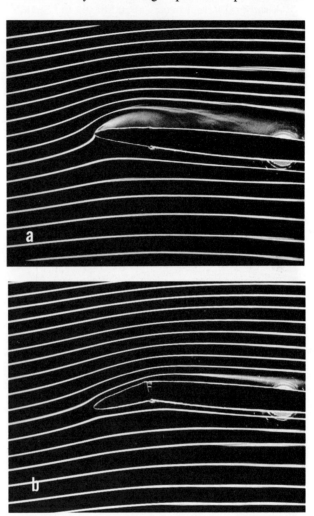

11. Thin profile with deflectable nose. (a) No deflection. Separation occurs at leading edge. (b) Nose deflected. Flow remains attached at leading edge.

12. Vortex generators on transport-airplane wing.

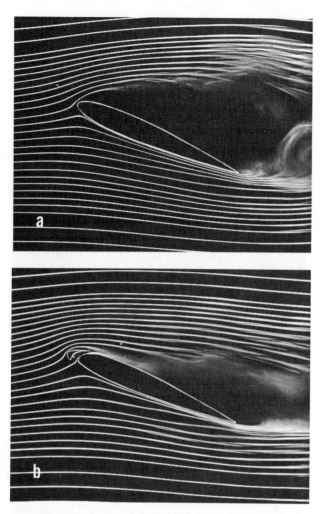

13. (a) Conventional airfoil at large angle of attack. (b) Leading-edge slot reduces extent of separation.

pressure on the upper surface. The same concept is used to decrease the extent of separation on deflected high-lift flaps. Fluid is led from the high-pressure region below the flap through a converging channel and ejected over the upper surface close to the point of minimum pressure (Fig. 14). This helps to overcome the strong adverse pressure gradients existing on the

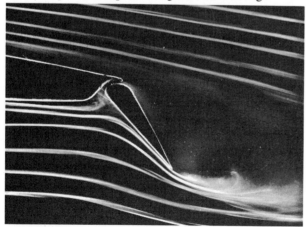

14. Flow near trailing-edge flap with slot.

upper surface of the flap. Multiple-slot arrangements, though more complicated, have proven to be particularly effective (Fig. 15).*

Still greater effectiveness can be obtained from blow-

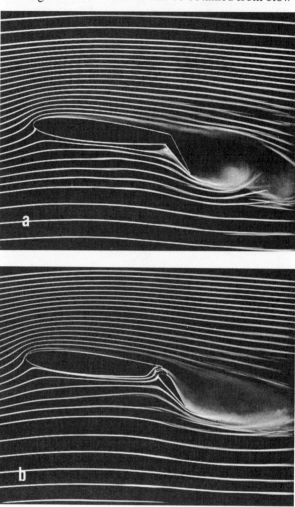

15. (a) Airfoil with deflected trailing-edge flap. (b) Multiple slots reduce degree of separation.

ing jets if they are produced by the direct application of power rather than by the limited pressure differentials obtainable on the body itself (Fig. 16).

If more suction or blowing is supplied than that necessary to prevent separation of the flow over a deflected flap, rather more lift is measured than would be expected from the predictions of potential theory. In the case of suction, the low pressure at the slot inlet can act like a concentrated sink, altering the potential flow in a manner that increases the lift. Blowing provides a component of direct momentum, plus a lift attributable to the pressure difference required to curve a blowing-jet sheet. This is the so-called *"jet-flap"* effect (Fig. 17).

---

*The smoke-emitting tubes upstream are more widely spaced near the top and bottom of the tunnel (cf. Fig. 1). Note that in Fig. 15 and in subsequent pictures some of the lower, widely-spaced smoke lines pass near the airfoils.

new separations are induced at other points. An example is a thin profile with deflected nose and trailing-edge flaps (Fig. 18). Because of severe adverse pressure gradients, major separations occur on the lower surface of the leading edge and over the trailing-edge flap, and a minor separation occurs at the break of the

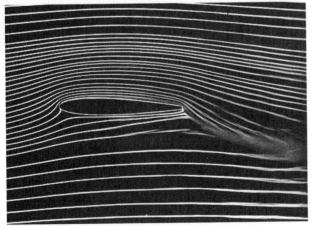

16. **Profile with deflected trailing-edge flap. (a) Separation occurs over upper flap surface. (b) Jet issuing from near the hinge blows over upper surface of flap, suppressing separation.**

17. **Profile equipped with a deflectable jet at trailing edge shows effects of high blowing quantity and large jet-flap deflection.**

Separation in More Than One Location

Care must be taken in applying the methods of boundary-layer control. Frequently, prevention of separation at one point may so alter the flow field that

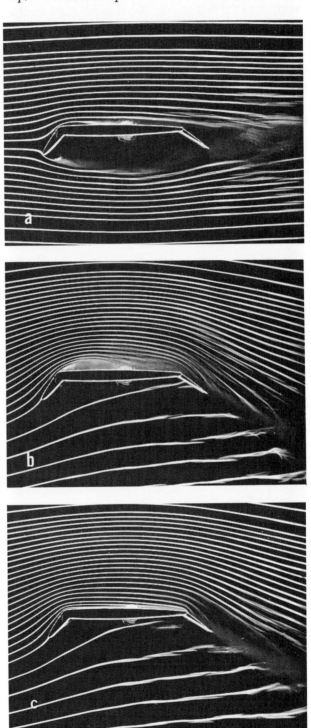

18. **Thin profile with deflected nose flap and trailing edge flap. (a) Separation occurs at three locations. (b) Blowing over trailing-edge flap. Separation is suppressed there, and on underside of leading-edge flap, but separation on upper surface is more severe. (c) Blowing also at break of leading-edge flap. Separation is eliminated.**

leading-edge flap, followed by reattachment of the flow (Fig. 18a). When blowing is applied over the trailing-edge flap alone, separation is suppressed at this point and the flow field is so altered that the flow is reattached to the underside of the leading-edge flap (Fig. 18b). Lift has been improved, but the separation at the break of the leading-edge flap is much more severe, and the flow is separated from most of the upper surface. Total reattachment is achieved only when some of the blowing air is diverted from the trailing-edge flap and blown over the knee of the deflected nose flap (Fig. 18c).

Summary

Although only airfoil applications have been considered, the techniques of boundary-layer control can readily be applied to diffusers, to bodies of revolution, and to fluid machinery. They may, in fact, be applied wherever transition or separation of the boundary layer affects performance.

References

Abernathy, F. H., "Fundamentals of Boundary Layers," (a NCFMF film)

Lachmann, G. V. (ed.), *Boundary Layer Control* (2 volumes), Pergamon Press, 1961

Pope, A. and Harper, J., *Low-Speed Wind Tunnel Testing*, Wiley, 1966

Schlichting, H., *Boundary Layer Theory*, McGraw-Hill, 1960

Secondary Flow

Edward S. Taylor
MASSACHUSETTS INSTITUTE OF TECHNOLOGY

Introduction

A number of different types of flow have been called "secondary flow." We shall deal here only with the most common type, that which occurs when fluid is made to follow a curved path.

A classic example is the teacup experiment. If one stirs one's tea so as to give it a generally circular motion, the tea leaves invariably gather at the center of the bottom of the cup (Fig. 1).

Clearly, tea leaves are denser than the liquid, since they are found at the bottom of the cup. There will thus be a tendency for the leaves to be centrifugally separated from the liquid, and one might expect to find them at the perimeter rather than at the center of the bottom. Some opposing phenomenon must cause them to seek the center.

A Model of the Teacup Experiment

In order to improve understanding of the phenomenon, a model, larger than the teacup, was made (Fig. 2). So that the fluid inside the circular cylinder could be viewed tangentially with little optical distortion, a transparent square box was placed around the cylinder and the space between was filled with water. The cylindrical container was rotated about its axis at constant velocity for a long time in order to insure that

1. Tea leaves at the bottom of a cup gather at the center.

initially, at least, the fluid motion would be simple rotation about the axis. In order to simulate the motion of the fluid in the teacup, we then stopped the rotation of the container. Except near the side wall and the bottom, the fluid continued to rotate as before. At these surfaces the velocity must be zero and near them the fluid is slowed by friction.

In developing a description of the actual flow, we

2. Apparatus modeling the teacup experiment.

first postulate a flow which we call "primary," assuming that each fluid element moves in a circle around the center of the container but that the elements near the bottom move more slowly because of friction. Clearly this flow satisfies the boundary conditions and satisfies continuity, but must be tested to see if it satisfies the momentum equation, which involves shear forces, gravity forces, pressure forces, and momentum.

Gravity forces produce a vertical pressure gradient, which affects the motion only insofar as it alters the shape of the free surface at the top of the liquid. In our experiment the rate of rotation is slow enough so that this surface remains nearly flat. We shall postulate that shear forces are negligibly small compared to pressure and inertia forces, and thus we may use Euler's equation to examine the momentum of the fluid.

For motion of a fluid element in a circle Euler's equation in the radial direction becomes

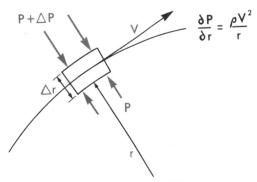

$$\frac{\partial P}{\partial r} = \frac{\rho V^2}{r}$$

For a given pressure gradient, $\frac{\partial P}{\partial r}$, the smaller the velocity, the smaller the radius the particle must follow. In our postulated primary flow there can be no vertical pressure gradients (other than that due to gravity, which has already been taken into account). Therefore, the pressure gradient toward the center, which keeps the fluid elements traveling in circles, must be the same at all depths. If we assume that the boundary layer occupies only a small part of the total

depth, then the main flow outside the boundary layer fixes the pressure gradient. Since the circumferential velocity of fluid in the boundary layer is low, its centrifugal acceleration is insufficient to balance this gradient. This flow is therefore forced inward toward the center, carrying the tea leaves with it.

In other words, the postulated primary flow does not satisfy the momentum equation in the radial direction everywhere, and a correction to this flow is necessary to give a reasonable approximation to the actual flow. It is this correction which we call *secondary flow*.

We can observe the secondary flow by watching a line of hydrogen bubbles produced by electrolysis from a vertical wire by a short pulse of current. The wire is positioned about one third of the radius in from the circumference and is shown in radial view in Fig. 3. The bubbles are carried by the fluid; therefore, the wire marks the initial position of a set of fluid particles, and a line of bubbles a later position. If the

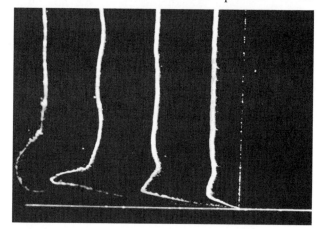

3. Radial view of vertical wire and bubble lines in teacup model.

vertical velocity of the bubbles is small, the line of bubbles represents a velocity profile with the wire as origin. By looking inward along a radius, we observe the tangential component of the velocity at any depth. Similarly, by looking in the tangential direction we can see the radial component of velocity at any depth. These two velocity profiles are shown in the composite photograph (Fig. 4). The tangential velocity shown at the left is seen to be nearly the same at all depths, except in the boundary layer where it goes to zero at the floor and shows an interesting overshoot near the outer edge of the boundary layer. This curious overshoot is a result of fluid in the radially-inward secondary flow being carried inward by its own radial momentum past the position of equilibrium where its tangential velocity matches that of the primary flow above. The right-hand bubble line (tangential view) shows the radial velocity components with the center of rotation toward the right. The radial velocity is essentially zero except in the boundary layer where there is considerable inward flow with again a few minor wiggles near the outer edge of the layer.

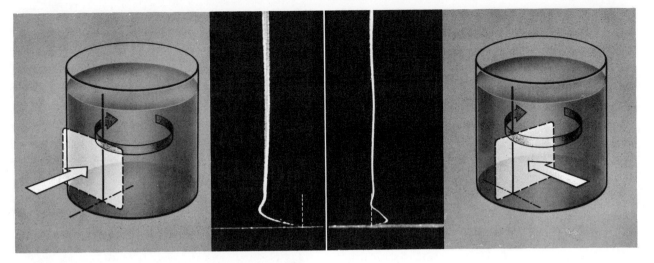

4. Tangential (left) and radial (right) velocity profiles.

Bödewadt (1, 2) gave a mathematical solution of a steady flow that is quite similar to the flow in our rotating tank. He postulated a semi-infinite body of fluid rotating uniformly about an axis perpendicular to a stationary bounding plane. Thus his solution corresponds approximately to what happens near the center of a very large and very deep tank wtih a stationary floor but with an outer wall which rotates with the fluid. Bödewadt's solution is shown graphically in Fig. 5.

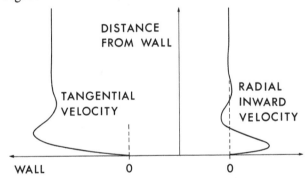

5. Computed tangential and radial velocity profiles (after Bödewadt).

While the experimental profile is slightly unsteady, it often shows the overshoot in tangential velocity predicted by Bödewadt's computation. The shape of the radial velocity profile also agrees qualitatively with his computation. We should not necessarily expect quantitative agreement, since Bödewadt's model is one of steady flow with a semi-infinite reservoir of rotating fluid to maintain the motion against friction, whereas our experiment involves a flow in a finite tank nonsteady in the sense that the entire flow is slowing down.

Steady Sink-vortex Flow

An example of a similar secondary flow occurs in a sink vortex (Figs. 6 and 7). Flow enters a circular tank tangentially and flows out a hole in the center of the bottom.

As might be expected, the slower-moving fluid in the boundary layer near the bottom is again forced inward by the radial pressure gradient established by the more rapidly moving fluid above. In fact, it appears that all of the flow out the drain hole comes from the boundary layer (Fig. 6) and the fluid above does indeed travel in circles (Fig. 7).

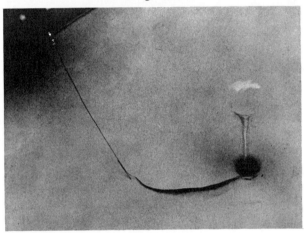

6. Dye marks the radially inward secondary flow near the floor of the sink-vortex apparatus.

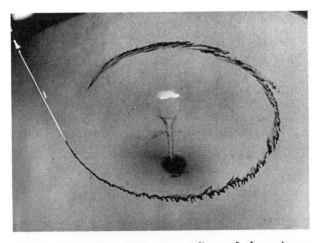

7. Dye marks the circular streamlines of the primary flow high above the floor.

8. Dye marks the radially inward secondary flow near the floor. The secondary flow gathers into a "tornado" under the suction tube, whose opening is just below the free surface.

Even if the flow is siphoned off near the top surface (Fig. 8), the fluid leaving comes from the bottom boundary layer, since this layer is the only region of lower-than-normal tangential velocity. The similarity of this latter flow to a tornado is evident (Fig. 9). It is also of interest to note the "burst" or sudden swelling of the vortex core in both Figs. 8 and 9. This phenomenon has also been observed in vortices arising from delta-wing airplanes.

Flow in a Channel with a Bend

Another model which exhibits secondary flow is a channel with a bend (Fig. 10). If the velocity of a vertical line of fluid elements entering the bend is the same at all depths (Fig. 11a) there will be no tendency for secondary flow in the bend (Fig. 11b). However, if we introduce a flow in which velocity varies with depth (Fig. 12a), in the bend, we can expect the slower-moving fluid to be swept inward, while the more rapidly moving fluid will be forced outward by the radial pressure gradient (Fig. 12b).

The concept of vorticity is useful in explaining and analyzing this flow. If we select a right-handed set of coordinates x, y, z, the corresponding components of

9. Tornado. (Wide World Photos.)

vorticity ω_x, ω_y, ω_z can be written in terms of the components of velocity u, v, w thus:

$$\omega_x = \frac{\partial w}{\partial y} - \frac{\partial v}{\partial z}$$

$$\omega_y = \frac{\partial u}{\partial z} - \frac{\partial w}{\partial x}$$

$$\omega_z = \frac{\partial v}{\partial x} - \frac{\partial u}{\partial y}$$

If we take x in the direction of flow, z vertically upward, and y in the transverse direction, we see that the initial vorticity of the flow shown in Fig. 12a will be

$$\omega_x = 0$$

$$\omega_y = \frac{\partial u}{\partial z} \approx \text{constant}$$

$$\omega_z = 0$$

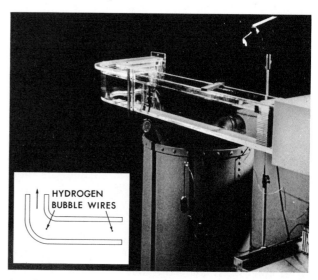

10. Water channel with a bend.

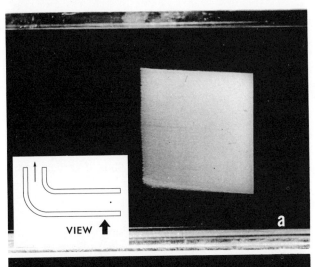

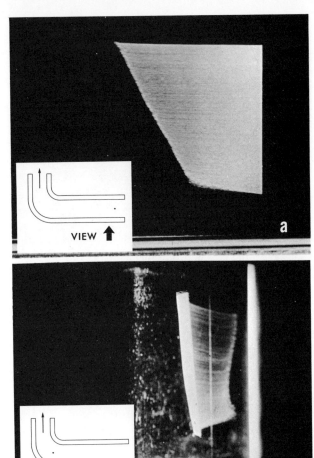

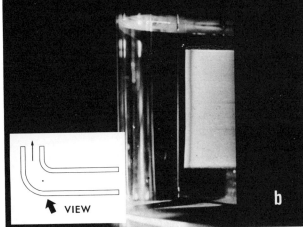

11. **Uniform flow.** (a) Straight section, (b) Tangential view in bend.

12. **Shear flow.** (a) Straight section. (b) Tangential view in bend shows that transverse velocities are induced by the radial pressure gradient.

Vortex lines (lines everywhere parallel to the vorticity vector) are, in this case, horizontal, and perpendicular to the flow direction.

The Helmholtz theorem states that in an inviscid flow, vortex lines are transported by the fluid. We shall assume that the fluid acts in an essentially frictionless manner (an assumption which will be validated later). Fig. 13 views the water channel from above and shows the displacement of a bubble cross produced upstream of the bend. One leg of the cross is a part of a streamline and the other a part of a vortex line. As the cross enters the bend the streamline leg must turn. The rate of turn is $\frac{\partial v}{\partial x}$. The rate of

13. **A cross composed of a streamline and vortex line is formed upstream (a) and enters the bend (b).**

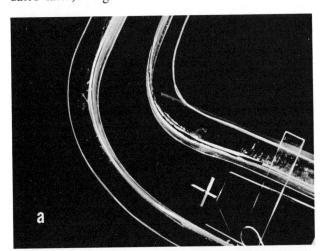

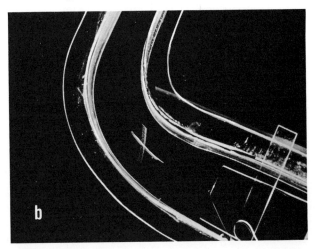

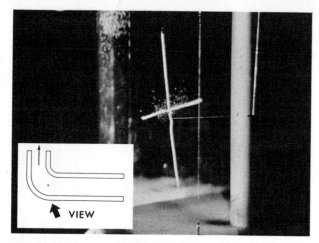

14. Tangential view of bend. Inclination of initially erect fluid cross shows streamwise vorticity.

turn of the vortex line is $\dfrac{\partial u}{\partial y}$. If the vertical vorticity, ω_z, is to remain zero

$$\frac{\partial v}{\partial x} = \frac{\partial u}{\partial y}$$

that is, the initial rate of turn of the vortex lines is equal and opposite to the initial rate of rotation of the streamline. The two lines, therefore, cease to be at right angles, and close together like scissors.

There is horizontal vorticity entering the bend in Fig. 13, and it is marked by a bubble line. The fact that this vortex line ceases to be perpendicular to the streamline indicates that a component of vorticity is appearing in the streamwise direction. An initially upright bubble cross viewed in the streamwise direction (Fig. 14) shows by its tilt additional evidence of this streamwise component of vorticity. It is evident from this picture that vertical velocities have been produced.

Notched Flow Around a Bend

A flow with a wake or defect in velocity at the center was created upstream of the bend. We will call this a "notched" flow (Fig. 15a). A tangential view in the bend (Fig. 15b) shows the familiar inward flow of the low-velocity region. The limited extent of the vorticity, and the proximity of two layers of vorticity with opposite sign tend to minimize vertical velocities.

If we assume that vertical velocities are zero everywhere, and that vertical vorticity remains zero throughout the flow, the components of vorticity in the bend become:

$$\omega_x = -\frac{\partial v}{\partial z}$$

$$\omega_y = \frac{\partial u}{\partial z}$$

$$\omega_z = 0$$

$$\vec{\omega} = \mathbf{i}\,\omega_x + \mathbf{j}\,\omega_y = -\mathbf{i}\,\frac{\partial v}{\partial z} + \mathbf{j}\,\frac{\partial u}{\partial z}$$

Fig. 16a is a view from above of an initially vertical bubble line in the notched flow in the bend, taken a short time after the bubble line was formed. We note that it lies in a vertical plane; that is, it appears from above as a nearly straight line. It can be verified that the direction of this line in the horizontal plane is

$$\mathbf{i}\,\frac{\partial u}{\partial z} + \mathbf{j}\,\frac{\partial v}{\partial z}$$

and that it is perpendicular to $\vec{\omega}$. Thus, if vertical components of velocity can be neglected, the plane determined by the velocity profile is perpendicular to the local vorticity.

We now have a means of checking whether friction is important: if a vortex line marked upstream coincides with the vorticity as indicated locally by the perpendicular to the projection of the velocity profile on the horizontal plane, then Helmholtz's theorem holds, an observation which is consistent with the absence of friction. Fig. 16b includes a vortex line sent down

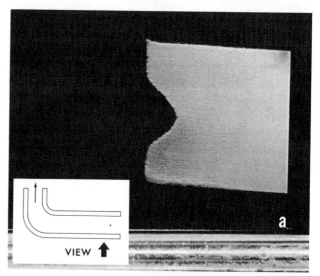

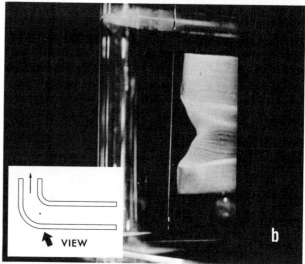

15. "Notched" flow. (a) Straight section. (b) Tangential view in bend.

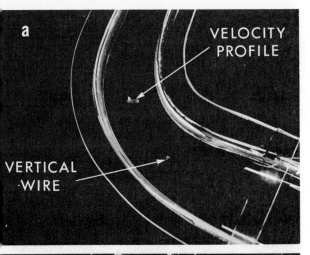

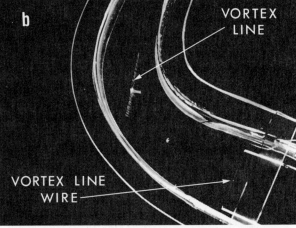

16. Notched flow. (a) Initally vertical bubble line remains in a vertical plane. (b) The plane of the velocity profile is perpendicular to a vortex line convected from upstream.

from upstream which is indeed nearly perpendicular to the velocity profile.

Boundary-layer Flow Around a Bend

Secondary flow due to the boundary layer on the bottom surface shows behavior similar to that of the notched flow. Here vertical velocities are suppressed by the floor of the channel.

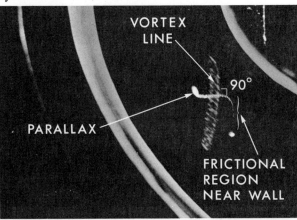

17. Intersection of velocity profile and vortex line. Flow in straight section has uniform velocity and a boundary layer on the floor.

In this case the bubble line is no longer a plane curve (Fig. 17). Looked at from above it appears as two approximately straight lines joined by a short curved section.* One of the straight portions is indeed perpendicular to the vortex line from upstream. This portion is in the outer part of the boundary layer, where local friction is unimportant; hence this part of the boundary layer behaves like the notched flow. The other straight portion is in a part of the boundary layer very near the wall, where friction is important. This part is not perpendicular to the vortex line, as might be expected. The curved portion is a transition region.

Transport Phenomena Associated with Secondary Flows

Secondary flow in a curved channel acts to replace the slow-moving fluid near the walls with faster-moving fluid, thereby greatly increasing viscous friction at the wall. Frictional losses are increased not only in the bend itself, but also downstream, as a result of the persistence of the secondary flow there.

Materials as well as momentum may be transported by secondary flow. The causes of the meandering of rivers are complex, but the phenomenon is undoubtedly associated with the transport of silt by secondary flows.

Heat transfer and momentum transfer follow similar laws, so that changes in the flow distribution which promote heat transfer generally cause increased friction, and vice versa. Heat exchangers are sometimes made with wavy passages, to promote secondary flow. The resultant heat transfer (and friction) can be more than double that of a comparable straight passage (3).

Turbines and compressors have blades which change the direction of the working fluid. There are boundary layers not only on the blades, but also on the hub and casing, and secondary flows are often important. In general, the fluid in the hub and casing boundary layers is turned more than the fluid outside the boundary layers, and impinges on the next set of blades at an increased angle of incidence.

Other Secondary Flows

The causes of some secondary flows are difficult to determine. For example, Fig. 18 shows a secondary flow arising from an oily film on the water surface. Such films can behave very much like solid walls.

Other secondary flows are very difficult to analyze. For example, when the flow in a bend has progressed around a considerable corner, a portion of it originating from the boundary layer rolls up into what is called a "passage vortex" (Fig. 19). Note the start of a passage vortex in Fig. 18.

---

*The apparent upward turn at the left-hand end of the line is due to parallax; the line of sight was along the wire and as the bubble line moved away from the wire, the camera looked at an angle to the line of bubbles.

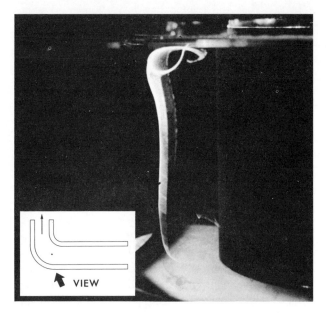

18. Secondary flow in a bend resulting from dirt film on free surface.

All of the pictures were taken with laminar flow to insure that dye or bubble lines would not be diffused by turbulence but would persist long enough to permit observation. Secondary flow phenomena, however, are not confined to laminar flows, but are present in the turbulent flows which often occur in pipes, channels, and ducts, as well as in rotating machinery.

Another more complex secondary flow is the horseshoe vortex formed when the boundary layer on the bottom of the channel meets an obstruction (Fig. 20). Fluid from the boundary layer cannot reach a stagnation point on the obstruction, since its total pressure is less than the static pressure at such a point. The static pressure in the center of a horseshoe vortex is low, and the low-energy fluid finds its way to this point. This particular flow pattern results in very high shear stresses on the bottom wall directly under the vortex and in front of the obstruction. (Note that the fluid near the bottom is flowing upstream at this point and the friction force on the wall is *upstream.*)

After a blizzard you may have noticed the way the snow is scooped out on the windward side of trees and

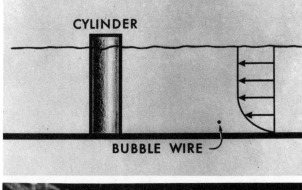

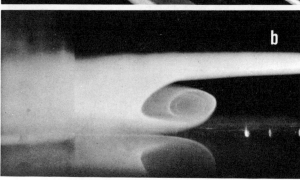

20. Horseshoe vortex. (a) Oblique view. (b) Cross section at plane of symmetry.

telephone poles. This is evidence of horseshoe vortices resulting when the boundary layer on the ground encounters obstructions. These beautiful snow patterns have counterparts underwater near bridge piers in streams. The resulting undermining of the earth in front of and around such piers can seriously affect the structure.

Many other flows can be analyzed by postulating a primary flow and deducing a secondary flow. The book *Boundary Layer Theory* by H. Schlichting gives several examples, such as turbulent flow in pipes of non-circular cross section (pp. 415-416) and oscillating disks, spheres, and cylinders (pp. 196-197).

References

1. Bödewadt, U. T. See also Schlichting pp. 176-180
2. Schlichting, H. *Boundary Layer Theory,* McGraw-Hill, 1960
3. Kays, W. M. and London, A. L. *Compact Heat Exchangers,* McGraw-Hill, 1958

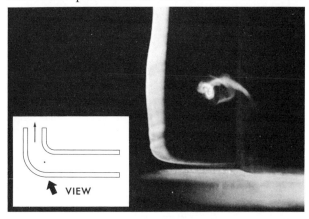

19. A "passage vortex" in a bend.

Waves in Fluids

Arthur E. Bryson
HARVARD UNIVERSITY

Introduction

Waves occur all around us. Most common, perhaps, are gravity waves on water and compressibility waves in air. Because waves in water are easy to observe, we will begin with them. Later we will see that many of the concepts we have developed with these gravity waves apply equally well to compressibility waves in gases.

A surface gravity wave is a region of increased or decreased depth that moves relative to the fluid. A large-amplitude wave is one that produces a change in depth large compared to mean depth as it goes by. When the change in depth is small compared to both mean depth and wave length, we call it a small-amplitude wave. Even when we cannot see the surface, we can detect the passage of a wave by the change in pressure at a fixed point underneath the surface. We can also detect it by the motion of the fluid particles. One way to see the motion of fluid particles is to introduce neutrally buoyant solid particles into water as shown in Fig. 1. The mean location of these solid particles is stabilized by using slightly buoyant particles (specific gravity approximately .999) and attaching a thread to them which extends to the bottom where a short length

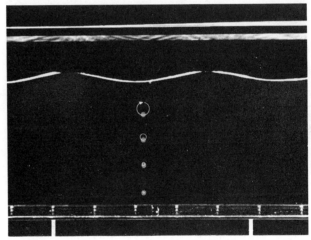

1. Circular particle paths typical of periodic waves where wave length is comparable to or less than depth ("deep water waves").

of miniature chain is attached; the particles lift one or two links of the fine chain and become neutrally buoyant.

Deep Water Waves

Superposed on Fig. 1 are the paths of the particles at various depths in a horizontal water channel when periodic waves of small amplitude go by. Here the wave length is comparable to the depth. In these waves the particles move in approximately circular paths. The diameter of the circles decreases with depth. The particles near the bottom hardly move at all. This is the pattern of fluid movement characteristic of waves on the surface of deep water, that is, where the wave length is comparable to or less than the depth.

Shallow Water Waves

Shallow water waves are waves with wave length long compared to the depth. Fig. 2 shows the characteristic pattern of particle motion for shallow water waves. Notice that the horizontal amplitude of the particle motions is nearly the same at all depths. We say that such waves are essentially one-dimensional, since the horizontal fluid motion depends only on time and horizontal distance from a reference point. Be-

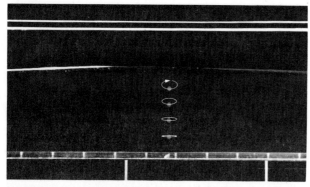

2. Elliptical particle paths typical of periodic waves where wave length is greater than depth ("shallow water waves").

cause they are simpler and yet display most of the important features of wave motion, we will deal with one-dimensional waves hereafter.

Step Waves

Not all waves are periodic. The upper photograph in Fig. 3 shows a single progressive, depth-increasing wave moving to the right in the horizontal water channel. The wave was generated by moving a "piston" at nearly constant velocity at the left end of the channel. The two arrows show the leading and trailing edges of the wave. Ahead of the wave (to the right), the fluid is at rest. Superposed on the photograph are the positions of a particle at equal time increments after passage of the step wave. The particle moves through a much shorter distance than the wave. Since the particle positions were recorded at equal time intervals, it is clear from their spacing that the wave is a region of acceleration. As the step wave continues to the right, particles behind it stop accelerating and move at a constant speed as shown in the lower photograph. Fluid acceleration occurs only during the time

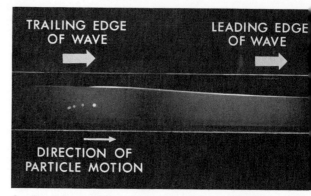

3. Particle position at equal time increments durin (above) and after (below) passage of a step wave that in creases depth ("compression wave"). Note forwar acceleration of particle during wave passage and constan forward velocity of particle after wave passage.

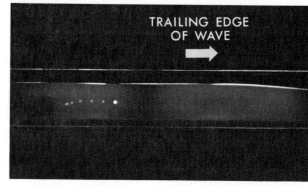

the depth is changing; i.e., *waves are accelerators.*

The depth in a given length of the channel can only increase if more fluid is moving into it than is moving out (a simple statement of the conservation of matter) This is apparent in the upper photograph of Fig. 3 where the fluid is stationary at the leading edge of the wave and moving to the right at the trailing edge.

Fig. 4 shows a depth-decreasing wave moving to the right. It was generated by "pulling a gate," allowing fluid to flow toward a shallower region at the left. More fluid is flowing out of the left side of a given length of the channel than is flowing in on the right side, causing the depth to decrease. The decrease

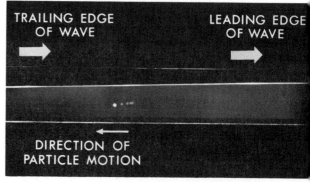

4. Particle position at equal time increments during passage of a step wave that decreases depth ("expansion wave"). Note backward acceleration of particles during passage.

in depth is accompanied by an acceleration of the fluid toward the left, i.e., in the *opposite* direction of the wave motion, as the multiple exposure in Fig. 4 shows. As the step wave moves on to the right, the fluid particles behind it continue to move to the left with constant speed.

The force that causes the fluid to accelerate is a force arising from differences in pressure. The pressure at the bottom of the channel increases as the wave of Fig. 3 goes by. For shallow water waves the vertical acceleration of the fluid is small compared with the gravitational acceleration. Thus, the vertical pressure variation is nearly hydrostatic, that is, the pressure is atmospheric at the surface and increases almost linearly with depth. The pressure difference in a horizon-

5. Pressure distribution at leading and trailing edges of a compression wave (superposed at left). Note that difference in pressures is constant at all depths.

tal direction is nearly the same at all depths, as shown in Fig. 5, where the pressure distributions some distance apart are superposed. This means that the horizontal acceleration of fluid particles is nearly the same at all depths. As the wave of Fig. 3 passes, the pressure on a fluid particle increases. For this reason, we may call this type of depth-increasing wave a *compression wave*. Note that in a compression wave the fluid acceleration is in the direction of the wave motion.

In the wave of Fig. 4 the fluid acceleration is in a direction opposite to the wave motion. This is so because the pressure is higher at the leading edge of the wave than at the trailing edge. As this depth-decreasing wave passes, the pressure on the fluid particle decreases; hence, we may call this type of wave an *expansion wave,* in contrast to the compression wave. In both cases, the horizontal pressure gradient arises because of the gravitational force acting on the fluid. Thus it is really gravity that causes these waves. Whenever the fluid surface is not horizontal, a horizontal pressure gradient is produced by gravity that accelerates the fluid horizontally toward the shallower depth.

Expansion Waves Flatten Out

As it advances, the expansion wave flattens out, the trailing edge falling behind the leading edge as shown

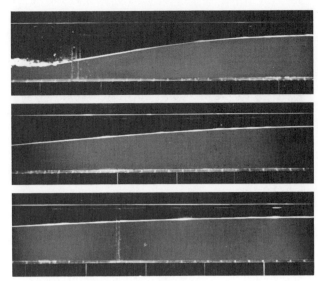

6. Camera moving with an expansion wave shows how it "flattens out" as it moves.

in Fig. 6. Differences in local fluid velocity and depth within the wave itself alter its shape as it advances. There are two reasons for this. Where the depth is greater, that portion of the wave travels faster, since the wave speed relative to the fluid is given by $\sqrt{gh}$ where h is the depth. The velocity of the leading edge of the expansion wave relative to the fluid is $\sqrt{gh_1}$, where h_1 is the depth at the leading edge. Similarly, the velocity of the trailing edge relative to the fluid is $\sqrt{gh_2}$, where h_2 is the depth at the trailing edge. Clearly the local wave speed at the shallower trailing edge is less than the local wave speed at the deeper leading

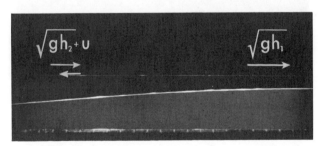

7. Speeds of leading and trailing edges of an expansion wave. Rearward drift speed at trailing edge further decreases trailing edge speed relative to leading edge.

edge. This is one reason why an expansion wave tends to spread out as it moves.

But there is another factor at work here. The local wave velocity given above is relative to the local fluid velocity. Now we have already seen that the fluid at the trailing edge of an expansion wave is moving away from the fluid at the leading edge with a velocity u (Fig. 4). So, relative to the leading edge, the slower speed of the trailing edge is rendered even slower by virtue of a local fluid velocity there in the opposite direction. This is shown in Fig. 7.

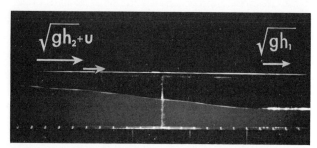

8. Speeds of leading and trailing edges of a compression wave. Forward drift speed at trailing edge further increases trailing edge speed relative to leading edge.

Compression Waves Steepen

The fact that the local wave velocity increases with depth also explains why the compression wave tends to steepen. The local wave velocity is higher at the deeper trailing edge of the wave than at the shallower leading edge. Furthermore, in compression waves the fluid at the trailing edge is moving toward the leading edge. This local fluid speed, u, adds to the increased local wave speed, $\sqrt{gh_2}$, at the trailing edge, causing the trailing edge to overtake the leading edge even more quickly. This is shown in Fig 8.

The Breaking of a Compression Wave

What happens when a compression wave becomes so steep that its slope is vertical? Anyone who has been to the beach knows the answer: the wave topples over as shown in Fig. 9. The leading edge of the toppled wave travels faster than it did before it toppled. Fur-

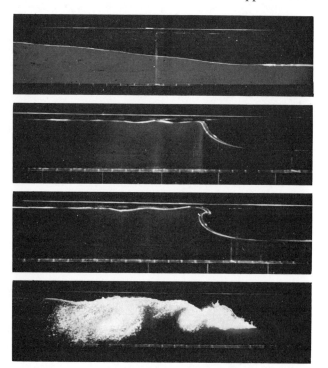

9. Camera moving with compression wave shows how it steepens and topples over to form a positive surge wave.

thermore, the toppled wave continues in its turbulent condition, does not change its shape, and travels at constant speed.

The Positive Surge Wave

The speed of such a toppled wave, which is called a *positive surge wave,* depends on both the depth ahead (h_1) and the depth behind the wave (h_2).

The larger the depth behind the wave, the faster the wave travels. In fact, it can be shown that the speed of a positive surge wave, V_s, is given by:

$$V_s = \sqrt{g\,\frac{(h_1 + h_2)}{2}\left(\frac{h_2}{h_1}\right)}$$

Small-amplitude waves ahead of the surge wave move at the speed, $\sqrt{gh_1}$, relative to the fluid. If you

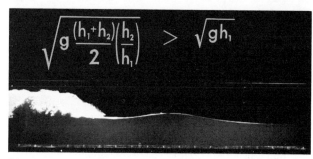

10. Positive surge wave overtaking a small-amplitude wave.

compare these latter two expressions, you will see that this local wave speed is always less than the speed of the surge wave. Thus the surge wave catches up to small-amplitude waves ahead of it as shown in Fig. 10. Small-amplitude waves behind the surge wave move relative to the fluid at a speed, $\sqrt{gh_2}$. However, be-

11. Small-amplitude wave overtaking a positive surge wave.

hind the surge wave the fluid is moving; there is a drift speed of the fluid $u = V_s\left(1 - \dfrac{h_1}{h_2}\right)$. The drift speed plus the local wave speed, $\sqrt{gh_2}$, is always greater than the speed of the surge wave (Fig. 11). Thus, small-amplitude waves behind the surge wave, and moving in the same direction as the surge wave, always catch up to it.

The Hydraulic Jump

As the positive surge wave moves down the channel, there is zero velocity ahead of it and a drift velocity behind it. In a reference frame moving with the surge, the fluid ahead appears to be moving into the wave front. The wave appears stationary and the fluid behind has a net velocity in the opposite direction to the drift speed. The flow is steady in this reference frame.

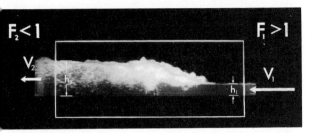

12. Control surface for analyzing conditions upstream and downstream of a hydraulic jump. Froude number ahead, F_1, is greater than one; Froude number behind, F_2, is less than one.

Exactly this situation occurs in a flume, as shown in Fig. 12. This is a stationary wave called a *hydraulic jump*. The structure of this turbulent wave is very complicated and difficult to analyze. However, if we consider the fluxes of mass and momentum into and out of a control surface, like the one in Fig. 12, the velocity V_2 and depth h_2 behind the hydraulic jump can be determined in terms of the velocity V_1 and depth h_1 ahead of it:

$$V_2 h_2 = V_1 h_1 \qquad \text{(conservation of mass)}$$

$$\frac{1}{2}\rho g\, h_2^2 + \rho V_2^2 h_2 = \frac{1}{2}\rho g h_1^2 + \rho V_1^2 h_1 \quad \text{(conservation of momentum)}$$

Solving these two relations for V_2 and h_2 gives:

$$\left.\begin{array}{c} \dfrac{h_2}{h_1} = \dfrac{\sqrt{1+8F_1^2}-1}{2} \\[2mm] \dfrac{V_2}{V_1} = \dfrac{2}{\sqrt{1+8F_1^2}-1} \end{array}\right\} \quad \text{where} \quad F_1 = \frac{V_1}{\sqrt{gh_1}}$$

The ratio of entering fluid speed V_1 to the local wave speed $\sqrt{gh_1}$ is called the Froude number ahead of the jump, F_1. It is always greater than unity, that is, the fluid speed entering the jump is greater than the local wave speed. We say the flow ahead is supercritical (or "shooting"). The Froude number behind the jump, F_2, is always less than unity — that is, the fluid speed behind is subcritical (or "tranquil"). This is easily demonstrated in a flume. A disturbance behind the jump moves upstream into the jump. A disturbance ahead is washed downstream into the jump. This is just another way of describing the effects shown in Figs. 10 and 11 in a different reference frame. Small-amplitude waves behind a positive surge wave catch up to it. Small-amplitude waves ahead are overtaken.

Dissipation of Energy in a Hydraulic Jump

Energy is dissipated in a hydraulic jump. Traversing a pitot tube downstream through the jump shows that the total head decreases. Mechanical energy is converted into internal energy. The jump in Fig. 12 is dissipating about one-third of a horsepower. Below big dams like the one in Fig. 13, hydraulic jumps may

13. Large hydraulic jump below a dam (U.S. Army Corps of Engineers).

14. Circular hydraulic jump on a plate below a kitchen water tap.

dissipate thousands, even millions, of horsepower. The temperature rise is hardly noticeable, however, since a total head loss of 778 feet is required to heat water one degree Fahrenheit.

It is straightforward to show that the loss in total head is given by

$$H_1 - H_2 = \frac{(h_2 - h_1)^3}{4h_1 h_2}$$

where H_1, H_2 represent total head ahead of and behind the jump, respectively.

You can see hydraulic jumps all around you. Fig. 14 shows one you can see in your kitchen sink. The water spreads out at a supercritical speed from under the tap, goes through a cylindrical hydraulic jump, and moves at subcritical speed to the edge of the dish.

Sound Waves and Shock Waves

Although gravity waves in liquids are easily seen, they are not the only common waves we encounter in fluids. Compressibility waves in gases are also very common; for instance, sound waves are small-amplitude compressibility waves. Large-amplitude compressibility waves are *shock waves*. Compressibility waves move at very high speeds; furthermore, they can usually be seen only with the aid of special optical devices (such as schlieren and shadowgraph systems). On rare occa-

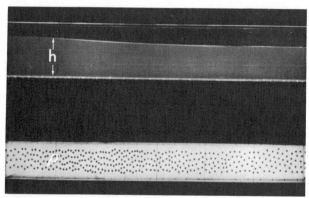

15. Analogy between depth h, in a surface gravity wave and particle density, ρ, in a compressibility wave.

sions, if the light is just right, shock waves may be seen with the unaided eye. However, the passage of a compressibility wave is easily detected by the changes in pressure associated with it; our ears, for instance, are very sensitive detectors of pressure changes.

As a model for a gas, the bottom half of Fig. 15 shows many particles floating on the surface of water. The number of particles per unit of surface area represents the density. As a compression wave comes along, the particles accelerate in the direction of the wave motion, and the density increases — that is, the particles move closer together (see Fig. 15).

Much of what we have discussed concerning one-dimensional surface gravity waves applies equally well to compressibility waves if we simply replace the depth of the fluid, h, by the density of the fluid, ρ, as shown in Fig. 15. In fact, a compressibility wave may be described as a region of increased or decreased density that moves relative to the fluid. Just as depth is proportional to the amount of fluid per unit length in an open channel, so density is proportional to the amount

of fluid per unit length in a closed channel. The net flow into a length of open channel increases the depth. The net flow into a length of closed channel increases the density. In a compressibility wave, the pressure depends on the density, just as, in a gravity wave, it depends on the depth.

Although surface gravity waves and compressibility waves behave quite similarly, they are propagated by entirely different physical mechanisms. The fluid acceleration in waves in gases is due to elastic forces associated with the compressibility of the fluid rather than to gravitational forces.

Steepening of Sound Waves into Shock Waves

Recall that the local wave speed of a surface gravity wave increases with depth; the speed of a small-amplitude isentropic compressibility wave increases with density. This increase of sound speed with density causes compression waves to steepen. When we say that a compression wave steepens, we mean that the density and velocity gradients in the wave increase as the wave propagates. What happens when these gradients become very high, that is, when the wave becomes very thin? The compressibility wave cannot topple over like the surface gravity wave; instead, the density and velocity gradients reach limiting high values in a thin region and then remain constant. In compressibility waves, these limiting high gradients are determined by the effects of viscosity and heat conduction which, up to this point, we have been able to ignore. The reason that viscous forces become significant here is that fluid particles going through a steep wave experience high rates of deformation and high temperature gradients; viscous forces and heat conduction thus become enormously large. The steepening tendency of inertia is therefore opposed by the diffusive actions of viscosity and heat conduction; the steep wave of stationary form that results is called a shock wave.

Standing Shock Waves

Standing shock waves can be produced in a supersonic wind tunnel (Fig. 16). The gas moves at supersonic speed ahead of the shock (Mach number M_1 greater than one) and at subsonic speed behind the shock (Mach number M_2 less than one). This is analogous to the stationary hydraulic jump in the flume (Fig. 12) with fluid moving at supercritical speed ahead of the jump and with subcritical speed behind it. Recall that the total pressure (total head) dropped in going through a hydraulic jump. In the same way, total pressure drops in going through a shock wave because some of the mechanical energy is changed into internal energy by the dissipative action

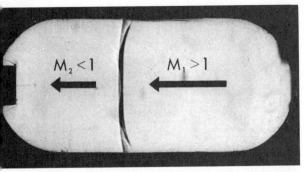

16. Stationary normal shock wave in a supersonic wind tunnel. Mach number ahead, M_1, is greater than one; Mach number behind, M_2, is less than one.

of viscosity and heat conduction. Unlike the hydraulic jump, a shock wave is extremely thin, on the order of a few mean free paths thick, about one-millionth of an inch at room temperature and pressure.

Moving Shock Waves

A moving shock wave is what you hear when there is an explosion; it is the thunder that accompanies lightning; it is the sonic boom you hear from a super-

17. Shock tube for producing moving normal shock waves in the laboratory.

18. Surge wave produced by pulling a gate in a water channel.

19. Circular trace on ground of a spherical shock wave produced by an explosion.

20. Shock tube for producing very strong normal shock waves.

sonic airplane. We can produce moving shock waves in the laboratory in a *shock tube,* like the one shown in Fig. 17. We put high-pressure gas on one side of a frangible diaphragm, and low-pressure gas on the other side. When the diaphragm is broken, a shock wave is created which moves at supersonic speed down the tube. By pulling a gate in the hydraulic channel, we produce a positive surge wave, as shown in Fig. 18; this is analogous to breaking the diaphragm to produce a shock wave in the tube. There is a wind behind a shock wave, just like the drift speed behind a positive surge wave. An explosion sends out a spherical shock wave followed by a strong wind, which is the air moving at the drift speed (Fig. 19).

We can produce very strong shock waves in the shock tube shown in Fig. 20. The high-pressure section is made of steel and the low-pressure section is made of pyrex tube. The diaphragm in the tube of Fig. 20 is

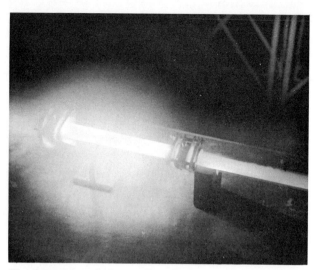

21. Ionization flash produced by passing strong shock wave through argon in a pyrex tube.

made of .050″ thick copper, because it must withstand very strong pressure forces. The temperature behind a very strong shock wave can become so high that the gas molecules are torn apart (dissociation), and even electrons stripped off the atoms (ionization). Ionized gases emit light, and hence this phenomenon is easily seen. Fig. 21 shows how the argon in the low-pressure section of the tube ionizes when a strong shock passes through it.

Diffraction of Shock Waves Over Obstacles

So far we have concentrated on one-dimensional waves; two and three-dimensional waves involve the same basic concepts but can become very much more complicated. Fig. 22 shows the shock pattern shortly after passage of a normal shock wave (moving toward right) over a sharp-edged vertical plate. Notice the vortices produced at the sharp edges, and the shock reflections and interactions.

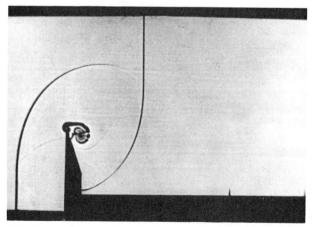

22. Shock pattern shortly after passage of normal shock wave (moving toward right) over sharp-edge vertical plate.

Summary

There are two main requirements to produce wave motion in fluids. First, it must be possible to have a *local accumulation of mass*, which is due to depth changes in the gravity wave and to density changes in the compressibility wave. Second, there must be *restoring forces*, pressure forces, which are due to gravity and depth gradient in the gravity wave and to density gradient in the compressibility wave.

References

1. Prandtl, L. *Essentials of Fluid Dynamics*, Hafner, New York, 1952, pp. 86-98, 271-277
2. Rouse, H. *Fluid Mechanics for Hydraulic Engineers*, Dover, New York, 1961, Chs. XIII, XV
3. Liepmann, H. W., and Roshko, A. *Elements of Gasdynamics*, Wiley, New York, 1957, pp. 57-84, 329-331
4. Shapiro, A. H. *The Dynamics & Thermodynamics of Compressible Fluid Flow*, Ronald Press, New York, 1953, pp. 112-158 of Vol.

Flow Instabilities

Erik L. Mollo-Christensen
MASSACHUSETTS INSTITUTE OF TECHNOLOGY

Introduction

Flow Instabilities play important roles in all branches of fluid mechanics.

The purpose of this film is to show several examples of flow instabilities and to point out some of their common features. The concepts of a critical condition and frequency selective amplification are illustrated in the context of an experimental investigation of a particular flow instability — namely, wind-generated waves.

These notes, in addition to serving as a reminder of the film, give some information on the techniques used in designing and performing the experiments, as well as some references for further study.

Three Examples of Flow Instability

A spectacular but poorly understood instability involving an interplay among viscous, surface tension, and buoyancy effects is illustrated in Fig. 1. A drop of

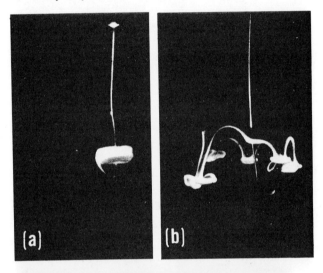

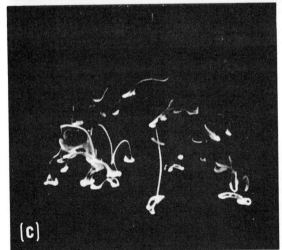

1. A drop of light cream falling into cold water creates an unstable flow pattern.

light cream is released from an eyedropper, one millimeter above a water surface. A vortex ring forms (Fig. 1a) then develops an instability and breaks up into smaller vortex rings (Fig. 1b). Each of the smaller rings goes through the same cycle, and a cascade of rings develops (Fig. 1c). Light cream (half cream, half milk) and ice water work quite well as fluids.

A familiar flow instability occurs in the buoyant smoke plume of a smoldering taper (Fig. 2). The smoke plume is steady near its source. Higher up, waves occur, and the plume becomes unstable and then turbulent. In a very still room, or if the plume is protected from room currents by glass walls, the plume may rise quite far before becoming turbulent.

A third instability, easily seen by using a kitchen faucet, is a low-speed capillary jet of water approaching an obstacle (Fig. 3). The jet may be stable and smooth, or unstable and appear to buckle, depending upon its

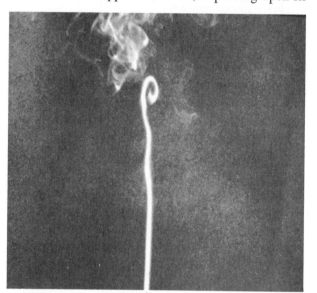

2. A smoke plume develops waves and then becomes turbulent.

3. A capillary jet of water shows a buckling instability.

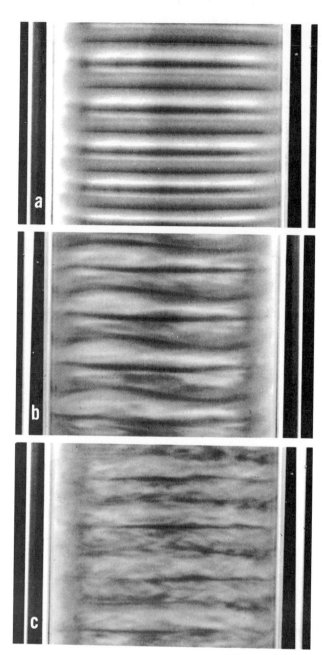

4. The flow in the circular space between two cylinders, the outer cylinder stationary, the inner rotating, is stable at low angular velocity, Ω. At higher Ω, Taylor vortices appear (a); at still higher Ω, waves develop (b); and ultimately, turbulence (c).

length. In Fig. 3 the jet is short, and capillary waves occur along its edge. This instability has been treated theoretically (Ref. 1, Article III, Ch. XII).

The Existence of a Critical Condition

The flow between concentric cylinders, produced by rotating the inner cylinder with the outer cylinder stationary (Fig. 4), has several critical speeds. At very low rotation speeds, the flow between the cylinders is a simple Couette flow and is stable. Instability, in the form of Taylor vortices (seen as bands from the side in ig. 4a), appears above a critical angular velocity, Ω, measured in terms of Taylor number, $\nu \Omega R^2/2$, where ? is the radius and ν the kinematic viscosity (Ref. 2).* At a second critical Taylor number, a second instability occurs (Fig. 4b) and the Taylor vortices develop circumferential waves. At still higher Taylor numbers (Fig. 4c), transition to turbulence occurs.

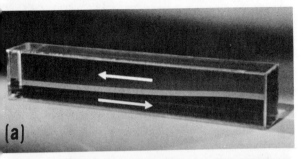

(a)

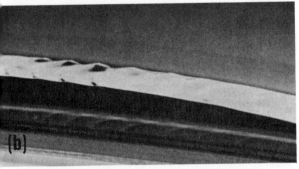

(b)

5. Two fluids of different density completely fill a tank. The shear flow created by tilting the tank (a) goes unstable and waves appear on the interface (b).

The flow of one fluid on top of another, heavier fluid can produce a shear instability. The closed rectangular tank of Fig. 5 contains two fluids, differing in density by 2 percent. The interface between the two fluids remains virtually horizontal when the tank is tilted very slowly. A sudden change in tank tilt causes internal waves. When the lower fluid flows to the right the upper fluid must flow to the left to fill up the space. (Fig. 5a) This produces a velocity shear at the interface between the two fluids. When the parameter $\rho \dfrac{(\Delta U)^2}{gl\Delta\rho}$ exceeds a critical value, the flow is unstable and small waves appear (Fig. 5b). Here l is a characteristic length, ΔU is the velocity difference between the upper and lower fluids, $\Delta\rho$ is the density difference, ρ is the average density, and g is gravity. This shear instability is often called Kelvin-Helmholtz instability. At still larger velocity shears, the waves break, resulting in turbulence. The particular fluids used in this experiment were: "Pine-X," an aliphatic hydrocarbon paint thinner, and "Doversol," a stove-fuel alcohol. The alcohol will weaken a plexiglass container in the course of less than a year.

* (2) Ch. 2. A more extensive account can be found in Ref. 1, Ch. VII.

Wind Blowing Over a Liquid Surface

The apparatus shown in Fig. 6 is used to study the generation of surface waves by wind. Air enters at the left, flows through the test section, and is sucked out by vacuum cleaners. The air speed may be varied and disturbances of different frequency may be introduced into the upstream end of the liquid surface by vertical oscillation of a horizontal rod placed across the tank. The liquid in the tank is a mixture of glycerine and

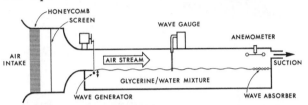

6. Wind tunnel for studying wave generation by wind.

(a)

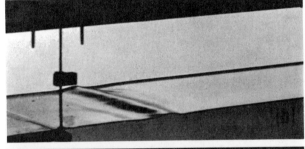

(b)

(c)

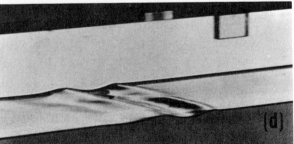

(d)

7. Just above critical wind speed, a disturbance introduced by the wavemaker grows as it travels downstream.

water, which causes capillary waves to be strongly damped, leaving just gravity waves unstable. At zero or low wind speeds a disturbance is damped as it travels away from the wavemaker. At wind speeds above the critical speed the same disturbances grow (Fig. 7). At air speeds well above critical speed, even with the wavemaker off, small accidental disturbances in the air stream are sufficient to make waves which grow

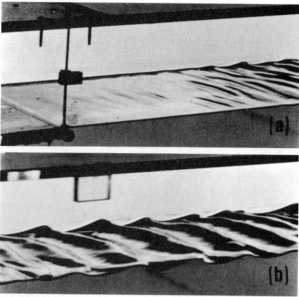

8. Well above critical wind speed, waves appear spontaneously and grow (a). A narrow range of wave lengths is predominant (b).

(Fig. 8a). Downstream these waves appear quasi-regular (Fig. 8b) because the wave length having the highest growth rate ultimately dominates.

The instability mechanism can be pictured as an amplifier (Fig. 9). The airstream is its energy supply, and its gain and frequency characteristics depend upon wind speed and other flow parameters. The input is disturbances in the liquid or the air which may be intentional or accidental. Surface waves are the output, and they may be damped or amplified. When the amplifica-

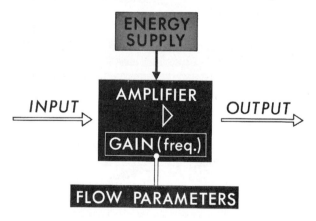

9. The mechanism of instability can be thought of as a frequency-selective amplifier.

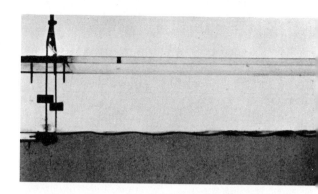

10. Wavemaker at constant frequency, wind above critical value, waves grow with distance.

tion ratio, the gain, exceeds unity we have instability. Fig. 10 shows the wavemaker producing sinusoidal disturbances with the wind speed above critical value. The waves grow with distance from the wave generator.

The essential features of the amplifier analogy are demonstrated in the generation of waves by wind (Ref. 3). We measure the amplifier gain as a function of the frequency of input disturbances and wind speed. In the experiment, waves of one frequency are being generated and the wind speed is increased continuously from a low value. The amplitude of the waves is

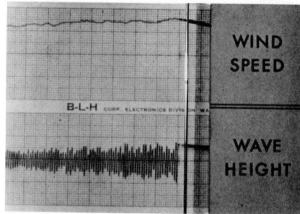

11. Recorder output.

measured downstream with a variable-capacity transducer. Fig. 11 shows a portion of a typical recording of downstream wave height vs. wind speed. The experiment is repeated at the same wave-generator amplitude but at different frequencies. When the recorder tracings are mounted on a chart of frequency vs. wind speed, a pattern becomes apparent (Fig. 12). The curve passing through the points where amplification first occurs is the neutral curve. To the left of this curve the waves are damped; the flow is stable. To the right of the curve, the waves are amplified and instability is present. To emphasize the fact that the amplifier has a limited band width, the experiment is performed the other way. Keeping the air speed constant, the frequency is slowly increased, starting at zero. The resulting record also fits the neutral curve

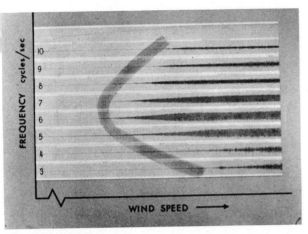

12. The neutral curve.

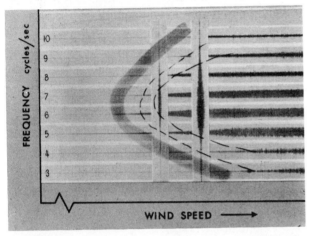

13. Varying the frequency at constant air speed shows the amplified frequency band. Contours of constant amplification have been added.

(Fig. 13). Crude contours of constant amplification rate are shown in Fig. 13. Without the wave generator, but with the wind speed above critical, accidental disturbances present will trigger the instability. The amplified waves occur first in the frequency range and at the wind speed corresponding to the nose of the neutral stability curve. A classical example of this experimental approach is given in Ref. 4.

We can develop a simplified physical explanation of this instability. The pressure pattern on the liquid surface depends on the speed of the wind relative to the waves. If the air were inviscid, pressure minima would occur at the wave crests and pressure maxima in

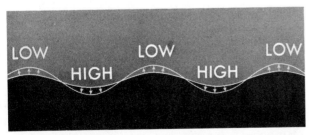

14. Potential flow would create high air pressure at the troughs and low pressure above the crests.

the troughs as shown in Fig. 14. The wave speed depends on wave length, liquid depth, gravity, and surface tension. Since waves of different wave lengths have different speeds of propagation (Fig. 15), they feel different relative winds. For all except the very

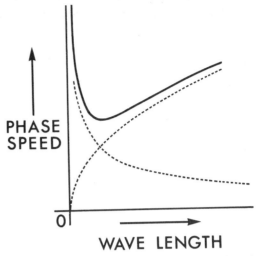

15. The wave phase speed depends upon wave length.

shortest waves, which are dominated by surface tension, the longer the wave the higher its phase speed and the less the relative wind for a given air speed. Thus, very long waves tend to be stable. Viscous damping in the liquid dominates the very shortest waves, making them stable. Therefore, only waves with an intermediate range of wave length may be unstable. The boundary layer in the air must be taken into account to obtain the final clue as to how the pressure forces do work on the water. The air velocity profile relative

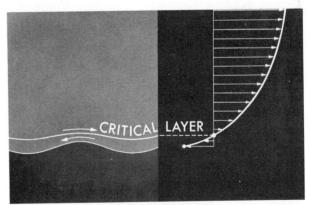

16. Air velocity profile and the critical layer in a coordinate system moving with a wave.

to moving waves is shown in Fig. 16. Across the *critical layer,* where the relative air velocity is zero, there is an exchange of vorticity as the air moves across the waves, resulting in a shift of phase of the pressure distribution (Fig. 17) so that high pressure occurs on the upwind face of the waves. This is how pressure does work on the liquid, putting energy into the waves.

In summary, the mechanism of flow instability be-

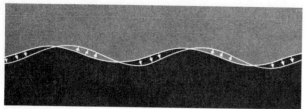

17. Approximate actual pressure distribution.

haves like a frequency-selective amplifier whose characteristics depend on flow parameters. The instability can be investigated experimentally by means of controlled disturbances, measuring the amplification as a function of flow parameters. The experiment may provide a valuable guide to an explanation of the physical mechanism underlying the instability.

A Density-Difference Surface-Tension Instability

A layer of heavy fluid above a lighter fluid will tend to fall down, exchanging places with the lighter fluid, unless there are restraining effects. Surface tension may provide this restraint in a container of finite size. Fig. 18 shows a container with one fluid placed above a lighter fluid where the bottom of the container is open. If the interface is disturbed, it returns to its equilibrium position. However, if a surface-active agent is added near the interface to lower the surface tension, instability results; the heavier fluid pours down, the lighter fluid pours up. The neutral-stability case lends itself to a simple mathematical analysis. If $\eta(x)$ is the vertical position of the interface (x being horizontal distance), the pressure perturbation at the surface is $g\Delta\rho \cdot \eta(x)$ where $\Delta\rho$ = density difference, g = gravity. This pressure difference must be held in check by surface tension: $\sigma \frac{d^2\eta}{dx^2} = g\Delta\rho\eta$. Solving for $\eta(x)$ with boundary condition $\eta = 0$ at $x = 0$ and

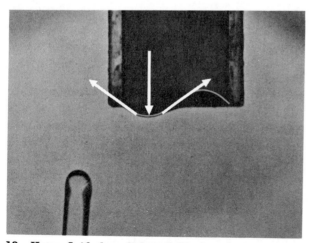

18. Heavy fluid above lighter fluid. A surface deflection downward increases the hydrostatic pressure, but surface tension pulls the bulge back up.

$x = l$, one finds neutral stability when $\frac{gl^2\Delta\rho}{\sigma} = (2\pi)^2$ after having imposed the constraint that the average deflection is zero $\left(\int_o^l \eta \, dx = 0\right)$.

A Convective Thermal Instability

The density of most fluids decreases with increasing temperature. Heating a fluid from below will therefore create an unstable density stratification, and the lighter fluid will tend to rise, producing thermal convection. The speed with which a fluid parcel rises depends upon its size, its excess buoyancy, and the fluid viscosity. As it rises, it encounters colder fluid and loses heat, and therefore buoyancy, to its surroundings. If it rises fast enough to get to colder (and therefore denser) surroundings faster than its own rate of increase in density, it will keep on going and we have instability (Fig.

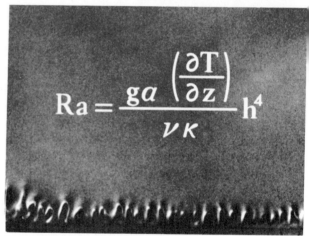

19. A thin layer of fluid confined between parallel vertical plates with a heating element near the bottom. Above a critical value of the Rayleigh number it goes unstable.

19). The critical parameter is the Rayleigh number $Ra = \frac{g\alpha\delta T/\delta z}{\kappa \nu}h^4$ where $\alpha = \frac{1}{\rho}\frac{\delta\rho}{\delta T}$, and z is distance, increasing downward. The stratification is given by $\frac{g}{\rho}\frac{\delta\rho}{\delta z} = g\alpha\delta T/\delta z$, and the larger this term is and the smaller the viscosity and the thermometric conductivity κ, the less stable the system will be. The smaller the depth, h, the more efficient are the restraining influences of viscosity and thermal conductivity. Instability occurs above a critical value of Rayleigh number. For an analytical treatment, see Ref. 1, Ch. 2.

A Thermal Instability Resulting in Convection Cells

When a shallow layer of fluid is heated from below and cooled from above, the instability discussed in the previous section results in a strikingly uniform array of convection cells called Benard cells. In Fig. 20 these

20. Above the critical Rayleigh number, Bénard cells appear in a layer of fluid.

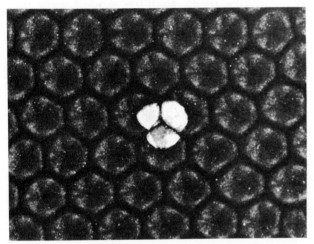

21. A drop of shiny fluid put at the corner between three cells will spread along the bottom and rise to the surface, showing the location of a cell as seen from below.

have been produced in a frying pan* with silicone oil. The flow is made visible with aluminum flakes and iron oxide. Each hexagonal cell is made up of six equilateral triangles (Fig. 21). The cell pattern as seen from below also consists of hexagons, but these are made up of two triangles from each of three adjacent hexagons as seen from above. Since the Rayleigh number, Ra $= \frac{g\alpha\Delta T h^3}{\kappa\nu}$, varies as the third power of the depth, h, the critical condition for instability is clearly demonstrated by tilting the frying pan (Fig. 22). No cells are visible until the depth reaches the critical value. At depths greater than critical, the cell size increases roughly proportional to depth. Cellular convection patterns may also be seen occasionally in the atmosphere where clouds may make them visible.

---

* Modified as follows: a layer of "cerrobend," a dense, low-melting-point (73°C) alloy was melted in the bottom of the pan. A flat aluminum plate was floated on the alloy and the alloy was allowed to solidify. This produced a level bottom surface at uniform temperature.

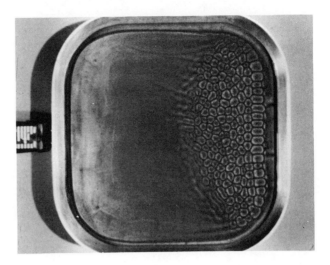

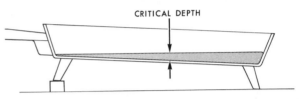

CRITICAL DEPTH

22. Varying the Rayleigh number by tilting the pan. The Rayleigh number is below critical on the left and supercritical on the right.

Shear Instability Without Stratification

Shear flows are especially unstable, without stabilizing density stratification. A jet of water into water (Fig. 23) creates shear layers which become unstable with a characteristic "most unstable" wave length.

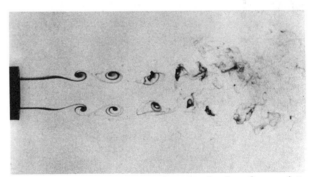

23. The shear layer surrounds a circular jet of water into water. It is unstable above a critical Reynolds number. The dye streamers are introduced at the top and bottom of the jet.

Non-linear effects and secondary instabilities appear which finally result in turbulence. The critical parameter for the initial instability is the Reynolds number based on shear velocity difference and shear layer thickness.

The Karman Vortex Street

A familiar example of flow instability is the flow around a cylinder. Below a critical value of Reynolds

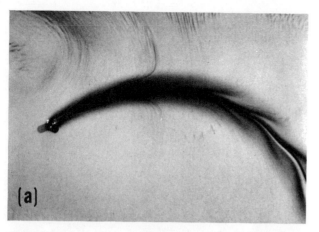

(a)

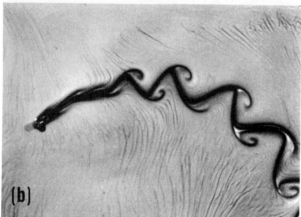

(b)

24. Flow around a cylinder. (a) Steady, stable flow with a slight wake instability. (b) Unstable flow.

Summary

Steady flows may become unsteady for certain ranges of the relevant flow parameters. The instability takes its energy from the mean flow or externally supplied heat and the mechanism can be thought of as a frequency selective amplifier whose characteristics are governed by flow parameters.

One can observe flow instabilities in the atmosphere both on the large and on the small scale, in smoke from chimneys, in rivers, in flickering flames, and in fact almost everywhere, along with their ultimate consequences: turbulence or random waves.

References

1. Chandrasekhar, *Hydrodynamic and Hydromagnetic Stability*, Oxford University Press, 1961
2. Lin, C. C., *Theory of Hydrodynamic Stability*, Cambridge University Press, 1955
3. Phillips, O. M., *The Dynamics of the Upper Ocean*, Cambridge University Press, 1966
4. Klebanoff, P. S. and Schubauer, G. B., NACA Report 1289, *Contributions on the Mechanics of Boundary-Layer Transition*, 1956

number, $\mathrm{Re} = \dfrac{Ud}{\nu}$, the flow is steady (Fig. 24a), although there is a weak instability in the wake. Above the critical Reynolds number (Fig. 24b), the steady flow around the cylinder becomes unstable, resulting in a periodic Karman vortex street in the wake. Just at critical Reynolds number the flow stability is precarious, and a small disturbance will cause the flow to become unstable. This instability causes telephone wires to "sing." The sound comes from small pressure fluctuations associated with the periodic vortex shedding.

Cavitation

Phillip Eisenberg
HYDRONAUTICS INCORPORATED

INTRODUCTION

Cavitation is defined as the process of formation of the vapor phase of a liquid when it is subjected to reduced pressures at constant ambient temperature. Thus, it is the process of boiling in a liquid as a result of pressure reduction rather than heat addition. However, the basic physical and thermodynamic processes are the same in both cases.

A liquid is said to *cavitate* when vapor bubbles form

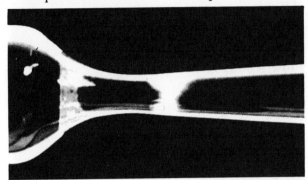

1. Patches of cavitation bubbles appear just downstream of the throat of a venturi tube.

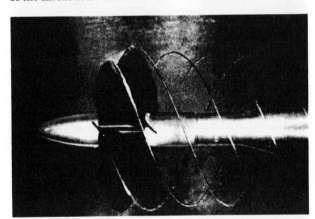

2. Cavitation in the cores of the tip vortices from the blades of a marine propeller forms a helical pattern. Cavitation on the faces of the blades is more easily seen in the close-up of Fig. 7.

and grow as a consequence of pressure reduction. When the phase transition results from hydrodynamic pressure changes, a two-phase flow composed of a liquid and its vapor is called a *cavitating flow*. Cavitating flow may be seen (and heard) as water flows through a glass venturi tube (Fig. 1), an experiment first exhibited by Osborne Reynolds in 1894. According to Bernoulli's equation, where the velocity is increased, the pressure is decreased. At sufficiently high flow rates, the liquid in the throat, where the velocity is highest and the pressure is lowest, begins to boil. The small bubbles formed there are filled with cold steam and other gases diffused from the liquid. Another example of cavitation occurs in the low-pressure regions on marine propellers at high rotation rates (Fig. 2).

TYPES OF CAVITATION

Flow About Hydrofoils

In the film, experiments are performed in a circulating water channel, where the water speed, absolute

3. As the flow speed is increased, cavitation first occurs at the intersection of the hydrofoil and the vertical supporting strut. With further increase in speed, cavitation on the blade itself occurs first in the low-pressure core of the laminar boundary-layer separation region.

4. At higher flow speeds, cavitation begins near the minimum-pressure line close to the leading edge.

5. The same flow as in Fig. 4, but here seen with stroboscopic lighting. The cavitation region is made up of individual bubbles.

pressure, and temperature can be varied independently of each other. A thick, symmetrical hydrofoil is suspended from a strut and set at an angle of attack to lift upward. With the ambient pressure well below atmospheric and held constant, the flow speed is increased until cavitation occurs. Cavitation starts first at the intersection of the strut and hydrofoil (Fig. 3), where the presence of the strut causes a greater pressure reduction than elsewhere on the foil. Cavitation on the foil itself first occurs in the low-pressure core of the laminar boundary layer separation region (Fig. 3). At higher speed, and thus higher Reynolds number, transition from laminar to turbulent boundary layer occurs, and cavitation begins near the minimum-pressure line close to the leading edge (Fig. 4). In Fig. 4, the cavitation region is seen under steady incandescent lighting. Under stroboscopic lighting, we see that the cavitating region is actually made up of individual bubbles (Fig. 5). In the film, high-speed motion pictures reveal that each bubble grows as long as it is in the low-pressure region. The bubbles collapse as they are swept downstream into the high-pressure region near the trailing edge. Individual bubble cavitation is characteristic of forms with gentle pressure gradients, such as those on the foil with a well-rounded leading edge (Fig. 5).

Supercavitating Hydrofoil

When the leading edge of a hydrofoil is sharp, cavitation begins at this edge (Fig. 6a). A continuous,

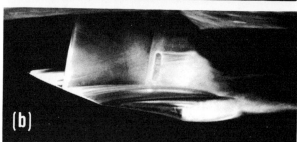

6. When the leading edge of a hydrofoil is sharp (a), cavitation begins first at this edge. When the angle of attack or the flow speed is increased, or the ambient pressure reduced, to the point where the cavitating region extends beyond the trailing edge (b), the flow is called "supercavitating." This type of hydrofoil is designed to operate efficiently in this condition, and is referred to as a *supercavitating* hydrofoil.

vapor-filled cavity is formed, rather than a mass of small individual bubbles. The cavity grows when the angle of attack is increased, when the ambient pressure is reduced, or when the water speed is increased. When the cavity extends beyond the trailing edge of the hydrofoil, the flow is called a "supercavitating" flow (Fig. 6b).

Flow About Propellers

On a marine propeller cavitation appears on each blade and in the cores of the tip vortices where the pressure is low (Figs. 2 & 7). Figure 8 shows a pro-

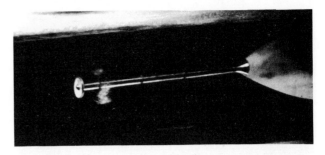

9. In the flow past a disc, cavitation appears in the zones of high turbulent shear at the edges of the separated wake.

7. A close-up of a marine propeller under stroboscopic lighting shows cavitation bubbles on the face of the blade and in the cores of the tip vortices.

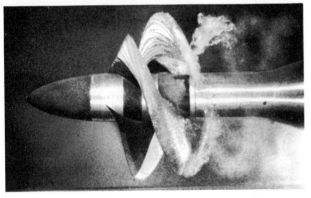

8. A marine propeller designed to operate under super-cavitating conditions has blade cross-sections similar to the hydrofoil of Fig. 6.

peller designed to operate under supercavitating conditions. Note the similarity of the cavity on each blade to that on the hydrofoil with the sharp leading edge.

Turbulent Shear Flow

Cavitation can also occur in turbulent shear flows because of the local pressure reduction in intense turbulent eddies. This phenomenon can be observed in the flow behind a disc with its axis of symmetry parallel to the flow direction. At high speeds, the pressure fluc-

tuations in the zone of high turbulent shear at the edge of the wake lead to cavitation (Fig. 9). With further increase in speed, the entire wake appears to be filled with vapor bubbles. Eventually, the flow becomes a true cavity flow (Fig. 17a), as distinguished from a cavitating wake flow. In the film, high-speed photography shows that the flow at the end of the cavity re-enters and moves upstream. The momentum flux in the re-entrant jet is equal to the pressure drag of the body. The jet kinetic energy is dissipated. A similar re-entrant flow can also be seen in the cavity trailing the supercavitating hydrofoil.

Water-Entry Cavity

A nonstationary type of cavity flow occurs when a solid body enters water at high speed (Fig. 10). The cavity follows the body into the water and eventually

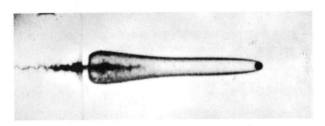

10. With the camera turned so the water surface appears vertical, a round body is shot into water. The cavity follows the body into the water and eventually pinches off at the rear, forming a re-entrant jet directed toward the body. (Courtesy of U.S. Naval Ordnance Laboratory)

closes off, producing re-entrant jets. Initially, the gas within the cavity attached to the body is air rather than vapor. Progressively, the air is left behind, and if the motion persists long enough, the cavity will contain vapor primarily.

Bubble Chambers

Still another example of cavitation occurs in bubble chambers used for studies of high-energy nuclear particles. In Fig. 11, a positron-electron pair has produced tracks of bubbles in liquid hydrogen which is at a pressure below the boiling point and is therefore unstable.

11. The tracks of a positron-electron pair through liquid hydrogen are made visible by cavitation bubbles.

We have seen that cavitation can take several forms: small, transient bubbles; large, more-or-less steady cavities; nonstationary cavities, and often a mixture of these types.

EFFECTS OF CAVITATION AND CAVITATING FLOWS

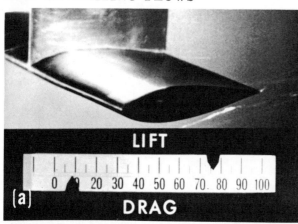

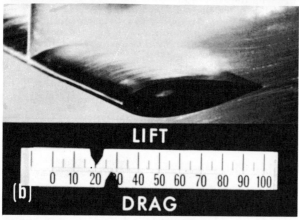

12. Force gauges measure lift and drag on a suspended hydrofoil. Prior to the inception of cavitation, reduction of the ambient pressure does not change lift or drag (a). When cavitation develops, further reduction in pressure causes a decrease in lift and an increase in drag (b).

Effect on Lift and Drag

Since hydrofoil sections make up so many different types of machines — pumps, turbines, propellers, propeller shaft struts, mixers — the effects of cavitation on such machines can be illustrated by studying the forces on the hydrofoil section itself. At constant flow speed, the lift and drag on a hydrofoil do not vary as the ambient pressure is lowered (Fig. 12a), until cavitation begins. As cavitation develops, the lift decreases and the drag rises (Fig. 12b). The flow becomes quite unsteady and often produces severe vibrations in hydraulic machines.

The loss in lift can be understood by examining the pressure distribution on the hydrofoil section (Fig.

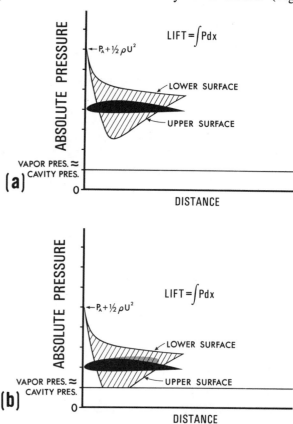

13. Prior to cavitation inception, reduction of absolute pressure does not affect the area between the curves (which is proportional to lift). When cavitation develops, the pressure on the upper surface cannot go below the vapor pressure, and the area (and lift) decreases with pressure reduction.

13a). The lift is proportional to the difference in the curves representing the pressure distributions on the upper and lower surfaces of the hydrofoil. As the ambient pressure is decreased, the pressures on the upper and lower surfaces decrease by exactly the same amount until the upper surface begins to cavitate. At this point, the pressure on the upper surface can no longer decrease below the cavity pressure, which is near the vapor pressure. Since the pressure on the lower surface does decrease, the lift drops (Fig. 13b). The shape of the pressure curve changes as cavitation develops; the resultant pressure distribution is such as to cause an increase in drag. Thus, hydrofoils and many hydraulic machines which are designed for efficient subcavitating operation lose efficiency when cavitation occurs. When cavitation is unavoidable and conditions are such that supercavitating flow can be assured, it is possible to use supercavitating hydrofoil profiles designed specifically to achieve high lift-drag ratios.

Noise Produced by Cavitation

Collapsing cavitation bubbles produce noise. In the film, this effect is demonstrated using water in a tube with a partial vacuum in the space above the water surface. By accelerating the tube downward, low-enough pressures can be produced to cause the liquid to cavitate. When the tube is brought to rest, the pressure gradient resulting from the acceleration is removed, the pressure returns to its original value, and the bubbles collapse. The noise produced is the result of shock waves generated upon bubble collapse.

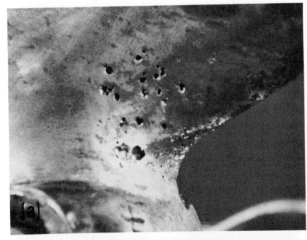

14. Cavitation damage to a marine propeller (a), and to a power dam spillway tunnel (b).

Cavitation Damage

The pressures associated with bubble collapse are high enough to cause failure of metals. Figure 14a shows cavitation damage on a propeller operated under cavitating conditions for a few days. Figure 14b shows cavitation damage in a spillway tunnel of a large power dam.

Cavitation can cause damage in a very short time. This effect is demonstrated by oscillating an aluminum specimen in the surface of water, using a magnetostriction oscillator. The specimen is driven vertically at a rate of 14,000 cycles per second. The total amplitude

15. Cavitation erosion (b) to a highly polished aluminum button (a), which was oscillated normal to the page at 14,000 cps for one minute.

is only .002 inches, but the pressure is changed from below vapor pressure as the button is accelerated upward to a high pressure as it is accelerated downward. After one minute, the highly polished surface has been eroded and considerable weight loss has occurred (Fig. 15).

DYNAMICS OF TRANSIENT CAVITATION BUBBLES

The life history of the small transient bubbles (Fig. 5) is measured in milliseconds. They grow during their passage through the low-pressure region, and then collapse as they enter the region of increasing pressure. If the bubbles have a relatively high initial

16. A cavitation bubble collapsing on a wall. A re-entrant jet is visible in the center of the bubble.

gas content, they will collapse and then rebound. If such cavitation bubbles remained spherical throughout their life history, extremely high pressures would be developed upon collapse (of the order of thousands of atmospheres). However, distortions occur because of Taylor instability, or if the bubble collapses in an unsymmetrical pressure field (in a gravity gradient or near a wall, for example). In the latter case, an internal jet is formed (Fig. 16), much as in the case of the water entry cavity. The velocity of the jet is very high; the impact on a surface can produce high stresses, and is another mechanism which may account for severe damage.

CAVITATION SCALING LAWS

The parameter which describes the conditions for cavitation similarity is the *cavitation number*,

$$\sigma = \frac{P_a - P_c}{\frac{1}{2}\,\rho\,U^2}$$

where P_a is the ambient absolute pressure, P_c is the cavity pressure, ρ is the mass density of the liquid, and U is a reference speed characteristic of the flow. It is the basis for scaling cavitation phenomena and for designing model experiments. The cavitation number at which cavitation begins is called the *critical cavitation number*. Above the critical cavitation number, no cavitation occurs; below the critical, it does occur. Operation with a cavitation number well below critical produces a very large cavitated region. In two-phase, one-component flow, the cavity pressure is just the vapor pressure. In a multicomponent flow, the cavity pressure is the sum of the partial pressures of the vapor of the liquid and of any gases that may have been introduced into the cavity. In fact, a cavity developed entirely with gas from an outside source behaves and appears very much like a cavity formed by means of the vaporization process discussed above. This is illustrated in Fig. 17, where a vapor cavity is compared with a cavity caused by introducing air into the wake of the disc. To form the vapor cavity, the water channel was operated at high speed and low pressure. The air cavity was formed by operating the channel at atmospheric pressure and a much lower speed to obtain approximately the same cavitation number.

The size of the physical system being studied does not appear in the cavitation number, but it is a factor in the Reynolds number. Consequently, as long as Reynolds-number effects are taken into account properly, cavitation similarity requires only that the cavitation number be the same for model and prototype. A model experiment can be made at lower speed than that at which the prototype operates if we simultane-

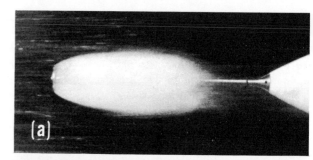

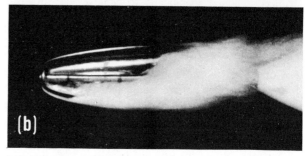

17. (a) shows a cavity behind a disc set normal to a high-speed, low-pressure flow. (b) shows a similar cavity formed by introducing air into the wake of the disc, at low flow speed and at atmospheric pressure.

ously reduce the pressure under which the model is operated. The operation of the cavitation tunnel is based on this principle.

INCEPTION OF CAVITATION

The Role of Nuclei

How does cavitation, or ordinary boiling for that matter, actually begin? Inception of cavitation in a multi-component liquid at *pressures near the vapor pressure* requires the presence of nuclei which contain minute amounts of vapor, gas, or both. Cavitation will occur only when these nuclei become unstable and grow when subjected to a pressure reduction. The conditions for such growth can be derived from an analysis of the static equilibrium conditions for a spherical nucleus (Fig. 18). The internal forces, produced by

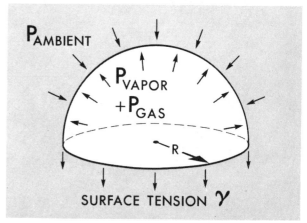

18. The forces on a cavitation bubble nucleus.

the partial pressures of the gas and vapor within the nucleus, must be balanced by the ambient pressure and the surface-tension pressure at the nucleus-liquid interface. Thus, the condition for static equilibrium is that the ambient pressure plus the surface-tension pressure equal the vapor pressure plus the gas pressure

$$P_A + \frac{2\gamma}{R} = P_V + \frac{\text{const.}}{R^3}.$$

This equation has been plotted in Fig. 19 for two gas contents. The pressure adjacent to the bubble has a

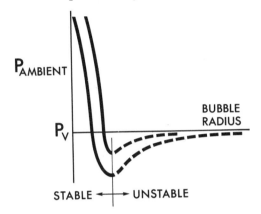

19. Plots of the conditions for static equilibrium of cavitation bubble nuclei. The upper curve is for a bubble with a larger gas content than the lower one.

minimum value which is below the vapor pressure of the liquid. As long as the ambient pressure is above this minimum, and the initial bubble radius is smaller than the radius associated with it, the nucleus is stable and tends to reach an equilibrium radius along the left-hand portion of the curve, where the slope is negative. If, however, the pressure drops below the critical value, the bubble becomes unstable and grows without bound. If a smaller gas content is available, even lower pressures are required.

Stable and Unstable Nuclei

Stable and unstable behavior of nuclei are shown in the film by observing nuclei that have radii just below and just above critical value. Small bubbles of the order of .005 inches in diameter are generated in a tube that is partially filled with water and evacuated to near vapor pressure. This is done by evolving hydrogen from one platinum wire and oxygen from a second wire by means of an electrical impulse generator. The bubbles are unequal in size, the hydrogen bubble being the larger. As they rise toward the free water surface, into regions of lower hydrostatic pressure, the larger hydrogen bubble reaches critical size and grows explosively, while the oxygen bubble does not (Fig. 20). Although the small oxygen bubbles expand as the pressure is reduced, they never reach a size large

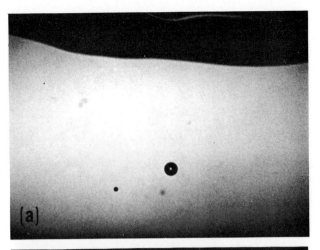

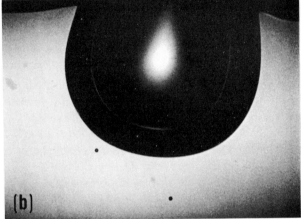

20. (a) A hydrogen bubble (on the right) and a smaller oxygen bubble (left) rise toward the free surface of water in a partially evacuated tube. (b) As a result of the lower hydrostatic pressure near the free surface, the larger hydrogen bubble reaches critical size and grows explosively, while the oxygen bubble does not. (Courtesy of Albert T. Ellis, University of California.)

enough to become unstable. On the other hand, some of the larger hydrogen bubbles, which move along a curve such as the one corresponding to larger gas content in Fig. 19, do expand to a size large enough to slip into the unstable region and grow explosively. Bubbles represented by the portion of the curve of Fig. 19 to the right of the minimum point are already unstable, and cannot exist unless stabilized by some external mechanism.

Effect of Nuclei Size and Content on Cavitation Inception

A stable nucleus can decrease in size and eventually disappear because its gas content diffuses into the surrounding fluid. This is shown in the film with a venturi experiment, by comparing the cavitation inception pressure for freshly drawn water with that for water which has rested undisturbed for some time (Fig. 21). Because the discharge is to the atmosphere, increasing the driving head increases the velocity and thereby de-

21. With freshly drawn water in the bucket, cavitation begins in the venturi throat at a head of about 2 feet. When the water in the bucket is allowed to rest undisturbed for some time, a higher head (about 3 feet) is required for onset of cavitation.

creases the static pressure at the throat. With freshly drawn tap water, cavitation begins at a head of about 2 feet, and continues to develop as the head is further increased. After leaving the water undisturbed for about an hour and then repeating the experiment, a much higher head is required for onset of cavitation — in this case, about three feet. During the settling period, some nuclei could rise to the surface and vent if they were very large, and some could decrease in size by diffusion of air into the surrounding water. Consequently, a greater pressure reduction was required to cause cavitation in that sample.

A Dilemma

The result of the experiment on nuclei size and content leads to a dilemma. Nuclei in mechanical equilib-

rium disappear through diffusion of the gas into the surrounding liquid. Unstable nuclei cannot persist. Yet we have seen that nuclei must be present for cavitation to occur at pressures near vapor pressure. How, then, can we account for the persistence of nuclei? We must conclude that some external mechanism is required, such an entrapment of gas in the crevices of solid boundaries or on dust particles, or by the accumulation of some foreign material on the nucleus gas-liquid interface which prevents diffusion.

CONCLUSION

We do not fully understand all of the mechanisms of nucleation, and many basic aspects of cavitation remain to be explained. Nevertheless, we do know a great deal about cavitation, and its effects on performance of machines can be predicted fairly well.

REFERENCES

1. "Cavitation," Section 12, *Handbook of Fluid Mechanics*, McGraw-Hill Book Co., Inc., 1961 (Section 12-I: "Mechanics of Cavitation," by Phillip Eisenberg; Section 12-II: "Supercavitating Flows," by Marshall P. Tulin).

2. Flynn, H. G., "Physics of Acoustic Cavitation in Liquids," Chapter 9, *Physical Acoustics — Principles and Methods*, Vol. 1, Pt. B, Academic Press, 1964.

3. *Cavitation and Hydraulic Machinery*, Editor: F. Numacki, Proc. International Assoc. Hydraulic Research Symposium, Tohuku University, Sendai, Japan 1963.

4. *Cavitation in Hydrodynamics*, Proc. Symp. at National Physical Lab., London: Her Majesty's Stationery Office, 1956.

Rarefied Gas Dynamics

F. C. Hurlbut and F. S. Sherman

UNIVERSITY OF CALIFORNIA

Introduction

In the upper atmosphere and in many applications of vacuum technology, the average distance which a molecule travels between collisions with other molecules can become comparable to the characteristic dimensions of flight vehicles or vacuum chambers. When this happens, the qualitative characteristics of flow fields, and the appropriate conceptual framework for their analysis, depend upon the Knudsen number.

$$\text{Kn} = \frac{\text{Molecular mean free path}}{\text{Characteristic length of body or chamber}} = \frac{\lambda}{d}$$

In the atmosphere, the mean free path increases from about 7×10^{-6} cm at sea level to about 80 meters at an elevation of 160 kilometers. The variation is roughly exponential with elevation.

In the film we show the evolution of a variety of flows as the Knudsen number assumes a wide range of values. These flows have greatly different appearances in the limits of very small and very large Knudsen number. Since the mechanics of collisions between gas molecules and solid surfaces assumes dominant importance when $\text{Kn} \gg 1$, we also show experiments which reveal important aspects of this "gas-solid interaction."

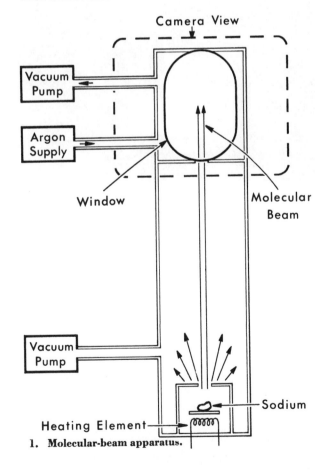

1. **Molecular-beam apparatus.**

A Molecular Beam and Its Scattering By Collisions

Figure 1 shows an apparatus in which an initially well-collimated beam of sodium atoms travels through a chamber in which we can maintain a controllable density of argon. We observe the sodium beam, and its attenuation by collisions with the argon "background gas," as a function of argon density.

Sodium is evaporated in a small electrically-heated oven. A portion of the evaporated atoms emerge through an orifice, and some of these pass through the second orifice to form a fairly well-collimated molecular beam in the upper chamber. Both chambers can be pumped down to pressures below 10^{-5} torr (mm of Hg).

When a narrow sheet of light from a sodium discharge lamp is passed through the axis of the chambers, the sodium becomes visible because of a very efficient (resonant) light-scattering process.[1] The intensity of scattered light is proportional to the density of illuminated sodium atoms. The argon has a negligible scattering cross section for this light, and remains invisible.

2. Illuminated sodium atoms in the emerging flow from the oven in the lower chamber.

Figure 2 is a scattered-light picture of the molecules effusing from the oven aperture. They move on straight-line paths away from the aperture, having a negligible number of collisions with other gas atoms at this low pressure. The mean free path is much longer than the interior dimensions of the chamber. Hence $Kn \gg 1$, and we say that a *free-molecule* flow is present.

Since the molecular paths diverge away from the aperture, the density decreases with distance from the gas source. However, with very fast film and long exposures,* we see the collimated sodium beam in the upper chamber. (Fig. 3a) The sharp edges of this beam indicate that the molecules do indeed follow straight lines between collisions.

*ASA 3000, 10-minute exposure

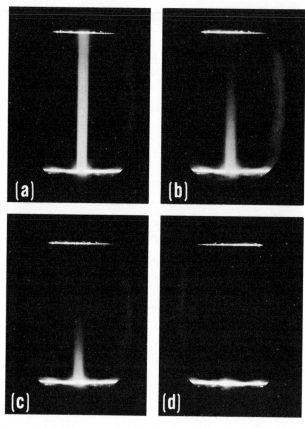

3. (a) Collimated sodium beam in the upper chamber. (b), (c), and (d) Increasing the density of argon in the chamber decreases the mean free path of the sodium atoms and reduces the visible penetration of the beam.

Next, argon is admitted to the upper chamber. As the density of argon is increased, the sodium beam fades out at the top (Fig. 3b), and becomes progressively shorter, (Figs. 3c and 3d). The mean free path of sodium atoms is inversely proportional to the density of the argon, and the visible penetration of the beam into the scattering chamber is roughly proportional to the sodium mean free path.

Collisions with the argon do not cause a diffused or fuzzy appearance of the sodium beam, because the scattered sodium atoms are dispersed into a very tenuous three-dimensional cloud, of which only a thin slice is illuminated.

When the sodium beam has virtually disappeared from the top chamber (Fig. 3d) the Knudsen number is very small. Sodium atoms still travel to the top of the chamber, but on often-interrupted and erratic paths, in a process which approaches ordinary diffusion.

Supersonic Flow Over a Blunt Body

The second experiment shows supersonic flow over a blunt body, with a 3000-fold variation of Knudsen number (Fig. 4).

A small washer is placed on the axis of a free jet of argon to which a trace of sodium vapor is added to

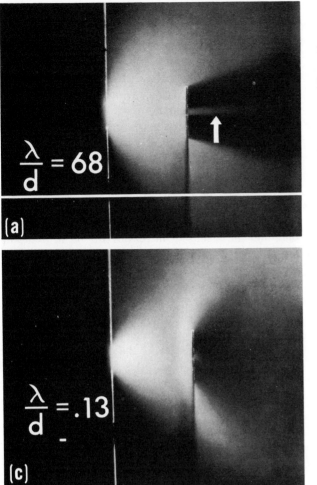
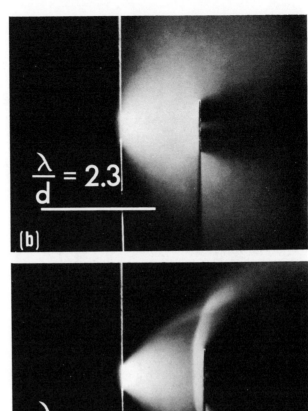

4. A washer in a free jet illustrates transition regimes between free-molecular and continuum flows. The characteristic dimension, d, in the Knudsen number is the diameter of the washer. The horizontal lines under the values of λ/d indicate the length of the mean free path, λ.* At the free-molecular extreme, the washer casts a sharp shadow and the hole forms a molecular beam. (See arrow.) (b) and (c) The shadow behind the washer becomes more diffuse and a compressed layer begins to form upstream. (d) At the continuum extreme, a bow shock wave is formed ahead of the washer.

allow flow visualization. The sodium light is introduced from above, in a thin sheet which passes through the axis of the jet. The washer is electrically heated to prevent condensation and collection of sodium. The purpose of the hole in the washer is to show what happens to molecules of the jet as they encounter molecules scattered back from the washer.

Figure 4a shows free-molecular flow for this configuration. By the time they reach the washer, mole-

cules of the jet are traveling nearly straight away from the center of the jet-forming orifice. They pass right through the tenuous cloud of molecules scattered back by the washer, to define a sharp aerodynamic shadow in the wake of the washer. A well-collimated molecular beam emerges from the hole in the washer.

As the Knudsen number is decreased by increasing the rate of argon flow, the cloud of molecules scattered back by the model becomes dense enough to scatter molecules approaching the model. The flow through the hole in the washer loses its collimation and hence its visibility, and the aerodynamic shadow broadens. (Fig. 4b)

Further decreases in the Knudsen number (Figs. 4c and 4d) lead gradually to the formation of a shock layer in front of the body, within which a molecule entering from upstream suffers many collisions with other molecules before eventually finding its way to the model surface, or directly into the wake via the hole or around the outer perimeter of the model.

In Fig. 4d we see, with Kn $<< 1$, a familiar hypersonic continuum flow. The Mach number at the model location is about nine. Further decrease in Knudsen number would make little noticeable change, except in the wake.

*The mean free paths represented by lines on the film are calculated from the formula

$$\lambda = \frac{1}{\sqrt{2}\,n\,\sigma^2}$$

The number density, n, is taken at the value which would be found on the jet centerline where the washer is located, if there were no washer. The collision cross-section used is $\sigma^2 = 36.5 \times 10^{-15}$ cm^2 (the viscosity cross-section of argon at the free-stream stagnation temperature). The sharp dark line below the washer is its optical shadow.

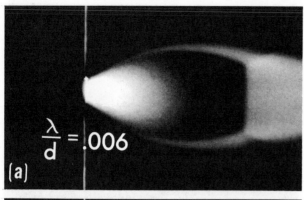

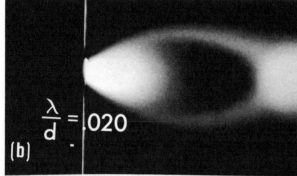

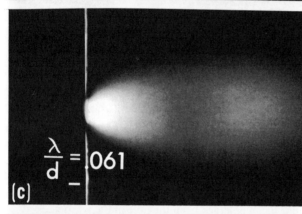

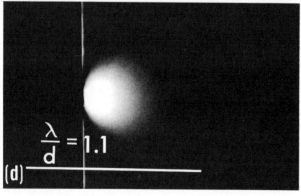

5. Transition from continuum to free molecular flow for the free jet alone. (a) At the continuum extreme a "barrel shock" exists from the nozzle exit to the "Mach disk." The characteristic dimension, d, is taken as the distance from the nozzle to the Mach disk for the continuum case. (b) and (c) Increase of mean free path causes shocks to become thick and diffuse. (d) At the free-molecular extreme, the shock system no longer exists. Further increase of the mean free path would not appreciably alter the flow.

Evolution of the Free Jet

The transition from continuum flow to free molecular flow from the same nozzle as in Fig. 4, but without the model, is shown in Fig. 5. With a small mean free path (Fig. 5a), the free jet gives rise to the familiar shock-wave pattern involving a barrel shock and a Mach disk. Similar patterns are often seen in rocket exhausts under appropriate continuum-flow conditions. These shock patterns are only clearly defined when the mean free path in the background gas is small compared to the distance from the nozzle to the Mach disk. Under these conditions the shock configuration depends only on the pressure ratio across the nozzle.

This pressure ratio is held constant at a hundred to one, and the results of increasing the mean free path by decreasing the argon flow are shown in Figs. 5b through 5d. When the mean free path is no longer negligible compared to the jet dimensions, the shocks become very thick and fuzzy (Fig. 5b). The barrel shocks spread out until they overlap and mix in with the Mach disk. In Fig. 5c, only a hazy, slightly bright spot remains to remind us of the shock system. In the free-molecule limit (Fig. 5d), the jet molecules no longer collide with the background molecules, but fly straight past them to the chamber walls. The background molecules, for their part, can fly right through the jet without collision.

Surface Interactions

To make quantitative predictions of aerodynamic forces or heat transfer in free-molecule flow, we must have at least some statistical knowledge of the outcome of collisions between gas molecules and solid surfaces.

A. *Sodium-Beam Experiment*

Figure 6 illustrates visually the important concept of diffuse reflection.

6. A polished aluminum oxide surface is positioned at 45° to the sodium molecular beam. The beam is not reflected as a light ray is from a mirror, but rather is scattered in all directions as the superposed arrows indicate. The density of the scattered molecules is too small to be recorded on the film.

A surface of polished aluminum oxide is mounted in the upper chamber of the sodium-beam apparatus, at an angle of 45° to the incident beam. The target is heated to the temperature of the beam oven, so that condensation of the sodium on the target is prevented.

Although the sodium beam appears to end at the target, the sodium atoms are actually re-emitted in every direction away from the spot of impact. Once again they are scattered into a cloud too tenuous to be seen. Clearly, the molecular beam is not reflected specularly from the target, as a light beam reflected from a mirror would be. More quantitative experiments confirm that nearly diffuse re-emission of molecules is typical at ordinary or even highly polished surfaces. However, highly directed scattering is sometimes observed, particularly when the surface is composed of atoms in a well-ordered array and is free of adsorbed gases.

B. *Momentum-Transfer Experiment*

The integrated effects of momentum transfer in gas-solid collisions can be studied directly by observing the torque exerted on a target surface subjected to the off-center impact of a molecular stream (Fig. 7). The

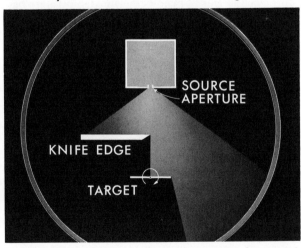

7. Plan view schematic of momentum-transfer experiment. A torque is applied about the center of suspension of the target.

apparatus consists of a gas source with slit aperture, a knife edge, and a target, all within a cylindrical glass vacuum chamber.

The target and a small mirror are attached to a vertical torsion wire. When enough gas is supplied to the source to produce a free-molecule effusion through the slit, the target is turned by the impact of molecules on its unsheltered half. A light beam reflects from the mirror to record the angle of target deflection on a large index circle.

The experiment is begun by indexing the light spot with the target oriented as shown in Fig. 7. Gas is admitted to the source, and source pressure and target deflection are noted. The top of the torsion wire is then twisted to return the target to its initial position.

The source pressure is doubled, and the new deflection noted. Figure 8 shows typical data of the torque required to hold the target in the position of Fig. 7,

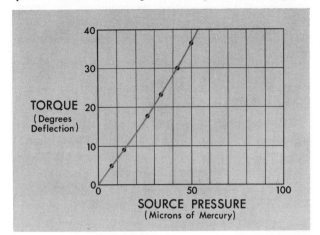

8. Typical results of momentum-transfer experiment of Fig. 7. For free molecular effusion from the source (up to about 30 microns of mercury), torque is proportional to source pressure. (Data points shown in the film were misplotted. Collisional effects near the source aperture cause the torque to increase at higher source pressures, as shown here.)

plotted versus source pressure. The departure from linearity at high pressures is due to departures from free-molecule flow in the vicinity of the source slit at high pressures.

Next, with source pressure held constant, the temperature of the target is increased by radiant heat transfer from a 1000-watt projection lamp. The resulting deflection of the target is like that produced by an increase in source pressure, because of the increased recoil momentum of molecules re-emitted from the hotter target.

The results of a series of experiments[2] of this type are shown in Fig. 9. Beams of various gases have exchanged momentum with an aluminum target over a range of target temperatures. The theoretical torque for the case in which the gas molecules are re-emitted

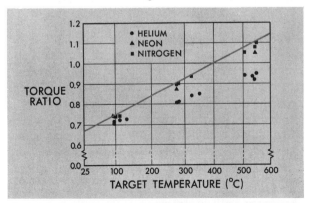

9. Increasing the target temperature (with source pressure held constant) causes an increase in the recoil momentum of the reflected molecules. Experimental results for nitrogen lie closer to the theoretical line for perfect accommodation than do the results for the helium and neon.

from the target as though they came out of a gas at equilibrium at the target temperature is shown as the solid line. This is called the case of *complete accommodation* of the re-emitted molecules to the target temperature. The experimental results for nitrogen lie close to this line, while those for helium and neon do not.

Mean Free Path as the Characteristic Length for Shock-Wave Structure

The shock wave in front of a model in a continuum hypersonic flow is examined in the wind tunnel shown in Fig. 10. The experiment also teaches something about heat transfer in rarefied flow.*

The model is a flat bar of aluminum mounted on a thin-walled stainless steel tube to reduce heat conduc-

10. Flow from the freely expanding jet at the left is visualized by fluorescent excitement by an electron beam at right angles to the flow (center). A bow shock wave is seen in front of the sting-mounted object at the right.

tion to the wind-tunnel shell. It is placed in a free jet of air where the Mach number is approximately seven. The flow is made visible by the fluorescence excited in air by a beam of electrons which is passed across the flow well upstream of the model.**

In the experiment the structure of the shock wave is indicated by the varying recovery temperature of a fine

---

* The idea for the experiment comes from Sherman (1955),[3] who gives theoretical details of the wire-temperature interpretation at points within the shock wave, and empirical data on how small the wire must be.

** Because the fluorescence excited by the electron beam[4] is rather dim, each frame of the pictures showing the wire moving through the shock required an exposure of about 15 seconds.

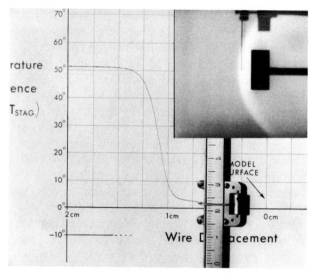

11. Wire temperature is plotted versus displacement as the thermocouple wire is traversed through the bow shock wave in front of the model (insert).

thermocouple wire which is traversed through the shock wave. The wire is small enough to be always in free molecule flow, and does not disturb the shock wave by its presence. Figure 11 shows the wire temperature being plotted by an *x-y* recorder for a flow stagnation

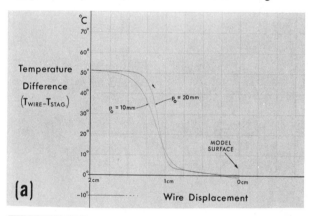

(a)

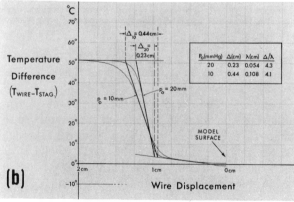

(b)

12. Comparison of the shock-wave profiles at stagnation pressures of 20 mm of Hg (20 torr) and 10 mm of Hg. (b) Determination of shock-wave thickness from asymptotes and tangents to the profiles. Results in table at upper right show that shock thickness is inversely proportional to the pressure, and hence proportional to the mean free path.

pressure of 20 torr. Also shown in the insert are the model, shock wave, and wire supports.

Figures 12a and 12b show the temperature traces for stagnation pressures of 20 torr and 10 torr, and the shock thicknesses deduced from these traces. When the radius of curvature of the shock wave is much greater than the shock thickness, the mean free path is the only physical length to which the shock thickness can be proportional. The experiment confirms this proportionality and shows that the factor of proportionality is about 4 in air at Mach number 7.

Notice that when the wire is in the undisturbed flow upstream of the shock wave, it is much hotter than the model. The essential difference between wire and model is that $Kn = 20$ for the wire, while $Kn = 0.03$ for the model. This decrease in equilibrium temperature with decreasing Knudsen number is directly connected with the collisions between back-scattered molecules and those of the jet, and accompanies the formation of a shock wave.

References

1. Vali, W., and Thomas, G. M. (1963), AIAA J. *1*, 2, 469-471.
2. Stickney, R. E., and Hurlbut, F. C. (1963), "Rarefied Gas Dynamics" J. A. Laurmann, editor, Vol. 1, 454-469, Academic Press.
3. Sherman, F. S. (1955), NACA TN 3298.
4. Muntz, E. P., and Marsden, D. J. (1963), "Rarefied Gas Dynamics" J. A. Laurmann, editor, Academic Press, Vol. II, 495-526.

Stratified Flow

Robert R. Long
THE JOHNS HOPKINS UNIVERSITY

Introduction

A stratified fluid is a fluid with density variations in the vertical direction. One example is a system of two superimposed fluids in a channel with the lighter fluid on top. In this case the density changes abruptly with height, as illustrated in Fig. 1. Layered systems of stratified fluids occur, for instance, where warm water lies above cold water or fresh water above salt water. An abrupt change also occurs at the interface between air and water. Both air and water are fluids, and together they may be thought of as a stratified fluid system. Fre-

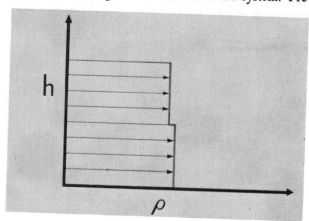

1. Density-height curve in a two-fluid system.

2. Smog. (Courtesy U.S. Public Health Service.)

quently, however, density varies continuously, as in the oceans and atmosphere. Density variations profoundly affect the motion of water and air. Wave phenomena in air flow over mountains and the occurrence of smog (Fig. 2) are examples of stratification effects in the atmosphere.

In the film we use laboratory demonstrations to illustrate the basic phenomena in stratified fluids. We emphasize fluid systems in which the density decreases with height. When such a system is disturbed, gravity

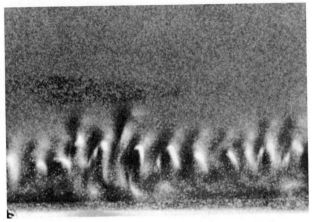

3. Convective motions in a fluid heated from below.

waves result, but gravity and friction eventually restore undisturbed conditions, and the system is judged to be stable. If the density increases with height, however, the fluid tends to be unstable. Figure 3 illustrates the cellular type of motion that results when a fluid is heated in its lower portions to produce an unstable distribution.

In our examination of fluids that are stably stratified in the vertical, we concentrate on flows over obstacles. Such flows reveal most of the fundamental phenomena that occur in stratified fluids. Moreover, such flows frequently occur in nature.

Surface Waves Produced by Flows over Obstacles

Figures 4-8 show water flow over a barrier in a channel. The flow relative to the obstacle is produced by moving it through the water. When the camera moves with the obstacle, we see a uniform flow from the left over a stationary barrier.

Phenomena in water flows depend on the dimensions of the obstacle relative to the depth and on the approach Froude number. A local Froude number F may be defined at any section as $F = U/\sqrt{gh}$, where U is the fluid velocity at the section, h is the local fluid depth, and g is the gravitational acceleration. Since $\sqrt{gh}$ is the speed of the fastest small-amplitude gravity

4. Subcritical flow of water over an obstacle; the local Froude number is less than one everywhere.

5. Hydraulic jump at a water surface. The approach Froude number is less than one.

wave, the Froude number compares the fluid speed to this wave speed. Where the Froude number is *less than one,* the flow is called *subcritical.* Where it is *greater than one,* the flow is called *supercritical.*

We first illustrate the change of flow patterns as the Froude number based on the approach conditions is increased from low to high values. Figure 4 shows a very low approach Froude number. The flow is subcritical everywhere. The free surface dips down slightly over the obstacle. In Fig. 5 the approach Froude number is higher, although still subcritical. When started from rest the upstream and downstream levels

6. Hydraulic jump at the base of a dam.

were the same. A blocking effect sets in, however, and causes a rise in the upstream level. When the motion becomes established, as in Fig. 5, the upstream subcritical flow near the obstacle draws down as it passes over the barrier, becoming supercritical in the lee. This supercritical flow changes abruptly to subcritical flow downstream as the fluid passes through a hydraulic jump. The hydraulic jump is an important phenomenon in nature. Figure 6 shows a jump at the base of a dam.

When we increase the approach Froude number to a value somewhat greater than one, we get an upstream hydraulic jump as in Fig. 7. The flow goes from supercritical to subcritical as it passes through the

7. Upstream hydraulic jump. The upstream Froude number is greater than one.

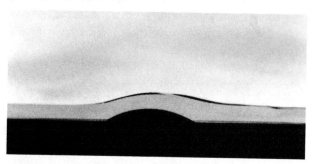

8. Supercritical flow. The approach Froude number is greater than one. The *local* Froude number exceeds one everywhere.

jump, then accelerates to supercritical as it passes over the obstacle crest.

When the approach Froude number is much greater than one, the fluid swells symmetrically over the obstacle (Fig. 8), and conditions are supercritical everywhere. Since the Froude number is much greater than one at all sections, the fluid speed everywhere exceeds the speed of the fastest waves, and all disturbances are swept downstream.

Flows of a Two-layer Fluid over Obstacles

Phenomena similar to those of water flows in a channel occur in liquids with internal density variatons. For example, waves, similar to waves on a water surface, can occur at the interface of two liquids with a slight density difference. The frequencies of the interfacial waves and the water waves are both proportional to the square root of the restoring force. The restoring force for water waves is the gravitational force g. But for the internal waves the restoring force is g multiplied by the small number $\Delta\rho/\rho$, where $\Delta\rho$ is the density difference between the liquids and ρ is the average density. As a result, the frequency of internal waves is usually much smaller than that of water waves. The slow motion of internal waves is a characteristic feature of the experiments of the film.

In the flow of a two-layered fluid over obstacles, the important Froude number is the internal Froude number of the approach flow, $F_i = U / \sqrt{g \dfrac{\Delta\rho}{\rho} h}$,

in which g is replaced by "modified gravity" $g\Delta\rho/\rho$, h is the total depth of the two fluids, and U is the common approach velocity of the two fluids. A critical value of F_i, as in water flows, corresponds to a fluid speed equal to the speed of long waves. However, the critical value is no longer one, since the wave speed also depends on the ratio of the depth of the lower fluid to the total depth. For two fluids, the flow patterns are determined by F_i, the ratio of the depths, and the height and length of the obstacle relative to the depth.

As these parameters are changed the flow patterns change. A long obstacle and a low value of the internal Froude number of the approach flow leads to subcritical flow at every section, and a slight draw-down of the interface over the barrier (Fig. 9). This long, gently-shaped barrier minimizes vertical accelerations of the fluid, and the pressure remains nearly hydrostatic, i.e., the pressure is proportional to the depth. However, for the same ratio of fluid depths and the same internal Froude number, a shorter model of the same height

9. Subcritical flow of two fluids over an obstacle. The flow condition is analogous to that in Fig. 4.

10. Weak lee waves in a two-fluid system.

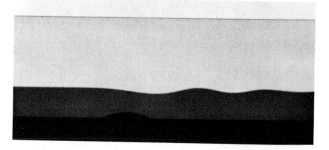

11. Strong lee waves in a two-fluid system.

results in weak lee waves (Fig. 10), as nonhydrostatic pressures become important. As we increase the Froude number by increasing the speed, these lee waves become stronger (Fig. 11). At a still higher Froude

12. Breaking waves in a two-fluid system.

13. Hydraulic jump in the lee of an obstacle in a two-fluid system.

14. Supercritical flow in a two-fluid system. Compare this with Fig. 8 for the water-air case.

15. Hydraulic drop in a two-fluid system. Note the increase of thickness of the upper fluid downstream.

number (Fig. 12), the waves show signs of breaking, as in the undular hydraulic jump on a water surface. Further increase in Froude number only results in a lengthening of these waves.

When we increase the size of the obstacle, a strong hydraulic jump occurs downstream (Fig. 13). The

upstream subcritical flow changes to supercritical as it passes over the barrier. Further downstream, it jumps to a new level, and the flow again becomes subcritical. As in water, if the approach Froude number is high enough, the flow is supercritical everywhere (Fig. 14). To this point, phenomena at the interface resemble those on a water surface, largely because there is a deep upper fluid in these cases, as is shown by the fact that when we make the lower fluid deeper than the upper, as in Fig. 15, the depth of the lower fluid abruptly decreases downstream of the obstacle in what might be called a "hydraulic drop."

A common feature of all the two-fluid experiments we have described is that the liquid-air surface remains quite undisturbed despite very large disturbances at the interface. We can explain this if we forget for a moment that there are two fluids in the channel, and notice that in all of these two-fluid experiments the *ordinary* Froude number is much less than one. In other words, conditions are very much subcritical as far as disturbances of the liquid-air interface are con-

16. Free-surface disturbance of a two-fluid system.

cerned. The free surface becomes disturbed if we move the obstacle fast enough to obtain an ordinary Froude number of order one (Fig. 16). In such cases the internal density variations are no longer important.

We can also disturb a two-layer system by moving an obstacle in the upper layer (Fig. 17). This experiment has an interesting application. It was noticed

17. Disturbance of a two-fluid interface by an obstacle moving in the upper fluid.

long ago that ocean-going vessels off the coast of Norway suddenly found themselves unable to maintain their accustomed speed as they moved past the mouth of a fjord. This "dead water" phenomenon was first explained by the Swedish oceanographer Ekman. He pointed out that the fresh water from the fjords flows out in a density current over the heavier water of the ocean, and forms a two-layer stratified fluid system. At certain speeds, much of the power from the ship's engines goes into the creation of waves at the fluid in-

18. **Disturbance of the interface of a two-fluid system by a model ship moving at the free surface.**

terface (Fig. 18), while the liquid-air surface is rather undisturbed. If the ship moves at a higher speed, the interfacial disturbances are much weaker and the dead-water phenomenon no longer exists.

Flows of a Continuously Stratified Fluid

The fundamental mode of oscillation in a two-fluid system involves a simple sine wave at the interface. A three-layer system, however, has two distinct modes (Figs. 19 and 20). Indeed, one mode is added for each layer. Carried to the limit, an infinite number of infinitely thin layers has an infinity of modes. This is equivalent to a fluid system in which the density varies continuously with height, as in the atmosphere and oceans.

To understand experiments with continuous density stratification, we consider the parameters which govern the flow of a continuously stratified flow over an obstacle. Again, the most important nondimensional number is the internal Froude number,

$$F_i = \frac{U}{\sqrt{g \frac{\Delta \rho}{\rho} h}},$$

where $\Delta \rho / \rho$ is now the fractional density difference from the top to the bottom of the channel, and h is the total depth. The flow patterns are influenced not only by the internal Froude number, but also by the characteristics of the obstacle relative to the depth.

Our experimental fluid is a saline solution in which the salt concentration, and therefore the density, de-

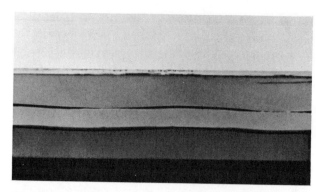

19. **Symmetric disturbance of a three-layer system.**

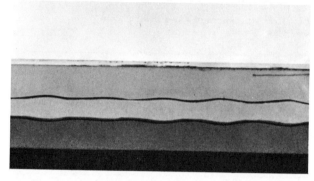

20. **Antisymmetric disturbance of a three-layer system.**

creases continuously with height. The tracer particles are neutrally buoyant polystyrene beads. The variation of the salt concentration, and therefore the density, is more or less linear with height.

In Fig. 21 the motion is analogous to the supercritical flow of water. The internal Froude number is high, and the fluid simply swells over the barrier as water does at supercritical speeds.

If we drop F_i below its first critical value of π^{-1}, simple sine waves appear in the lee of the obstacle (Fig. 22). These internal waves are not possible in a homogeneous fluid. To see how they can arise in a stratified fluid, consider the vorticity equation for a frictionless, nonhomogeneous fluid,

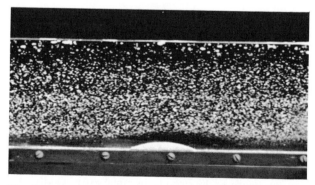

21. **Supercritical flow of a continuously stratified fluid over an obstacle. The uneven distribution of tracer particles in this and subsequent pictures does not reflect the density distribution. The density distribution depends only on salt concentration and is nearly linear with height.**

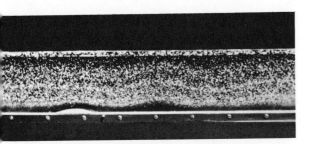

22. Flow of a continuously stratified fluid over an obstacle at an internal Froude number of less than $1/\pi$.

$$\frac{d\omega}{dt} = (\omega \cdot \nabla)\, \mathbf{V} + \frac{1}{\rho^2}(\nabla \rho \times \nabla p).$$

ω is the vector vorticity, $\mathbf{V}$ is the velocity, and p is the pressure. $d\omega/dt$ is the rate of increase of vorticity of a moving parcel of fluid. The last term shows that the vorticity can be generated through the interaction of the pressure and density fields. This equation can be solved under certain circumstances to yield flow patterns corresponding to the experiments we have just seen (Fig. 23). We can make use of this pattern to show the relationship between the generation of vorticity and the wave motion. The density varies from point to point in the fluid, but because the motion is

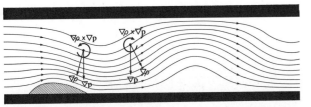

23. Theoretical flow pattern of a continuously stratified fluid over an obstacle.

steady, the density is constant along a given streamline. The density gradient $\nabla \rho$ is everywhere normal to the streamlines. Despite the motion, the constant pressure surfaces are nearly horizontal, i.e., the pressure gradient ∇p is vertical. As the fluid descends in the lee of the obstacle, the last term in the vorticity equation, $\nabla \rho \times \nabla p$, yields a vorticity generation, or fluid rotation, in a counterclockwise sense (Fig. 23). This rotation permits the descending fluid to turn upward toward its original level. On the way up, vorticity of the opposite sense is generated. The fluid can then turn back into the next phase of the wave.

Of course, these waves can also be explained from a buoyancy-force viewpoint. The descending fluid particle finds itself in a heavier environment and is forced back up as it moves along. It overshoots, then descends again, and so on.

When the barrier is small, the waves have small amplitudes, as in Fig. 22. If we increase the height of the barrier, keeping F_i the same, the amplitude of the waves gets much larger, as shown in Figure 24a.

In the theoretical flow pattern (Figure 24b) corresponding to the experiment in Figure 24a, closed streamlines appear near the free surface above the obstacle. In the experiment, such regions appear to be unstable and break into turbulent eddies. The loss of energy in this turbulence causes the wave amplitude downstream to be less than that indicated by the corresponding theory.

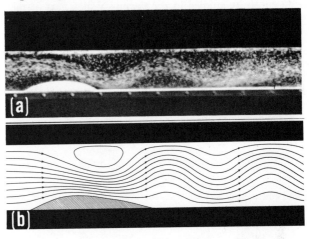

24. Experimental (a) and theoretical (b) flow of a continuously stratified fluid over an obstacle.

In going from the experiment of Fig. 22 to that of Fig. 24, we raised the height of the model but kept F_i the same. A *decrease* of F_i reduces the length scale of the fluid motion, and, in addition, the wave structure is no longer a simple sine wave.

When the Froude number is low, but the obstacle is large, the flow pattern again has a small scale, but is of a very different character (Fig. 25). The approach flow consists of a number of jets moving rapidly toward the obstacle, sandwiched between layers of fluid that are stagnant with respect to the obstacle. In these

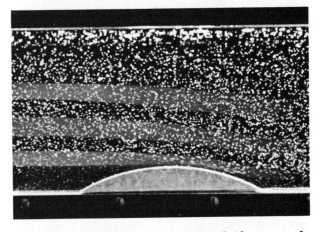

25. Jet-patterns in flow of a stratified fluid over an obstacle. Notice that the fluid is stagnant with regard to the obstacle in a number of layers, with jets sandwiched between. (The jets are delineated by the lightened areas.) The first stagnant layer extends from the bottom to the level of the obstacle crest.

flows, there is always a stagnation layer extending from the bottom of the channel to the level of the crest. At low Froude numbers, the fluid has insufficient kinetic energy to move up over the obstacle from this region. This illustrates the tendency of a slowly moving parcel of stratified fluid to maintain its original level.

An example of a natural flow of a stratified fluid is the flow of air over a mountain ridge. Such flows have been carefully studied in the Owens Valley east of the Sierra Nevada in California. A typical streamline pat-

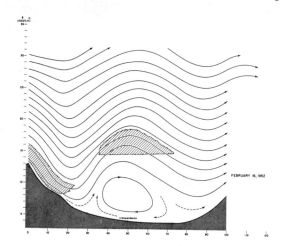

26. Streamline pattern over the Owens Valley. The vertical scale is exaggerated in this line drawing.

tern constructed from balloon and glider observations is shown in Fig. 26.

We can model this flow in the laboratory by creating a model of the terrain in this region and simulating the density stratification of the atmosphere by the density variation in salt water. Figure 27 shows an experimental photograph in which the Froude number is the

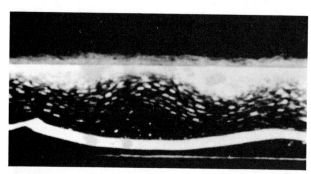

27. Flow over a model of the Sierra Nevada Mountains simulating the observed flow in Fig. 26.

same as that in the atmosphere on the day the observations were made in Fig. 26. A single large wave appears over the valley. When the wind is weaker the number of waves over the valley increases. When the flow in the model was adjusted downward to give a corresponding Froude number, the same increase of wave number appeared in the experiment.

Effects of Stratification on Diffusion

Density stratification has an important effect on diffusion in fluids. For example, smoke coming from a chimney diffuses turbulently if the atmosphere is not

28. Diffusion of smoke in conditions of neutral stability.

stably stratified, as seen in Fig. 28. When the lower air is stable, as in the morning or early evening, the smoke comes out and flattens into a long, thin layer (Fig. 29). Strong stratifications, or inversions as

29. Diffusion of smoke when the atmosphere is stably stratified.

they are sometimes called, confine contaminants to the lower layers of the atmosphere, and cause many of our air-pollution problems.

Conclusion

Thus, stratification gives rise to forces that generate internal waves, inhibit turbulent diffusion, and create strong velocity gradients and jets. These phenomena have far-reaching effects on the motion of air and water in the atmosphere, oceans, lakes, and reservoirs.

References

1. Long, R. R., The Motion of Fluids with Density Stratification. *Jour. Geophys. Res., 64,* p. 2151, 1959.
2. Rouse, H., *Fluid Mechanics for Hydraulic Engineers,* Dover Publications, Chs. 13, 16, 1961.
3. Scorer, R. S., *Natural Aerodynamics,* Pergamon Press, London, Chs. 3, 9, 1958.
4. Yih, C. -S., *Dynamics of Nonhomogeneous Fluids,* Macmillan Co., N. Y., 1965.

Rotating Flows

Dave Fultz
UNIVERSITY OF CHICAGO

Introduction

Rotating fluids occur in a wide variety of technical contexts and in geophysics, particularly in the atmosphere and the oceans (Figs. 1 and 2). The phenomena involved are so varied that, in order to illustrate some of the principal features, we shall confine our attention to homogeneous fluids and to motions that do not deviate greatly from a rigid-body rotation. It is then advantageous to use a coordinate frame which rotates

1. Cloud patterns seen from a satellite show rotational effects in the earth's atmosphere. (Courtesy V.E. Suomi, University of Wisconsin, and T. Fujita, University of Chicago.)

2. The whirlpool galaxy in Canes Venatici. (Courtesy Mt. Wilson and Palomar Observatories.)

at an appropriate rate Ω with respect to inertial space. The inviscid fluid equations of motions take the form*

$$\overbrace{\frac{\delta \mathbf{U}}{\delta t} + \mathbf{U} \cdot \nabla \mathbf{U}}^{\text{Relative acceleration}}$$

$$= \underbrace{-\frac{1}{\rho} \nabla p}_{\text{Press. grad.}} \underbrace{- \nabla \phi_g}_{\text{Gravity}} \underbrace{- \nabla \left(\frac{\Omega^2 r^2}{2}\right)}_{\text{Centrifugal}} \underbrace{- 2\,\Omega \times \mathbf{U}}_{\text{Coriolis}}$$

where ϕ_g is the gravitational potential, the velocity vector $\mathbf{U}$ is measured relative to the rotating frame,

*The first four terms alone would constitute the complete equation of motion in a non-rotating, *inertial,* system. In the *non-inertial* rotating frame two additional terms are present.

and the other symbols are customary. The last two terms are parts of the absolute acceleration which are customarily shifted to the right-hand side and interpreted as forces. $- \nabla (\Omega^2 r^2/2)$ is the centrifugal force which, since it is a function of relative position only, can be combined with the gravity term to give an apparent gravitational force. $- 2 \Omega \times U$ is the Coriolis force, which is responsible for many of the unfamiliar features of rotating flows. A careful discussion of the origin and status of these terms is given in Ref. (1).

The Rossby Number and the Frequency Ratio

Two dimensionless parameters are especially important in the examples we shall consider — they characterize the relative importance of fluid relative accelerations as against Coriolis forces. If σ is a characteristic frequency, U a characteristic (relative) fluid velocity, and L a characteristic length, these parameters are the Rossby number, $U/L\Omega$, and a frequency ratio σ/Ω. The Rossby number has order of magnitude of the ratio of convective acceleration to the Coriolis force, $(|U \cdot \nabla U| \div |2\Omega \times U|)$; and the frequency ratio has order of magnitude of the ratio of local acceleration to the Coriolis force, $\left(\left| \dfrac{\partial U}{\partial t} \right| \div \left| 2\Omega \times U \right| \right)$.

If the Rossby number is much less than one, the relative convective acceleration is small, and if the frequency ratio is low, the local acceleration is small. Under these conditions there is a very close balance between the vertical component of the pressure gradient and apparent gravity (hydrostatic balance). More importantly, there is a close balance between the *horizontal component* of the pressure gradient and the horizontal component of the Coriolis force. Such flows are called geostrophic. They occur commonly in the atmosphere and oceans, where the Rossby number is small mainly because the scale of the motion is large. Full discussions of the theoretical features of geostrophic flows and of additional dimensionless parameters (such as those associated with the viscosity of real fluids) are given in Ref. (2).

Elementary Coriolis Effects

The experiments in the film were carried out on a vertical-axis turntable driven by a very stable variable-speed motor and transmission (Fig. 3). The rotation axis must be vertical within a few seconds of arc when a free surface is present on the water. The top and side cameras rotated with the fluid containers, and all sequences where the rotation was not zero are identified by an indicator arrow. All rotations are counterclockwise seen from above.

One of the initial and subtle surprises is that direct

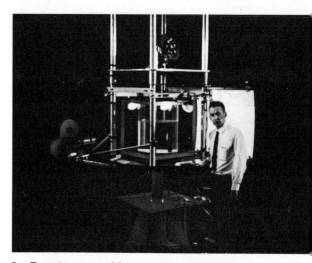

3. Rotating turntable carrying cylindrical tank inside a square tank to reduce optical refraction effects. Top and side cameras rotate with the tanks.

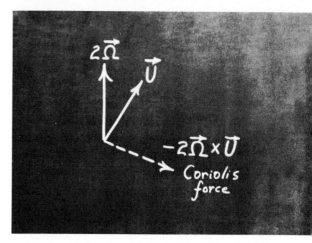

4. The Coriolis force is directed to the right of the rotational and relative velocity vectors.

reactions of the fluid to the Coriolis force (in the sense of motions deflecting to the right, toward $- 2\Omega \times U$) (Fig. 4) are obvious mainly for high Rossby numbers, where the Coriolis forces are relatively small. A small three-dimensional fluid body projected horizontally from a vortex-ring generator travels straight across a diameter of a cylindrical tank when the rotation is zero (Fig. 5a). In the same situation, but with the tank and fluid rotating, the vortex ring curves sharply to the right (Fig. 5b); the curvature increases as the horizontal velocity decreases. The Rossby number is of order one or more throughout (Ref. 3).

The effects of Coriolis forces are more complicated in an extended fluid body. When the rotation is zero, the lowest-frequency sloshing mode of surface gravity waves in a layer of fluid in a circular cylinder has the property that the free surface merely tilts back and forth (Fig. 6). Particles on the surface move horizontally in linear harmonic motion perpendicular to a nodal diameter. When the fluid layer is rotating (at

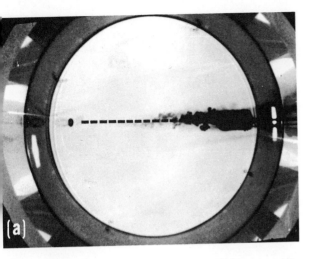

5a. A small parcel of dyed fluid, projected from a vortex-ring generator, travels straight across a non-rotating tank.

5b. When the tank and fluid are rotating counterclockwise at a moderate rate, the projected vortex ring curves sharply to the right (clockwise).

2.0 sec.$^{-1}$), there are two fundamental modes, now progressive, with crests moving either with, or opposed to, the sense of rotation (as seen in the rotating frame) (Fig. 7). These progressive modes stem basically from the transverse deflections due to the Coriolis forces, but the kinetic effects are substantially modified by the

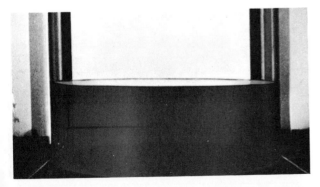

6. Free surface tilt in the lowest sloshing mode of a gravity wave generated by a vertically-oscillating disk positioned near the rim on the left side. The tank is not rotating. Wave frequency = 8.4 sec.$^{-1}$.

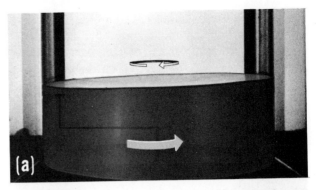

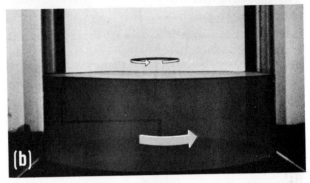

7. When the fluid layer of Fig. 6 is rotating counterclockwise, a disk frequency 15% higher than the non-rotating fundamental mode produces a progressive wave which travels clockwise around the tank (a). A second progressive mode, at a frequency 13% lower than the non-rotating mode, travels counterclockwise (b).

pressure fields. Thus the "negative" mode whose angular phase speed is clockwise has a 15% higher frequency than the non-rotating mode, while the "positive" mode has 13% lower frequency (Ref. 4). Moreover, while the horizontal trajectories of particles seen from above are circles curving to the right (in the same sense as the Coriolis force) for the negative mode, they are circles described to the left, exactly contrary to elementary expectation, for the positive mode.*

Low-Rossby-Number Motions Around Spheres

Some of the most striking instances of low-Rossby-number geostrophic flows were demonstrated by G. I. Taylor about 1920 (see Ref. 2 for discussion and sources), using various cases of flow around an immersed sphere. These and some later cases are considerably illuminated by thinking in terms of the vortex tubes of the absolute motion. In a low-Rossby-number, low-frequency-ratio, homogeneous flow, the inviscid Helmholtz vorticity equation contains (in the lowest-order approximation) only the term

$$(2\,\Omega \cdot \triangledown)\,\mathbf{U} \approx 0.$$

*The sequences in the film using aluminum powder on the top surface also show some unrelated motions resulting from local flow around the generating disk that is located below the free surface.

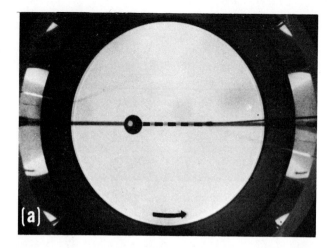

(a)

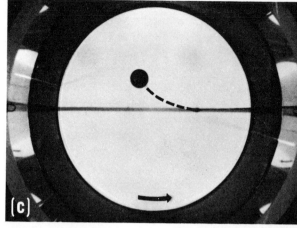

(c)

(b)

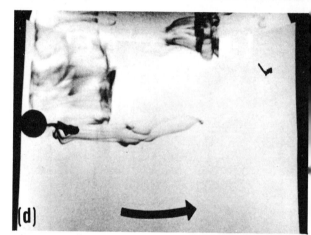

(d)

8. (a) A neutrally buoyant sphere, suspended from a long string so that it is free to deflect, travels straight across a diameter when pulled at a very slow speed in rotating fluid. (b) In a side view, dye streaks above the sphere show that a column of fluid is carried with the sphere at low Rossby number. (c) When the towing speed is increased so that the Rossby number is no longer small, the sphere deflects to the right. (d) A side view shows that the dyed column is left behind when the sphere is accelerated to higher Rossby number.

This is the Taylor-Proudman theorem (Ref. 2). It is equivalent to the statement that the relative velocity field does not vary in the direction of the rotation axis (since $2\Omega \cdot \nabla$ is a differentiation operation in that direction), and that the flow tends to be two-dimensional in planes perpendicular to the rotation axis. The absolute vortex tubes tend to remain parallel to the axial direction. They resist both bending and shrinking or stretching.

A small, nearly neutrally buoyant sphere was suspended on a fine thread and translated horizontally at a very slow speed. The sphere moved along a straight path, instead of deflecting to the right (Fig. 8a). It moved straight because a pressure field developed which exactly compensated the Coriolis force on the sphere. Dye trails above the sphere show that a Taylor column was carried with the sphere (Fig. 8b); in this manner the flow becomes approximately two-dimensional, in accord with the Taylor-Proudman theorem. When the towing speed is increased to a higher Rossby number, the flow becomes more three-dimensional, is

no longer geostrophic, and the sphere deflects to the right (Fig. 8c). When the sphere is accelerated, the dye column is left behind (Fig. 8d).

9. A sphere is towed slowly parallel to the rotation axis. Ink shows the Taylor column above the sphere, and the helical motion in the column below the sphere where there is a strong counterclockwise spin.

When the sphere is moved slowly upward parallel to the rotation axis, dye trails again show a Taylor column both above (forward wake) and below the sphere (Fig. 9). In the fluid ahead of the sphere, the compression of the absolute vortex tubes generates clockwise relative spin, while behind the sphere, the stretching generates counterclockwise relative spin. The corresponding pressure fields yield increased pressure in front and decreased pressure behind, so that the pressure drag on the sphere can be increased by several orders of magnitude. Thus the terminal velocity of a positively buoyant sphere is large when the rotation is zero, and becomes very small when the rotation is high enough to make the Rossby number small.

Taylor Walls

Even disorganized velocity fields can be radically affected by vortex tube stability when the Rossby number is small. Figures 10a and 10b show a turbulent field produced by injecting dyed water into a resting body of fluid. In Fig. 10c, the fluid was initially rotating rigidly and the dye was injected in an identical manner. The initial three-dimensional motions are similar to Fig. 10a, but they are rapidly converted (Fig. 10c) to a nearly two-dimensional flow as the velocities decay and the Rossby number decreases. The dye eventually becomes distributed in vertical Taylor walls (Fig. 10d). While the system is highly stable to the initial three-dimensional motions and rapidly converts them into two-dimensional geostrophic flows, it is nearly neutral to the latter, which undergo only a very slow viscous decay.

Inertia Oscillations and Rossby Waves

The rotational stiffness of the axial vortex tubes (or, alternatively, the resistance to changes of fluid circuit length in Kelvin's circulation theorem) make possible modes of oscillation that do not exist in non-rotating

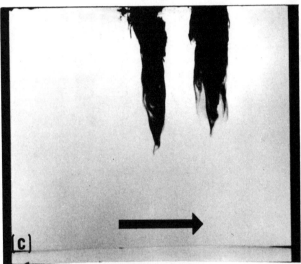

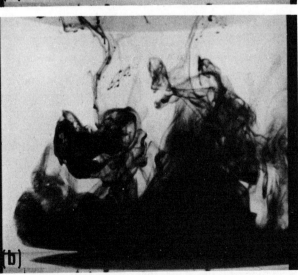

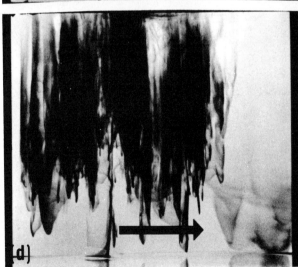

10. When dye is injected into non-rotating water (a), it spreads in a typical turbulent field (b). With identical ink injection into rotating water, the initial motions are rapidly converted to nearly two-dimensional motions (c), and after a time, the ink is found distributed in vertical Taylor walls (d).

fluid bodies. A small neutrally buoyant sphere, free to move on a vertical thread in rotating fluid, will oscillate up and down when given an impulse. This is a response to an inertia oscillation in the fluid.

In a rotating circular cylinder, normal modes of this type of oscillation can be generated by oscillating a small disk up and down along the axis at the proper frequency and position (Refs. 2 and 5). When there is

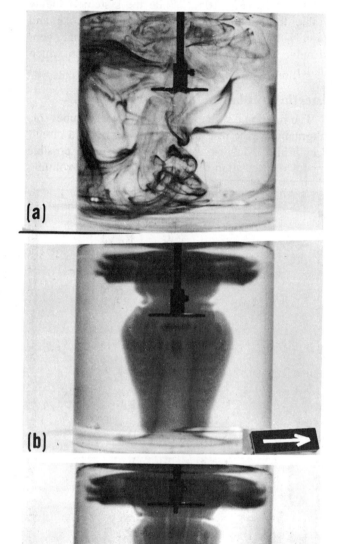

11. When there is no rotation, ink injected near a vertically oscillating disk shows only a disorganized flow (a). With rotation, the ink remains in a column which oscillates in an inertial mode with a nodal plane at the midpoint (b). (c) Ink column 180 degrees out-of-phase with (b).

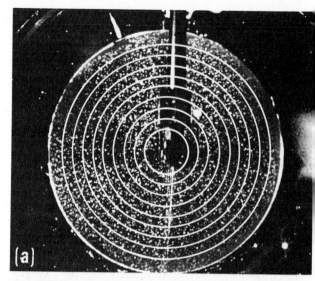

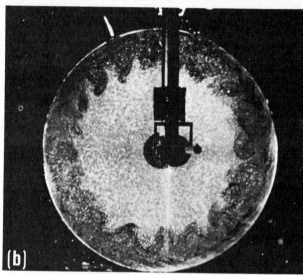

12. (a) A centrally placed disk oscillating vertically in a relatively shallow circular body of water produces rings (delineated by the overlaid lines) which oscillate in a tangential direction. The direction of motion is opposite for adjacent rings. (b) When the amplitude of the disk is increased beyond a certain critical value, the ink near the circumference shows that the motion becomes unstable.

no rotation, ink injected near the disk shows only a disorganized flow developing from vortex rings shed off the edge of the disk (Fig. 11a). With the cylinder rotating, the ink forms a stable Taylor column as the disk oscillates, in this case at a frequency ratio of 1.25. This frequency gives the mode with a nodal plane at middepth for the particular depth-radius ratio (2.00) of this cylinder. In contrast to a two-dimensional geostrophic flow, the oscillating ink column undergoes substantial expansions and contractions (and corresponding vorticity changes) as shown in Figs. 11b and 11c at two phases differing by half a period. Inertia oscillations of the same general character have been shown to be very common in the oceans and large lakes (Ref. 6).

In a much flatter circular body of water than that of Fig. 11, and with an excitation frequency ratio of only 1/2, the normal mode obtained is one with a large radial wave number (10 or so nodal circles), shown in plan view in Fig. 12a. An ink streak placed in the fluid along a circle within one cell of the normal mode expands, contracts, and simultaneously oscillates in the tangential direction during the inertia oscillation (Ref. 2, 5). When the amplitude of the generating disk is increased beyond a certain point, the flow becomes unstable to a regular set of waves in the azimuthal direction, Fig. 12b); these waves amplify and often become complete rows of oscillating vortices (Ref. 7).

A similar class of very low-frequency waves known as Rossby waves (Ref. 2) are very important in the motions of the atmosphere and oceans. This type of wave arises where there is a *variation* in the depth of a fluid such that axial vortex tubes are forced to shrink or stretch in a low-Rossby-number motion. Our example of a Rossby wave is produced in the annular cylinder shown in Fig. 13a. A conical bottom pro-

duces radial depth variation. A low, smooth radial mountain on the cone forces the wave perturbations when counterclockwise zonal flow is produced by slowly reducing the rotation of the container from an initial high value. With a Rossby number of about .045, a train of 5 sinusoidal stationary waves is produced (Fig. 13b). For larger characteristic values of the Rossby number, the stationary wave number decreases (Ref. 2).

Ekman Layers and Free Shear Layers

Effects of viscosity can be very substantial in rotating systems. The Taylor-Proudman theorem deals with inviscid fluids, for example, and does not hold in boun-

14. Ekman boundary layer produced by increasing the rotation rate of the floor under a rotating cylinder of water. The vertical streaks left by falling dye crystals show that the main body of water continues to rotate. The spiral streaks on the floor show that only in a very thin boundary layer are there appreciable radial velocity components.

dary-layer regions for any Rossby number. Figure 14 shows one of the simplest types of rotating boundary layers, called an *Ekman layer* (Ref. 2). The water and tank were initially in rigid-body rotation, and the tank speed was then increased to a slightly higher rotation rate. The dye streaks in Fig. 14 show the strong outward spiraling motions in the Ekman layer; the fluid in the inviscid region above is in clockwise relative motion that is almost purely tangential. A downward flux over the central region is required to supply the radial flux in the Ekman layer. This stretches the vortex tubes in the interior, and has the effect of increasing the angular velocity of the fluid to the new value for the tank in a much shorter time than would be expected from ordinary viscous diffusion mechanisms.

If the currents in the inviscid region vary from one place to another, the transverse Ekman layer fluxes also vary, requiring exchanges of fluid between the viscous and the inviscid regions that may be quite localized. In Fig. 15 the cylinder has a central disk

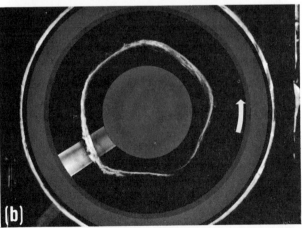

13. (a) Side view of annular cylinder combination with a conical bottom. The ratio of the outer to inner depth is about 5/3. (b) In plan view, a dye trace shows a train of 5 stationary Rossby waves excited in the annulus when there is tangential flow past a low ridge on the conical bottom. The dye trace was released at about middepth.

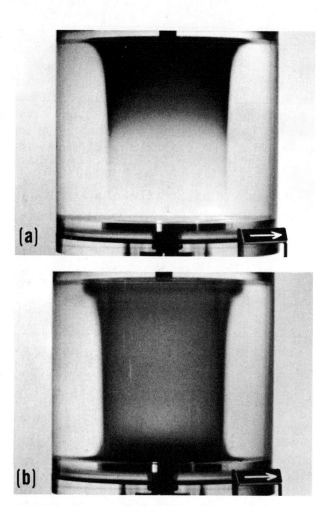

15. Side view of a cylindrical free shear layer moving downward from the solid lid of a rapidly rotating circular cylinder 130 sec.⁻¹) toward the edge of a more slowly rotating disk in the base. The Rossby number of the relative disk rotation is about 10^{-3} (a). (b) When the dye reaches the disk in the base, it flows inward along the disk and then upward in a slow flow parallel to the axis of rotation.

which is flush with the clear plastic bottom. The disk is rotated at a slightly lower speed than the cylinder. The interior fluid out to the disk radius adjusts to a rotation speed halfway between that of the disk and that of the upper lid. The accommodation to the no-slip boundary condition occurs in an Ekman layer on the disk, and in another Ekman layer under the lid. The fluid in the outside ring rotates at nearly the same speed as the cylinder. Ink is injected through a central hole in the solid lid, which is in contact with the fluid. The ink spreads rapidly outward in the top Ekman layer to the same radius as the disk, and stops there, even though the lid is a single rigid sheet. It then descends in a thin cylindrical Taylor sheet with a hollow core (Fig. 15a), a so-called free shear layer, to the base disk edge. There it flows inward in the disk Ekman layer and eventually completes the circuit by a slow upward flow parallel to the axis (Fig. 15b).

References

1. Webster A.G. (1942): *The Dynamics of Particles and of Rigid, Elastic and Fluid Bodies*, 3rd Ed., New York: Stechert, pp. 316ff.

2. Greenspan, H.P. (1968): *The Theory of Rotating Fluids*, Cambridge University Press, p. 327.

3. Taylor, Sir Geoffrey (1921): Experiments with rotating fluids, *Proc. Roy. Soc. London* (A), *100*, 114-121.

4. Lamb, H. (1932): *Hydrodynamics*, Cambridge University Press, pp. 317ff.

5. Chandrasekhar, S. (1961): *Hydrodynamic and Hydromagnetic Stability*, Oxford University Press, pp. 284ff.

6. Webster, F. (1968): Observations of inertial-period motions in the deep sea, *Rev. Geophys.*, *6*, 473–490.

7. Fultz, D. and Murty, T.S. (1968): Effects of the radial law of depth on the instability of inertia oscillations in rotating fluids, *J. Atmos. Sci.*, *25*, 779–788.

Aerodynamic Generation of Sound

John E. Ffowcs Williams
and
M. James Lighthill
IMPERIAL COLLEGE OF SCIENCE & TECHNOLOGY
UNIVERSITY OF LONDON

Introduction

A steady stream of air can produce sound (Fig. 1) — sometimes very beautiful sound, sometimes very ugly. Yet sound is a vibration of the air. How can it be produced by a steady air flow? In this film we study why, and how much, a flow, even though produced by a constant pressure difference, can generate the pressure fluctuations that we hear as sound — sound that may be amplified by the presence of obstacles in a stream, and still more powerfully by the action of resonators like the pipes of an organ, or the bubbles whose resonant vibration we shall find causes the "babbling of a brook."

A high-speed jet without resonators can be quite noisy enough. We especially want to study the sound a jet makes so that aircraft and their engines can be designed to keep ground noise levels within prescribed limits when the correct take-off precedures are adopted. Also, in the design of rocket launching sites an important issue is the sound, generated again by a

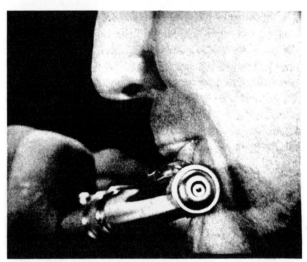

1. A steady stream of air produces sound.

flow designed to be as steady as can be achieved. Admittedly, fluctuations are already present in the rocket combustion chamber or the aircraft jet pipe.

But jet noise that is extremely comparable, if frequencies and intensity are properly scaled on jet diameter, can be produced in the laboratory with an air jet practically devoid of fluctuations in the emerging flow. This raises two questions. First, how do new and quite intense fluctuations arise as the flow emerges into the atmosphere? Secondly, what fraction of the energy radiates away as sound?

The answer to the first question lies in the words "flow instability." A vortex layer is formed between the jet and the outside atmosphere, and free vortex layers like this can be exceedingly unstable. There is a fast exponential growth of disturbances of a certain type as they travel downstream, so that any small fluctuations present at the orifice, if their frequency is less than a certain value, may have a large positive growth rate and quickly become very big. Now that is an effect of inertia, and it is counteracted by the viscosity of the fluid. The ratio of inertial to viscous effects is the important quantity known as Reynolds number. At high Reynolds numbers disturbances in a wide range of frequencies grow very fast and interact with one another, and we call the result "turbulence." At lower Reynolds numbers the effect of viscosity is to reduce the growth rate of small perturbations so that most disturbances do not grow at all. Then it is only disturbances in quite a modest range of frequencies that have a significant growth rate. So, under those circumstances the disturbance that appears tends to be relatively regular (Fig. 2).

3. An edge placed near the exit of a jet produces an audible whistle.

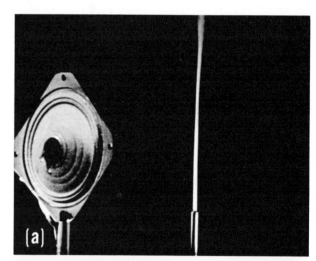

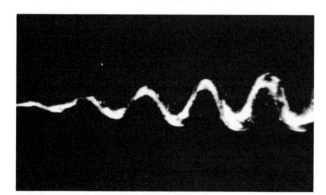

2. Disturbances in the wake of a wire or thin plate can grow to form a "vortex street."

We can make them quite regular by an interesting feedback mechanism using an edge and a jet (Fig. 3). The edge casts off an eddy each time the local angle of attack changes sign. The eddy reacts on the nozzle to create a disturbance at the right moment in the cycle, which in turn is amplified by the jet shear. An extremely regular cycle results from this jet-edge interaction. A similar experiment performed with a cylindrical jet is seen in Fig. 4. At low Reynolds number disturbances grow too slowly to be visible close to the jet. But if a loudspeaker artificially introduces dis-

4. (a) With the loudspeaker off, a low-Reynolds-number jet of smoke remains laminar for an appreciable distance. (b) The loudspeaker is emitting sound at a critical frequency. Disturbances in the jet grow quickly.

turbances of the right frequency, they quickly grow to something very substantial.

But it is only at these low Reynolds numbers that jets are in this way acoustically sensitive. At higher

Reynolds numbers external sound has no influence on the jet, which quickly assumes a turbulence condition because disturbances in a wide range of frequencies grow fast. At the very high Reynolds numbers that arise in engineering applications, the sound of a jet is just a minor by-product of the turbulence. The turbulence itself is unaffected by its presence.

The wake of a wire in the wind is another free vortex layer which, like a jet, is extremely unstable. At low Reynolds numbers (around 100 based on wire diameter) the flow is extremely regular, taking the form of Karman's vortex street, which reacts back on the wire causing it to shed just the right amount of vorticity in each cycle to maintain the street. At higher Reynolds numbers more and more frequencies are important and the flow soon looks quite chaotic.

We have seen, then, that air flowing steadily out of an orifice, or blowing steadily past a wire, may develop fluctuations; at low Reynolds numbers regular ones, at high Reynolds numbers turbulent ones. From this, sound may result in two ways. First, resonators if present may be excited by the pressure fluctuations in the flow and may radiate sound rather efficiently. A beer bottle is a so-called Helmholtz resonator, where the inertia of air flowing in and out of the nozzle and the stiffness due to air in the bottle resisting compression define a frequency, and sound radiation is the main loss term. Also, pipes of variable lengths are used as resonators in musical wind instruments based on the jet-edge method of excitation.

Sound Radiation

We shall now concentrate on how flow fluctuations, like the turbulence in a jet, generate sound without the action of resonators. How much of the energy of the fluctuations succeeds in escaping a sound? The key word here is "escaping." Radiated sound means pressure fluctuations whose energy propagates away from the source with the sound speed: 350 meters per second. Of course, within the flow there are other pressure variations balancing the local fluid accelerations; these do not propagate at all.

Actually the ear, or a microphone, can register these latter pressure fluctuations exactly as if they were sound, but they are called "pseudo-sound," — only "pseudo" because they do not propagate like sound. For example, the microphone placed very near the jet (Fig. 5) registers pressure fluctuations, with a minimum each time one of the mixing-region eddies passes the microphone and maxima in between. But those are pseudo-sound because the minima and maxima travel at the speed of the eddies, and not at the much larger sound speed. These fluctuations are substantial only very close to the jet. If the microphone is moved up and down near the jet, there are enor-

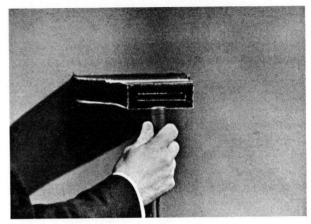

5. A microphone near a jet hears "pseudo-sound."

mous changes in the sound heard. Farther out the sound of the fluctuations disappears altogether, but in that region we can with the same microphone find the real sound if we increase the speed of the jet. But the real sound shows much less variation of amplitude with distance. There is some falling off with distance, actually like the inverse square of distance from the nozzle, as the energy is radiated outward all the time.

To summarize, then: at small distances from the orifice there is an intense level of pressure fluctuations known as "pseudo-sound." Here the real sound is at a very much reduced level from the pseudo-sound. Farther out, the pressure intensity decays like the inverse-square law. The reason why radiated sound is so much less than pseudo-sound emerges after we have looked at mechanisms of sound generation.

Mechanisms of Sound Generation

The mechanism that generates a simple inverse-square law of intensity at all distances is a simple source. In Fig. 6 a simple source is observed in a

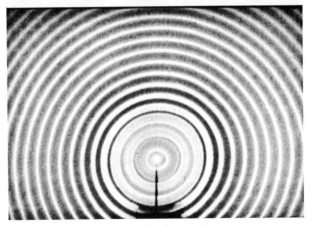

6. Shallow-water ripples from a simple source.

water tank by viewing the shadows of the waves produced on a screen. The depth of the water was chosen to represent sound as well as is possible in the sense that waves in the ripple tank travel at an effectively constant speed. This speed is 20-odd centimeters a second, and propagation from a source is clearly visible. Sound waves travel at speeds over a thousand times faster.

Air pressure fluctuations at a distance r from a sound source are given by: $\Delta p = \dot{q}(t - r/c)/4\pi r$. The variation in inverse proportion to r means that the intensity or energy flow (which varies with the square of Δp) obeys an inverse-square law. The pressure fluctuations faithfully follow those of the source itself except for a time lag r/c which represents the time taken for waves to travel a distance r at the sound speed c. From a source, as its name implies, there is a rate of mass outflow $q(t)$ in grams per second. But it is only rate of change of mass outflow that produces waves, and it is this rate of change $\dot{q}$, often called the source strength, which is mimicked by the pressure fluctuations at a time r/c later. Any foreign body in a fluid, whose volume V pulsates as a function of time, acts like a source, and the mass outflow from the body is equal to the density of the fluid times the rate of change of volume of the body, $\dot{V}$. So the source strength in that case would be proportional to $\ddot{V}$. In fact, the waves in our ripple tank are produced by a foreign body that is pushed in and out of the water (Fig. 7).

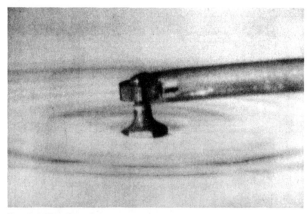

7. **Oscillating plunger acting as source.**

A quite different method of sound generation occurs when two equal and opposite sources are close together (Fig. 8). At any one moment the sum of their strengths is zero. This contrasts with the field of a single plunger (Fig. 6). Although the amplitude of motion near the source is somewhat increased, far less wave energy is propagating away from the source. (Compare the near field with the far field.) In fact, in particular directions (east and west) contributions from the positive and the negative sources almost cancel.

Those contributions do not cancel exactly in most directions, however, because the distance r from the positive source is not exactly equal to the correspond-

8. **(a) Two plungers close together oscillating 180 degrees out of phase produce the directional wave field of (b).**

ing distance from the negative source. For example, in a direction that makes an angle θ with the line of length l joining the sources, the difference between those distances is approximately $l \cos \theta$. Now, from that difference in the value of r, we can deduce, by differentiating the expression for the simple source field with respect to r, that the field of the pair of sources is:

$$\Delta p \approx \frac{\dot{q}(t - r/c)}{4\pi r \cdot r} l \cos \theta + \frac{\ddot{q}(t - r/c)}{4\pi r c} l \cos \theta.$$

Actually, it is the differences in the amplitude factor r^{-1} that produce the near field term, and the differences in the time lag r/c that produce the radiation, or far field, term. At distances r which are large compared with the distance l between the sources, the pressure fluctuations, in both near and far fields, depend not separately on the strength $\dot{q}$ of the sources and on l, but only on their product $(\dot{q}l)$. The fluctuations would be the same if we had sources that were twice as strong but twice as close together. We call them a dipole sound field, of strength $\dot{q}$ times l.

Now we can see that when l is small enough the dipole field is going to be very much smaller than the simple source field. The near field term is smaller than the simple source field by a factor of at least

(l/r). The far field term depends on the frequency, the ratio of $\ddot{q}$ to $\dot{q}$. The speed of sound divided by the frequency is the quantity λ, the radiated wavelength. The far field term then is smaller than the simple source term by the ratio $(2\pi l/\lambda)$. When this ratio is small, we say that the source region is compact. The source field is compact relative to the wavelength. Now, aerodynamic sound sources are often compact, and we have our first glimpse here into the dual structure of the pressure field. At distances greater than a wavelength or so, the pressure falls off like $1/r$, like a simple source. Closer to the source region than a wavelength or so, the near field term is dominant. The field is then relatively intense, being induced predominantly by the nearest simple source.

Force Associated with Dipole Radiation

In our ripple tank there is a local back-and-forth movement in the neighborhood of the sources, and locked in this local movement there is a fluctuating momentum. Now, no net momentum can be produced by simple sources, so it follows that in the external dipole field there must be an equal and opposite momentum. We can find out how much by considering the mass flow across different discs strung on the axis joining the two sources. We do not know how much that is, but it is obvious that it must drop suddenly by an amount $q(t)$ at the source where the mass outflow is $-q(t)$, and rise suddenly by $q(t)$ at the source where the mass outflow is $+q(t)$ (Fig. 9). It follows from

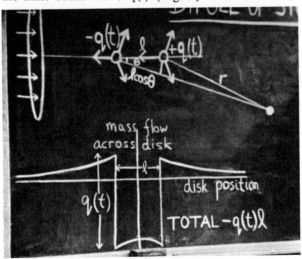

9. **Mass flux as a function of position for a dipole source.**

this that when the two sources come closer to make a dipole the total momentum in the region between the sources would be equal to the area under the curve of mass flux versus position (the mass flux being equal to momentum per unit length). That is, the momentum would be $-q(t)$ times the source separation distance, l. There must on the other hand be an equal and opposite momentum in the external dipole field,

and this therefore must be $+q(t)$ times l. Now rate of change of momentum equals force, and it follows that the strength of the dipole, $\dot{q}(t)\,l$, is exactly the force with which the dipole is acting on the external fluid, producing a rate of change of its momentum. Actually it would be the same not only in magnitude but also in direction if we take the direction of the dipole as pointing from the negative toward the positive source.

So we can say that a dipole whose strength in magnitude and direction is $\mathbf{F}$ is equivalent to the action on the external fluid of a force of exactly the same strength.

10. **Wave field from a group of three out-of-phase sources whose strengths do not add up to zero.**

In Fig. 10 we see the field from a group of sources whose strengths do not add up to zero, and they are not in phase. Yet they generate a rather symmetrical wave field. This illustrates the important principle that any group of compact sources radiates like a simple source at some central point if the sum of their strengths is not practically zero. That is because the difference between the two sets of source arrangements is simply three dipoles, each composed of a negative source at the center and a positive source at one of the peripheral points (Fig. 11). These dipoles are negligi-

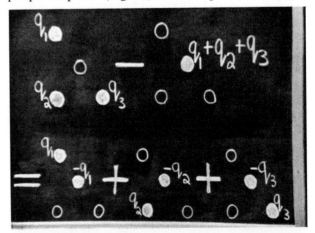

11. **The difference between a compact arrangement of three sources and a simple source having the strengths of the sum of the individual sources is three dipole sources.**

ble when the source is compact. The same conclusion would be reached if in addition to the sources there was an externally induced dipole resulting from the action of a force.

So, in sound-generation problems we always ask first whether the sum of all the source strengths is significantly different from zero; when that is so, then the total source strength is what dominates the external sound field. For example, when a body is changing in shape as well as in volume, we can represent that fact acoustically by sources distributed round its outside which correspond to displacements of the surface, and by dipoles corresponding to forces. All those would be important in the near field, but in the far field it is only the total source strength that matters, and that would depend on the total rate of change of volume of the body, since only that can affect mass outflow.

Thus, in the ripple tank an irregularly shaped plunger generates a far field similar to that of a simple source, essentially because the difference is the sum of dipole far fields that is reduced in magnitude by the factor $(2\pi l/\lambda)$. Similarly, if a bubble in water is vibrating, only its volume change is important in producing sound. Vibrating bubbles have a characteristic frequency depending on the compressibility of the gas. For air bubbles, these regular vibrations of bubble volume have a frequency of about 600 cycles per second divided by the diameter in centimeters, and make quite a musical note. Flows with bubbles produce sound mainly because the bubbles are resonators of various frequencies whose volumes fluctuate in response to pressure fluctuations. So, here again, pseudo-sound makes itself heard by exciting resonators. A simple-source sound field is not so amplified. When bubbles are introduced the radiated sound changes little, because a simple source has a far field comparable in intensity to that of its near field, which is exciting the bubble. But a dipole of the same frequency whose natural far field is weak in relation to its near field has its radiation greatly amplified when the same lot of bubbles respond to its near field.

But in many important cases of aerodynamic sound generation the total source strength is zero, because foreign bodies in the fluid are not changing their volume and the rate of introduction of new fluid is zero or, as with an air jet, practically constant. When the total source strength is zero, the equality between the source arrangements pictured in Fig. 11 remains true, but the second term on the left amounts to nothing. The equality states therefore that the whole sound field is equivalent to a sum of dipole wave fields . . . each corresponding to the action of a force on the fluid. Now, all those effective forces plus any real forces that may be present combine together to form a resultant force, and to a good approximation the far field is that

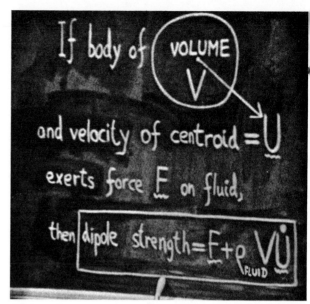

12.

of a single corresponding dipole. Actually, its strength is equal to the total force exerted on the fluid, plus a correction if any bodies are moving through the fluid (Fig. 12). This correction is equal to the rate of change of momentum of the fluid supposed to have been displaced by the body. Actually this correction is often unimportant for air with its low density, but it is used in propeller-noise theory. Each little element of a propeller blade can be represented by a dipole consisting of the force with which it is acting on the fluid, plus the displaced mass of air times its centrifugal acceleration.

We can illustrate this displaced inertia term in the ripple tank. If we allow a solid body to vibrate just at the amplitude with which the water moves, it makes no waves since the force between it and the fluid exactly balances the rate of change of displaced momentum. But if we hold an identical body at rest,

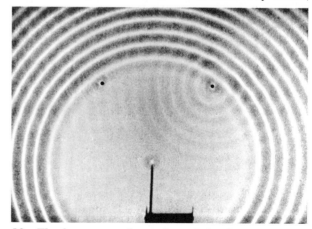

13. **The last waves of a packet of simple source waves pass two objects in the field. At the upper left a freely floating body does not generate waves. A similar object at the upper right is held at rest and produces a scattered wave field.**

which needs a different fluctuating force, we observe the dipole field associated with that force, which is what would usually be described as a scattered wave field. We illustrate the case when both bodies are seen together — one scattering and the other not (Fig. 13).

The dipole strength for a rod moved rapidly through the air is simply the force with which the rod acts upon the fluid. The correction term due to acceleration is unimportant. We can illustrate this motion and the waves it generates in the ripple tank. The dipole axis is seen in Fig. 14 to be normal to the direction of

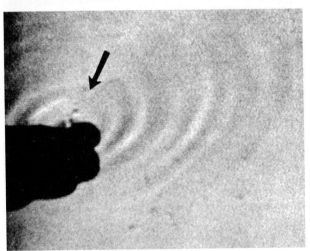

14. A body moving along the direction indicated generates a dipole wave field whose axis is normal to the direction of motion.

motion. The waves are induced by fluctuations of lift force. At higher Reynolds numbers, as we have already seen, lift fluctuations occur over a broad frequency range — they generate acoustic noise. This is an important part of the sound field of aero-engine compressors. Finally, we will leave the subject of dipoles by noting that the wind roars mainly owing to the forces with which solid obstacles resist it. Their geometry is very various and so a broad spectrum usually results.

Quadrupole Radiation

We introduce our next main subject with an experiment; two rods placed in the ripple tank are vibrated horizontally. They act on the fluid with identical fluctuating forces, in phase, and you can see the single dipole field associated with their resultant (Fig. 15a). Note the strength of the wave shadows in the top left-hand corner of Fig. 15a. In Fig. 15b the two rods are being moved in exact anti-phase so that the total dipole strength is zero; the strength of the wave shadows in the top left-hand corner is considerably reduced. This experiment is particularly important because it shows how much smaller the far field is when the resultant total external force on the system

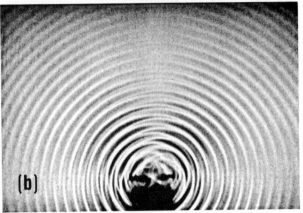

15. (a) Two oscillating rods moving in phase (north-south direction) produce a dipole wave field with intensity minima along an east-west axis. (b) When the rods move 180 degrees out of phase, a quadrupole wave field is produced. The intensity minima are in directions east, north, west, south with maxima in between.

is zero. So when we have air turbulence that is separated from foreign bodies that can act on the air with forces, we shall get radiation of this type with a reduced far field. This wave pattern has a name derived from the fact that two equal and opposite dipoles can be thought of as altogether four sources; the combination is called a *quadrupole*. Notice in Fig. 15b that there are four directions in which the far field is strongest.

We have seen earlier that the field of a dipole is smaller than that of an individual source by the factor $(2\pi l/\lambda)$, because two sources opposing each other, separated by a length l which is small compared to $(\lambda/2\pi)$, almost cancel. Now a quadrupole has two dipoles opposing each other, so again we repeat this cancelled factor and we get $\left(\dfrac{2\pi l}{\lambda}\right)^2$ for the efficiency of the quadrupole versus the individual source (Fig. 16). A common example of the quadrupole is a tuning fork, where the two tines vibrate in anti-phase, inducing equal and opposite forces. Much less energy is heard than when one of the tines is baffled so that the field of one of the elementary sources is heard.

These ideas explain why in a turbulent jet the

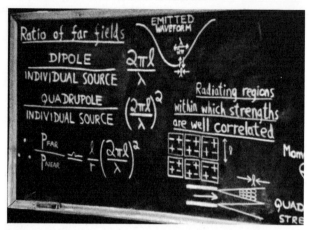

16:

pseudo-sounds, or near-field pressure fluctuations, are so large in relation to the radiated sound or far-field pressure fluctuations. There is no variable rate of introduction of new fluid, so the total source strength is zero. There is no force exerted on the fluid by any foreign body, so the total dipole strength is zero. Therefore we can say that the radiation is of quadrupole type, and far-field pressures to near-field pressures are not simply in the ratio l/r as for a source, but carry the additional factor $\left(\frac{2\pi l}{\lambda}\right)^2$.

We see already why it is so difficult to reduce the sound of a jet. Any idea that the turbulence in the jet may be modified by inserting solid obstacles will fail. This is because a solid obstacle will induce a dipole field, which is a much more efficient radiator than the quadrupoles which are in the jet alone.

Quadrupole Strength

To find the strength of the quadrupoles we use the fact that the main difference between laws governing how velocity components fluctuate in a turbulent jet and in a simple acoustic medium is that momentum transfer (in and out of a fluid element) is not accomplished simply by pressure. There is additional transport of, for example, the component of fluid momentum ρu by motions at right angles to the velocity v. The resulting momentum transport ρuv can be thought of as a force with which the fluid element acts on the external medium (Fig. 17). It consists of two equal and opposing forces, so it is effectively a quadrupole. Actually the quadrupole strength per unit volume, for quadrupoles of this orientation, may be proved to be ρuv. But what volume does a single quadrupole occupy? The answer lies in the statistical nature of turbulent fluctuations. There is a tremendous interference between waves generated from uncorrelated regions. But from any small region in which the fluctuations are well correlated the waves are relatively well-ordered (Fig. 18). This suggests viewing

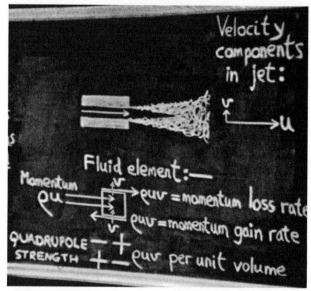

17.

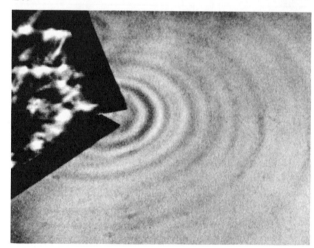

18. **Well-ordered waves emanate from a slit in a barrier separating a turbulent flow from tranquil water.**

the turbulence as made up of distinct elements of scale l within which fluctuations are well correlated. Contributions from different regions are uncorrelated. Pressure amplitudes from regions in which correlation is high add linearly, but the mean square pressures add linearly from uncorrelated regions. Measurements in jets have indicated the shape and size of correlated regions or eddies. Measurements also show that the ratio $(2\pi l/\lambda)$ is of the order of the fluctuating velocity in the turbulence divided by the speed of sound. In low-speed jets this ratio is very small.

Sound Output from a Jet

The general formula for quadrupole radiation can be used to estimate the total sound output from a jet of speed U. For modest speeds, $(2\pi l/\lambda)$ is about $(1/7)$ (U/c), because the fluctuation velocities just referred to are about 1/7 of the jet speed. So $(2\pi l/\lambda)^2$ is approximately the square of the speed measured on a

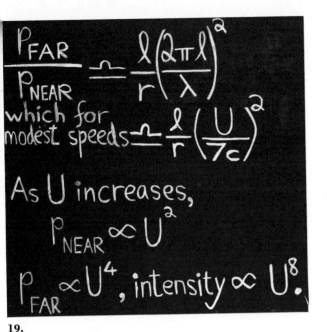

$$\frac{P_{FAR}}{P_{NEAR}} \simeq \frac{\ell}{r}\left(\frac{2\pi\ell}{\lambda}\right)^2$$

which for modest speeds $\simeq \frac{\ell}{r}\left(\frac{U}{7c}\right)^2$

As U increases,

$$P_{NEAR} \propto U^2$$

$$P_{FAR} \propto U^4, \text{ intensity} \propto U^8.$$

19.

scale of 7 times the atmospheric sound speed. But the near-field pressures themselves vary like U^2, so the far-field pressures go like U^4, and the intensity of sound radiation therefore like U^8 (Fig. 19). This rather crude argument gives a result in surprisingly good agreement with measurements of total acoustic power output for the more moderate values of the jet speed. But there are two corrections that we need to make to it in order to understand noise at the higher jet speeds. First, net turbulence production tends to be somewhat reduced at the higher Mach numbers. That would make power output increase somewhat more slowly at the speeds around $U/c = 1$, more like a U^6 law. But the second correction works the other way and restores the U^8 dependence up to a value of jet speed around 1.5 times the atmospheric sound speed (Fig. 20).

This second correction is due to the fact that turbu-

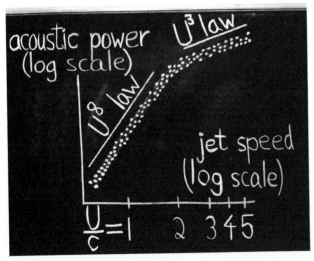

20.

lent eddies are moving at about half the jet speed relative to the atmosphere into which they radiate. Now, the best-known property of moving sound sources is the Doppler effect, whereby sound coming toward us seems to have a higher pitch than the same sound going away from us. This is illustrated in the ripple tank in Fig. 21. The wavelength ahead of the source is reduced, and the wavelength behind the source increased.

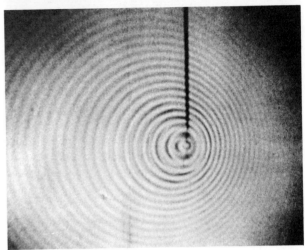

21. A source moving to the right produces shorter wavelengths ahead and longer wavelengths behind.

If an eddy is moving with speed U_e, the wavelength of sound emitted forward is reduced by a factor $(1 - U_e/c)$, essentially because during the time between emission of crests the sound has traveled a dis-

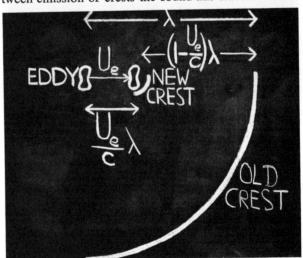

22.

tance λ while the eddy has moved on a distance $(U_e/c)\lambda$ (Fig. 22). For emission in other directions making an angle θ with the direction of motion of the eddy, the same argument applies with U_e replaced by the component in that direction, $U_e \cos\theta$. So the wavelength in that case is $(1 - \frac{U_e}{c}\cos\theta)\lambda$ (Fig. 23).

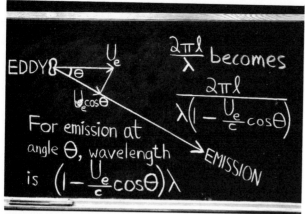

23.

This eddy Doppler factor changes the compactness ratio, $2\pi l/\lambda$, which now becomes $(2\pi l/\lambda)$ divided by the Doppler factor. This now is not so small in the forward directions. So we see here a tendency for preferential forward emission, such as is particularly clearly heard for jets with U/c around 1 to 2. For higher values of U/c preferential emission is at its extreme in the so-called Mach wave direction of supersonically moving eddies. Then $\dfrac{U_e}{c}\cos\theta$ is equal to 1 and the wavelength evidently reduces without

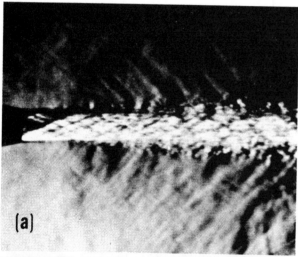

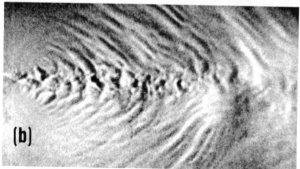

24. **(a) Shadowgraph visualization of a supersonic jet shows Mach waves emanating at about 45 degrees from the jet axis. (b) Ripple-tank visualization of waves radiating from a supercritical jet.**

limit. Then there can be no quadrupole cancellation and the sound is just simple source. It's as if the eddy were generating its own supersonic bang. A shadowgraph picture shows that the sound field of a properly expanded supersonic jet is simply a collection of Mach waves (Fig. 24a). A supercritical jet in a ripple tank produces similar waves (Fig. 24b).

At high speed, then, we have important consequences stemming from the breakdown of the compactness condition. The sound is then directional and of simple-source type — the pressure falling off like l/r all the way from the turbulence where it varies like U^2. The mean square pressure then varies like U^4.

Detailed arguments on these lines show that the high-speed form of the radiation curve follows a U^3 law and the transition occurs at about $U/c = 2$, where the eddies first become supersonic (Fig. 20). This agrees rather well with the observations. Another way of plotting the results is in terms of the acoustic efficiency, which is acoustic power output divided by jet

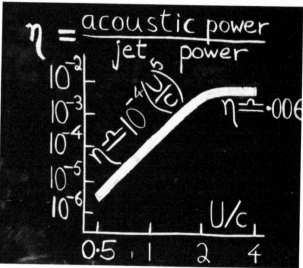

25.

power (Fig. 25). At modest values of the jet speed this takes quite low values, around 10^{-5}, but at higher speeds when the supersonic-bang mode takes over, the curve levels off to just under 10^{-2}, essentially because the inefficient quadrupole radiation has been replaced by efficient simple-source radiation. More detailed study of data in various frequency bands confirms the view of a turbulent jet as exciting the external acoustic medium in the same way as would clusters of eddy-sized quadrupoles that are traveling in the direction of the jet and whose strengths reflect turbulent fluctuations in the momentum transport.

Jet Noise Reduction

Of course, jet-engine powers have consistently increased while at the same time there has been an incentive to get the acoustic power down as far as possible. This has attracted engineers to come ever further down the curve of Fig. 25, with turbo-fan engines at low jet velocities achieving their increased power with large jet diameters. The turbo-fan is attractive for many reasons, of course, but the noise argument was one of the most compelling. This could at best be a long-term solution, however, so other means of jet noise reduction were vigorously sought. This was difficult, but nonetheless important improvements have been achieved using devices which for given jet speed reduce the ratio $(2\pi l/\lambda)$ by reducing the turbulent intensity.

The idea is to get the external atmospheric air moving in the same direction as the jet gases so as to reduce the total shear that is producing turbulent fluctuations. The multi-tube design of jet exits (Fig. 26) does this, and has a second important advantage in that the annoyingly intense directional peak from any one of those small jets gets flattened because its sound gets scattered by the presence of all the other jets.

More recent research on aircraft noise reduction has concentrated on compressor noise, which has become relatively more important for the big turbo-fan engines. But that is only one of several fascinating contemporary applications of the basic knowledge outlined on aerodynamic sound generation.

26. **Multi-tube jet exit device.**

Special scenes in the film were provided by the Pratt & Whitney Aircraft Division of United Aircraft Corporation, the National Aeronautics and Space Administration, and R. W. Webster, Imperial College of Science & Technology, London.

References

1. **Lighthill, M. J.**
    On Sound Generated Aerodynamically
    i. *General Theory,* Proc. Roy. Soc., A. 211, 566, 1952
    ii. *Turbulence as a Source of Sound,* Proc. Roy. Soc., A. 222, 1954

2. **Lighthill, M. J.,** *The Bakerian Lecture, Sound Generated Aerodynamically,* Proc. Roy. Soc., A. 267, 1329, 1962.

3. **Ffowcs Williams, J. E.,** *Flow Noise,* Annual Review of Fluid Mechanics, Annual Reviews Inc., 1969.

Magnetohydrodynamics

J. A. Shercliff
WARWICK UNIVERSITY, ENGLAND

Introduction

Magnetohydrodynamics concerns fluid motion under magnetic fields. Fluids are not, in general, magnetic but if they can conduct electric currents $\mathbf{j}$, these can interact with a magnetic field $\mathbf{B}$ to produce forces $\mathbf{j} \times \mathbf{B}$ per unit volume which profoundly affect the motion. Figure 1 shows a simple example of the effect of the $\mathbf{j} \times \mathbf{B}$ force.

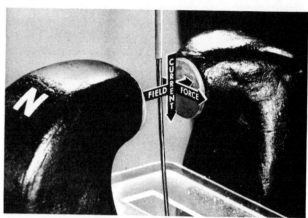

1. A vertical mercury jet carrying a vertical electric current $\mathbf{j}$ in a horizontal magnetic field $\mathbf{B}$ is deflected by the $\mathbf{j} \times \mathbf{B}$ force.

2. Solar flares.

The most abundant conducting fluid in the universe is ionized gas, or plasma, because all stars are composed of it. Plasmas can support phenomena besides those which are described by the magnetohydrodynamic approximation, which treats the fluid as a continuum.

Magnetic fields are common in cosmic phenomena, and Fig. 2 shows the strong effect of magnetic fields on solar flares. Plasmas also occur terrestrially, but on earth we also have conducting liquids, either electrolytic or metallic.

Magnetic Pumping

The $\mathbf{j} \times \mathbf{B}$ forces can be made to pump a conducting fluid. In Fig. 3, a vertical current is being passed through mercury in a duct in the horizontal field of a magnet.

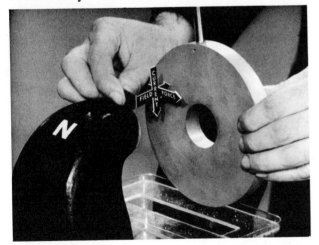

3. A simple electromagnetic pump.

The $\mathbf{j} \times \mathbf{B}$ acts to the left and pumps the mercury clockwise round the circuit, as is revealed by the "mercury-fall" in the right-hand side of the picture. The manometer above the magnet reveals the pressure rise across the pump.

Induced Currents

In the pump the current is *imposed*, but currents may be *induced* by motion of the conductor itself. In Fig. 4

4. When the aluminum ring moves into the magnetic field, induced currents produce a $\mathbf{j} \times \mathbf{B}$ force.

a ring entering a magnetic field links an increasing amount of magnetic flux. A current is induced around the ring, and $\mathbf{j} \times \mathbf{B}$ force strongly opposes the motion.

The Distribution of Magnetic Force and Its Effect on the Fluid's Vorticity

In a solid the distribution of $\mathbf{j} \times \mathbf{B}$ does not matter so much as the *overall* force, because the rigid solid sustains any force distribution. But a *fluid* subjected to $\mathbf{j} \times \mathbf{B}$ forces cannot in general sustain them without being continuously deformed. Figure 5 shows a top view of mercury being stirred by forces due to the cur-

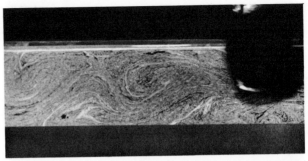

5. A mercury surface stirred by a moving magnet which is seen leaving the picture to the right. The other pole is under the trough, so the magnetic field is normal to the surface. The magnet poles do not touch the mercury.

rents induced by a moving magnet which provides a vertical field. Powder floating on the mercury reveals its motion. The effect of $\mathbf{j} \times \mathbf{B}$ forces on a liquid is more subtle than on a solid. We must examine how the force affects each individual fluid element.

A transparent electrolyte enables dye lines to be used to reveal the motion. Copper sulphate solution is

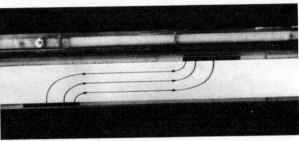

6a. Current distribution between electrodes at edges of flat duct containing electrolyte. The magnetic field is uniform and normal to the paper.

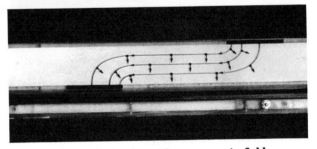

6b. $\mathbf{j} \times \mathbf{B}$ distribution in uniform magnetic field.

at rest in a flat box when the two-dimensional current shown in Fig. 6a is passed between the electrodes. If a uniform magnetic field is applied in the direction of viewing, dye lines in the fluid show that it barely moves, though the $\mathbf{j} \times \mathbf{B}$ distribution is complicated (Fig. 6b) and causes pressure changes which may be detected by the manometer shown in Fig. 6c.

Here $\mathbf{j} \times \mathbf{B}$ is being balanced by pressure gradients, i.e. $\mathbf{j} \times \mathbf{B} = \text{grad } p$ ($p = $ pressure). But curl grad $\equiv 0$ and so curl $\mathbf{j} \times \mathbf{B} = 0$, i.e. the magnetic field is an *irrotational* vector, incapable of making fluid elements spin, even though the force is far from uniform.

6c. Dye lines in the duct. The upper fluid in the manometer is also dyed. The manometer shows the pressure difference across the duct.

If the same **j**, **B** and **j** × **B** distributions* are applied with the fluid in motion at a fixed flow rate along the duct, the flow pattern is quite unchanged when the current is turned on. Again the pressure distribution is able to balance the **j** × **B** forces because these are still irrotational.

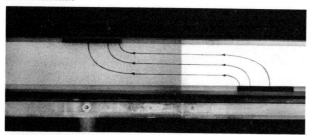

7a. Current distribution crossing the edge of the magnetic field. The field normal to the paper is uniform in the darker region on the left and falls to zero in the right-hand region.

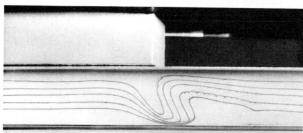

7b. Distorted dyelines in the fluid, which was at rest with the dyelines straight when the current shown in Fig. 7a was turned on one second earlier.

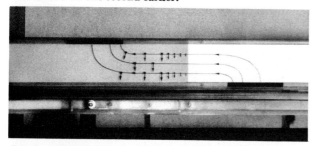

7c. j × B distribution, which is rotational because of the falling off of the magnetic field.

---

* Electrolytic conductivity is so low that the current distribution is not affected by the fluid motion, because the induced e.m.f.'s are far smaller than the ohmic potential differences.

These experiments are very different if repeated at the edge of the magnetic field. When the current flows through fluid at rest between the electrodes across the edge of the field as in Fig. 7a, the fluid is stirred. Figure 7b shows its state after one second. The **j** × **B** force now spins the fluid elements, i.e. it *creates vorticity*, which is counterclockwise here. Figure 7c shows the new **j** × **B** distribution with curl **j** × **B** ≠ 0 in the edge region. But curl grad $p \equiv 0$ and so grad $p \neq$ **j** × **B** here. The pressure gradient now cannot balance **j** × **B**, and the fluid cannot stay still.

If the fluid moving along the duct encounters the same **j** × **B** forces in the edge region, they give it counterclockwise vorticity with the result that the velocity profile is deformed.

In this electrolyte experiment, vorticity is generated by *imposed* currents, but in the mercury experiment shown in Fig. 5 it is generated by *induced* currents.

To sum up this section: to understand the effect of **j** × **B** on a fluid, we must consider also the unknown pressure field which can balance *irrotational* forces. Only to the extent that the **j** × **B** force is *rotational*, i.e. tending to alter the fluids vorticity, can it elude the pressure gradient and affect the motion. Thus the discussion must be in terms of vorticity.

Vorticity Suppression

If a metal loop is spun about a diameter perpendicular to a magnetic field, it links a changing magnetic flux (Fig. 8). As a result, induced currents produce **j** × **B** forces which damp out the motion. For small in-

8. Metal loop in a magnetic field. When the loop is spun about an axis perpendicular to the paper, a clockwise torque acts to oppose its motion.

clinations of the plane of the loop to the field, the opposing torque is proportional to angular velocity. But if the spindle is *parallel* to the field, there is no change of flux linked and no damping of the motion.

Similarly, vorticity of fluid elements about axes perpendicular to a magnetic field should be suppressed by

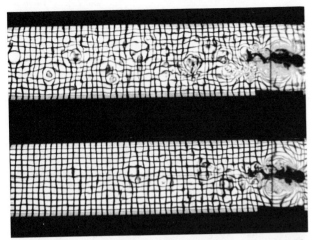

9. Damping of vorticity by a transverse magnetic field. The vortices in the mercury in the wake of the towed paddle decay more quickly when a transverse **B**-field is present (lower picture) than without (upper picture).

induced $\mathbf{j} \times \mathbf{B}$ forces.* Figure 9 shows two top views of the surface of mercury in a shallow trough. In the lower picture the arrows indicate the magnetic field. The mercury surface is being used to reflect an illuminated pattern. A moving paddle is shedding vortices in its wake. These vortices, with vertical vorticity, are visible because they create dimples which distort the reflection. Figure 9 presents a comparison between cases without the magnetic field (upper picture) and with the field (lower picture). The vortices are seen to decay much more swiftly when the field is present. A stronger field prevents the vortices ever being shed. So $\mathbf{j} \times \mathbf{B}$ forces can suppress vorticity as well as generate it.

Vorticity Redistribution

As another example† of rotational $\mathbf{j} \times \mathbf{B}$ forces altering vorticity, consider the circular motion of mercury between two concentric cylinders, the outer one fixed, the inner one rotating about its vertical axis. In the absence of magnetic effects and secondary flow, the fluid velocity falls off monotonically from its value at the inner cylinder to zero at the outer cylinder, and the vorticity is opposed to the rotation of the inner cylinder (Fig. 10). Now add a radial magnetic field between the cylinders and consider an imaginary loop initially lying in a plane through the axis. If the loop moves with the fluid, it rotates in such a way as to start linking magnetic flux. Thus induced currents must produce forces which change the motion so that the loop moves without linking magnetic flux, i.e. it stays

---

* This is a slightly oversimplified argument, because vorticity is the sum of the angular velocities of perpendicular planes in a fluid element, but closer investigation does not invalidate the argument in the cases chosen.

---

† This experiment is described and analyzed more fully by Heiser and Shercliff reference.

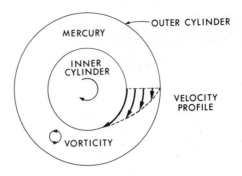

10. Plan view of cylinders and velocity profile without magnetic effects.

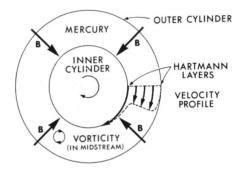

11. Velocity profile between cylinders, with the radial field turned on.

in planes through the axis and the mercury rotates like a solid with uniform vorticity and velocity proportional to radius, as in Fig. 11. The vorticity is now in the *same* sense as that of the inner cylinder. The uniform angular velocity of the fluid becomes exactly one-fifth of that of the inner cylinder when the field comes on, a consequence of the ratio of the cylinders' radii being 2:1 (see the Heiser-Shercliff reference).

The mercury is now "slipping" over both cylinders, the slip taking place across Hartman boundary layers (see Fig. 11) containing intense vorticity of opposite

12a. The vorticity indicator, which is free to rotate at the end of its restraining arm, about to be inserted into the mercury. The annular mercury surface is visible round the central, rotatable cylinder, which is painted black and white.

12b. Top view of the mercury and cylinders, with vorticity indicator in use.

sign. The *total* vorticity is unaltered by the magnetic field and the $\mathbf{j} \times \mathbf{B}$ forces have merely *redistributed* it.

The behavior of the vorticity outside the boundary layers when the field comes on is revealed by the vorticity indicator shown in Figs. 12a and 12b. The cruciform paddles take a correct mean for the fluid's angular velocity, which is indicated by the black-and-white disc. The freely rotating arm keeps the indicator vertical in midstream.

Perturbation of the Magnetic Field

So far we have considered only half of magnetohydrodynamics — the effect of the magnetic field on the flow. The currents in the fluid must also affect the field to some extent. Figure 13 shows the experiment in which a conducting loop swings through a magnet gap. The current induced in the loop affects the field, as is revealed by the twitching of the iron nails on the side of the magnet pole.

Field perturbation is weak in the experiments so far,

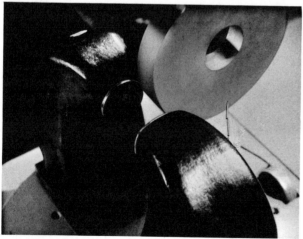

13. The currents induced in the metal loop affect the field and cause the iron nails (on the right of the nearer magnetic pole) to move.

and has been ignored. When the fields due to the currents induced in the fluid are too big to be ignored, the nature of magnetohydrodynamics changes drastically.

Consider a conducting loop rotating so that its plane becomes increasingly inclined to a magnetic field. Figure 14 shows how the field due to the induced current

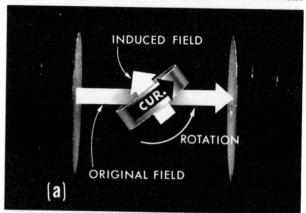

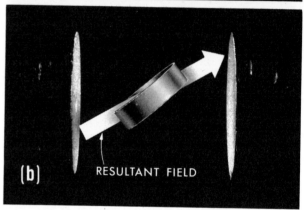

14. Currents induced in the rotating loop perturb the field so as to reduce the change of flux linked. (a) Field components. (b) Resultant field.

makes the total field less inclined to the loop than before. Thus the field is perturbed so that the rise of flux linked by the loop is reduced.

If the loop stops rotating, its resistance causes the induced current to decay and the field relaxes back to its original form in a "magnetic relaxation time" L/R that depends on the inductance L and the resistance R of the loop. It gets *longer* if we reduce the resistance. The extent to which a rotating loop perturbs the original field depends on how L/R compares with the time for a revolution.

When L/R is small, the field continually relaxes back to its original form and is hardly perturbed. Then the magnetic forces on the loop are *dissipative*; there is a torque only when the loop is moving, inducing the current.

But when L/R is large because the conductivity is high, the field is deformed strongly, and when L/R is very large, there is virtually no change of flux linked.

With perfect conductivity, the induced e.m.f. becomes negligible and spontaneous currents alter the field so that the flux linked never changes.

Modeling Perfect Conductivity by the Use of a Feedback

We may show how a perfectly conducting loop behaves by energizing a coil from a power source to which feedback is applied from a search coil in such a way that the flux linked is constant.* Figure 15 shows such a coil, pivoted in a horizontal magnetic field.

15. Artificial "perfectly-conducting" loop perturbing a magnetic field. When the coil is tilted slightly and released, it oscillates about the pivot. The compass needle in the center shows the direction of the magnetic field; it oscillates with the coil, showing that the magnetic field lines are distorted so that the coil never links any flux.

When the coil is disturbed, it oscillates as though subject to an elastic, torsional restraint. There cannot be dissipation when there is "perfect" conductivity. The torque is now proportional (for small angles) to the tilt, *not* to angular velocity (as in the low conductivity case). The time for an oscillation is much less than the magnetic relaxation time and the magnetic forces act in a pseudo-elastic, non-dissipative manner. There are corresponding effects in a fluid conductor.

High Conductivity Behavior in Fluids; the Alfvén Wave

For any loop drawn in the fluid, the magnetic relaxation time is of order $\mu_o \sigma l^2$, which is big if σ, the conductivity, is large or l, the scale, is large (as it is in astrophysics).

In the experiment shown in Fig. 9, $\mu_o \sigma l^2$ for loops drawn in the vortices is short compared with their rotation time, and the magnetic force is dissipative, not elastic.

To see the effect of "elastic" $\mathbf{j} \times \mathbf{B}$ forces in a highly

* For further details of this experiment see reference 2.

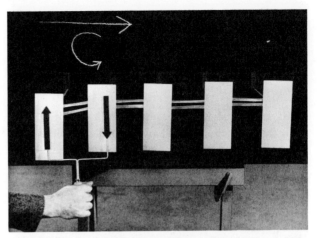

16. Schematic model of rectilinear shear motion. The curved arrow represents the vorticity, the horizontal arrow the field. The sides of white rectangular loops lying in a plane normal to the page are visible between the white cards. The vertical arrows on the left represent the magnetic torque on the first loop.

conducting fluid, consider the simplest motion that contains vorticity, namely a rectilinear, shear motion. In Fig. 16 the white rectangles represent layers of fluid with left-hand layer moving downward and counterclockwise vorticity between it and the next layer.

To understand the magnetic effects when there is an imposed horizontal field, consider rectangular loops lying between the layers, initially linking no magnetic flux. When the first layers move down, the induced currents produce elastic forces on the first loop, as shown in Fig. 16, tending to twist the loop back in line. The force on the second layer causes it to accelerate downward in turn, and then the third layer begins to feel the downward force, and so on.

17. Working model of Alfvén wave.

Figure 17 shows a working version of this apparatus,* with the rectangles mounted on pivoted rods. Flexible loops, attached to the rods, simulate perfect conductivity by means of feedback. The imposed magnetic field is again horizontal.

When the first rectangle is displaced downward, a downward motion propagates from rod to rod as a wave. In a fluid this corresponds to the propagation of

* See Melcher reference for fuller details.

the vorticity between the layers. Thus, when the rotational magnetic forces behave elastically because the conductivity is high, vorticity propagates along the field lines as a wave, known as an *Alfven wave*.

An Alfven Wave Experiment in Liquid Metal

Alfven waves can be produced in liquids if $\mu_o \sigma l^2$ is made large by choosing a highly conducting fluid such as NaK (sodium-potassium eutectic) and by making l,

18. Experiments on Alfven waves in NaK. The large magnet is at the rear, the stainless-steel tank (without lid) on the right, and the outer copper electrode in the foreground. Instrumentation is on the left.

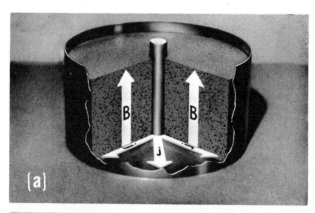

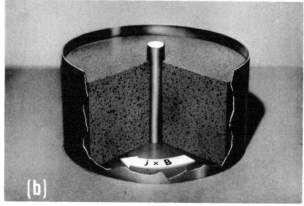

19. (a) Initial current and (b) $\mathbf{J} \times \mathbf{B}$ distribution in Alfven wave experiment.

the scale, as large as possible, and at the same time making the characteristic time (the transit time of the waves) small by making them travel fast in a strong magnetic field.

Figure 18 shows the apparatus. The stainless steel

20. Model of Alfven wave deforming magnetic field.

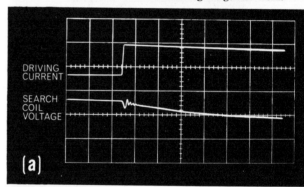

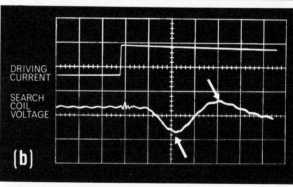

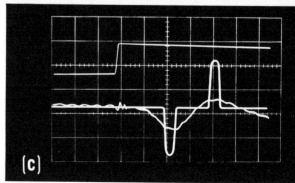

21. Oscilloscope traces from Alfven wave experiment. (a) Field off. (b) Field on. (c) Perfect conductor case.

tank on the right contains the NaK. The motion is excited by passing a current radially between the central rod and the outer wall, both made of copper (Fig. 19a). The imposed magnetic field is vertical.

The current at first flows across the bottom of the NaK and the resulting $\mathbf{j} \times \mathbf{B}$ force (Fig. 19b) makes just the bottom layer of fluid swirl. There is horizontal vorticity between the moving layer and the stationary fluid above. Immediately this vorticity propagates away along the vertical field lines as an Alfvén wave. The layer of radial current travels with it, and this deforms the magnetic field, as in Fig. 20. A search coil in the middle of the NaK detects the magnetic perturbation due to the traveling current layer, and the signal is displayed on an oscilloscope.

Figure 21 shows the results; the upper trace shows the sudden onset of the driving current, and the lower trace shows the voltage in the search coil. In Fig. 21a, the vertical magnetic field is absent, no waves occur and the lower trace shows only some stray signals, because the NaK shields the search coil itself. But in Fig. 21b, with the field on, the lower trace reveals the wave passing the search coil (the first arrow) and passing it again after reflection (the second arrow).

With a perfect conductor, the signals would be as in Fig. 21c, in which the pulse width is fixed by the finite width of the search coil. The resistance of the NaK makes the first pulse broader and weaker, and the second pulse after reflection is even broader and weaker. Nevertheless there is clear evidence of propagation of current and vorticity by the Alfvén mechanism in a real fluid.

Summary

In studying the effect of magnetic forces on a conducting liquid, it is fruitful to concentrate on their rotationality, or effect on vorticity, because then the unknown pressure gradient can be left out of consideration. When perturbation of the field by the currents is small, the magnetic force tends to be dissipative, altering the fluid's vorticity in various ways, but when the fluid conductivity is high enough for strong perturbation of the field to occur, the magnetic force is pseudo-elastic and vorticity propagates in Alfvén waves.

References

1. Heiser, W. H., and Shercliff, J. A. (1965), "A Simple Demonstration of the Hartmann Layer." *J. Fluid Mech. Vol. 22, p. 701.*

2. Melcher, J. R., and Warren, E. P. (1965), "Demonstration of Magnetic Flux Constraints and a Lumped Parameter Alfvén Wave." *IEEE Transactions on Education, Vol. E-8.*

Part II

8-mm Silent Film Loops

This section presents brief summaries of 133 short silent films of phenomena in fluid mechanics. Each of these films deals with a *single topic:* a particular experiment or a particular phenomenon.

Each summary first describes the apparatus used and the conditions and parameters of the experiments, and then proceeds to descriptions of what is to be observed, and the interpretations thereof.

The descriptions contained in Part II cover all the loops issued by the NCFMF. They are arranged in numerical order, but this bears no relation to topic or difficulty of topic. For the convenience of the reader, the loops are categorized according to topic in the contents list immediately following.

Subject listing of film loops

Boundary Layers and Turbulence

FM-3
Shear Deformation of Viscous Fluids

FM-6
Boundary Layer Formation

FM-88
Laminar Boundary Layers

FM-89
Turbulent Boundary Layers

FM-2
Structure of the Turbulent Boundary Layer

FM-1
Some Regimes of Boundary Layer Transition

FM-92
Stages of Boundary-layer Instability and Transition

FM-23
Tollmien-Schlichting Waves

FM-8
The Occurrence of Turbulence

FM-134
Laminar and Turbulent Pipe Flow

FM-135
Averages and Transport in Turbulence

FM-136
Structure of Turbulence

FM-137
Effects of Density Stratification on Turbulence

Cavitation

FM-125
Examples of Cavitation

FM-126
Cavitation on Hydrofoils

FM-127
Cavitation Bubble Dynamics

FM-128
Cavity Flows

Compressible Flows

FM-129
Compressible Flow Through Convergent-Divergent Nozzle

FM-130
Starting of Supersonic Wind Tunnel with Variable-throat Diffuser

FM-29
Supersonic Zones on Airfoils in Subsonic Flow

FM-28
Transonic Flow Past a Symmetric Airfoil

FM-90
Supersonic Flow Past Diamond Airfoil

Instabilities

FM-146
Examples of Flow Instability. Part I

FM-147
Examples of Flow Instability. Part II

FM-148
Experimental Study of a Flow Instability

FM-31
Instabilities in Circular Couette Flow

FM-32
Examples of Turbulent Flow between Concentric Rotating Cylinders

FM-46
Current-induced Instabilities of a Mercury Jet

FM-82
Water Jet Instability in Electric Field

FM-25
Low-speed Jets: Stability and Mixing

Internal Flows

FM-7
Propagating Stall in Airfoil Cascade

FM-15
Incompressible Flow through Area Contractions and Expansions

FM-16
Flow from a Reservoir to a Duct

FM-17
Flow Patterns in Venturis, Nozzles and Orifices

FM-49
Flow Regimes in Subsonic Diffusers

FM-65
Wide Angle Diffuser With Suction

FM-67
Flow through Right-angle Bends

FM-68
Flow through Ported Chambers

FM-69
Flow through Tee-Elbow

FM-19
Secondary Flow in a Bend

FM-18
Secondary Flow in a Teacup

Kinematics of Flow

FM-47
Pathlines, Streaklines, Streamlines and Timelines in Steady Flow

FM-48
Pathlines, Streaklines, Streamlines in Unsteady Flow

FM-107
Deformation in Fluids Illustrated by Rectilinear Shear Flow

FM-21
Techniques of Visualization for Low Speed Flows. Part I

FM-22
Techniques of Visualization for Low Speed Flows. Part II

FM-39
Interpretation of Flow Using Wall Tufts

Low-Reynolds-Number Flows

FM-112
Examples of Low-Reynolds-Number Flows

FM-113
Hydrodynamic Lubrication

FM-114
Sedimentation at Low Reynolds Number

FM-115
Kinematic Reversibility of Low-Reynolds-Number Flows

FM-116
Swimming Propulsion at Low Reynolds Number

Non-Newtonian Fluids

FM-122
Non-linear Shear Stress Behavior in Steady Flows

FM-123
Normal-stress Effects in Viscoelastic Fluids

FM-124
Memory Effects in Viscoelastic Fluids

Potential Flows

FM-80
Hele-Shaw Analog to Potential Flows. Part I. Sources and Sinks in Uniform Flow

FM-81
Hele-Shaw Analog to Potential Flows. Part II. Sources and Sinks

Pressure Fields

FM-33
Stagnation Pressure

FM-38
Streamwise Pressure Gradient in Inviscid Flow

FM-37
Streamline Curvature and Normal Pressure Gradient

FM-34
The Coanda Effect

FM-35
Radial Flow between Parallel Discs
FM-36
Venturi Passage

Rarefied Gas Dynamics
FM-103
Effect of Knudsen Number on Flow Past a Blunt Body
FM-104
Effect of Knudsen Number on a Jet

Rotating Flows
FM-138
Taylor Column in Rotating Flows (at Low Rossby Number)
FM-43
Buoyancy-induced Waves in Rotating Fluid
FM-44
Elastoid-inertia Oscillations in Rotating Fluid

Surface Tension
FM-72
Examples of Surface Tension
FM-73
Surface Tension and Contact Angles
FM-74
Formation of Bubbles
FM-75
Surface Tension and Curved Surfaces
FM-76
Breakup of Liquid into Drops
FM-77
Motions Caused by Composition Gradients along Liquid Surfaces
FM-78
Motions Caused by Electrical and Chemical Effects on Liquid Surfaces
FM-79
Motions Caused by Temperature Gradients along Liquid Surfaces
FM-83
Induced $J \times B$ Forces in Solids and Liquids
FM-84
An MHD Pump
FM-86
The Suppression of Vorticity by MHD Forces
FM-87
The Hartmann Layer

Vorticity
FM-13
The Bathtub Vortex
FM-20
The Horseshoe Vortex
FM-14A
Visualization of Vorticity with Vorticity Meter. Part I
FM-14B
Visualization of Vorticity with Vorticity Meter. Part II
FM-26
Tornadoes in Nature and the Laboratory
FM-70
The Sink Vortex

Wave Phenomena
FM-109
Source Moving at Speeds Below and Above Wave Speeds
FM-105
Ripple-tank Radiation Patterns of Source, Dipole, and Quadrupole
FM-108
Small-amplitude Waves
FM-102
Passage of a Shock Wave through a Circular Orifice
FM-99
Passage of Shock Waves over Bodies
FM-100
Passage of Shock Waves through Constrictions
FM-101
Reflections of Shock Waves

Some Regimes of Boundary Layer Transition

APPARATUS: In the first apparatus water flows in an open channel at approximately one half foot per second. The flow is made visible by injection of dye through a slit in the floor of the channel. Both plan and side views of the boundary layer are shown. In the second apparatus a sheet of water 1/8" deep flows over a level glass plate; turbulent spots are made visible by reflecting light from the water surface. Some phenomena are shown by both techniques.

SEQUENCE: (1) Schematic of channel apparatus and camera viewpoints are shown. Location and action of dye slit are indicated. (2) Laminar flow over the slit is shown. (3) The Reynolds number of the flow is increased, by moving downstream in the flow; large-scale, steady three-dimensional streaks then are seen in the flow. This is the second stage of transition. The first stage, Tollmien-Schlichting waves, is not seen, since the amplitude of these waves is small. (See references 1, 2 for further information.) (4) The Reynolds number is again increased, by moving further downstream; breakdown of the streaks into spots is shown by dye in plan and side views. (5) Turbulent spots are shown with the reflected light technique in the second apparatus. The increase in the number of spots with increase in Reynolds number (due to raising velocity) is seen. As the sequence continues the spots grow in size and number; finally they cover the entire flow at the right edge of the picture. (6) First apparatus again employed. The Reynolds number is again increased; this scene is much farther downstream, where the boundary layer has been turbulent for a considerable distance. The fine-scale, unsteady streaks characterizing the wall layers of the fully-turbulent shear layer are seen using dye injection, verifying that transition to turbulent flow is complete. In this last sequence, dye injection makes visible only the innermost layer of the turbulent boundary layers. A comparison of all four zones of the fully turbulent layers is shown in FM-2.

QUESTION: The transition from a fully-laminar boundary layer to a fully-turbulent one on a plate can pass through as many as four stages or regimes in between the laminar and turbulent states. Describe in order each of these regimes.

REFERENCES:

1. Klebanoff, P. S. and K. D. Tidstrom, "Evolution of Amplified Waves Leading to Transition in a Boundary Layer with Zero Pressure Gradient", NASA TN-D-195. September 1959.

2. Meyer, K. A. and S. J. Kline, "A Visual Study of the Flow Model in the Later Stages of Laminar-Turbulent Transition on a Flat Plate", Stanford University Report MD-7, December 1961.

3. Emmons, H. W., "The Laminar-Turbulent Transition in a Boundary Layer", Part 1, Jour. Aero. Sci., Vol. 18, No. 7, pp. 490-498, July 1951.

4. Emmons, H. W. and A. E. Bryson, "The Laminar-Turbulent Transition in a Boundary Layer", Part II, Proc. First U. S. National Congress Appl. Mech., pp. 859-868, 1952.

CREDITS: Experiments and film by H. W. Emmons, S. J. Kline, K. A. Meyer; Consultant, S. J. Kline, Stanford University; Director, A. H. Shapiro, M.I.T.; Editor, E. Carini, E.S.I.; Art Director, R. P. Larkin, E.S.I.

Structure of the Turbulent Boundary Layer

APPARATUS AND PROCEDURE: Water flows in a straight open channel at about 0.4 ft./sec. The boundary layer flow on the side wall is made visible by pulsed "timelines" of hydrogen bubbles generated by a .001" platinum wire parallel to the wall and normal to the flow. (For details see ref. 5.) The timelines are produced at 120 cycles per second. The major regimes of turbulent boundary layer structure are shown by moving the wire successively closer to the wall.

SEQUENCES: (1) Schematic views of the apparatus. (2) The velocity versus space coordinates to be used and the wire locations relative to them are displayed. The velocity distribution of the turbulent boundary layer is shown relative to these coordinates, and the use of a semi-log scale to expand the important inner layers is indicated. (3) Each of the five major regions is next shown, starting with the inviscid or outer flow and moving successively inward. In each case the region to be seen is first indicated schematically by a shaded zone on the semi-log coordinates, and is followed by the experimental observation. (i) Inviscid Flow: outside the boundary layer. In this flow the boundary layer on the channel wall is still developing; consequently there is an inviscid region of flow in the center of the channel which has not been affected by the vorticity and turbulence generated in the boundary layer. The timelines are unwarped, and maintain their positions relative to each other as they move downstream in the inviscid region. (ii) Wake Region: the outermost region of the boundary layer. Isolated eddies move through the flow. In this sequence the eddies occur mostly in the upper portion of the frame; this is merely a statistical accident depending on the particular frames chosen. This sequence shows the outer part of the wake region; eddies occupy considerably less than half the plane of view on the average. (Successive views moving inward in this region show a monotonically increasing fraction of the flow occupied by eddies. The distribution of eddy fraction is Gaussian; see references 1, 2, 4) (iii) Fully Turbulent Region: also called the log region. The entire flow is filled with eddies; the eddy motions appear smaller and more rapid than in the wake region. (iv) Buffer Region: the next to innermost layer. The flow is more organized, but still very unsteady. A more regular, streaky structure and very strong transverse fluctuations begin to appear in the flow. (v) Wall Layer: the innermost region. This region is seldom observable in applications because it is so thin; in the apparatus used, low velocities and large dimensions have been specially arranged to make it visible. The wall layers have a relatively regular streaky structure. The streaks consist of alternating lower and higher speed fluid. In this flow structure the bubbles concentrate in the low speed streaks. These low speed streaks oscillate and then "erupt", giving birth to strong eddies quite near the wall.

REFERENCES:
1. Klebanoff, P. S., "Characteristics of Turbulence in a Boundary Layer with Zero Pressure Gradient", NACA Report 1247, 1955.
2. Runstadler, P. W., S. J. Kline, and W. C. Reynolds, "An Experimental Investigation of the Flow Structure of the Turbulent Boundary Layer", Stanford University Report MD-8, June 1963.
3. Runstadler, P. W., S. J. Kline, and W. C. Reynolds, "A Visual Study of the Flow Structure in the Fully Developed Turbulent Boundary Layer on a Flat Plate", ESL film R-7, 1961, available from Engineering Societies Library, 345 East 47th St., NYC 17, N. Y.
4. Schraub, F. A., S. J. Kline, and W. C. Reynolds, "Structure Studies of the Turbulent Boundary Layer with and without Longitudinal Pressure Gradients", Stanford University Report MD-12, 1965.
5. Schraub, F. A., S. J. Kline, J. Henry, P. W. Runstadler, Jr., A. Littell, "Use of Hydrogen Bubbles for Quantitative Determination of Time-Dependent Velocity Fields in Low-Speed Water Flows", Journal of Basic Engineering, Trans. ASME, Series D, vol. 87, 1965.

CREDITS: Experiments and film by P. W. Runstadler, S. J. Kline, W. C. Reynolds; Consultant, S. J. Kline, Stanford University; Director, A. H. Shapiro, M.I.T.; Editor, E. Carini, E.S.I.; Art Director, R. P. Larkin, E.S.I.

FM-3
Shear Deformation of Viscous Fluids

PURPOSE: To illustrate concept of shear deformation in fluid.

Part I

APPARATUS & PROCEDURE: A rectangular tank is filled with glycerine. By means of a hypodermic needle filled with colored glycerine, a square grid of straight lines is marked on the surface of the glycerine. A circular cylinder is moved through the tank.

SEQUENCES: As the cylinder moves through the glycerine, the initially square fluid particles are deformed by the motion into non-square shapes. The deformations are large, and they are very different for particles ahead of and behind the cylinder. The time rate of change in the angle formed by a pair of mutually-perpendicular fluid lines is a measure of the local rate of shear deformation. In a viscous fluid, such shear deformations are accompanied by viscous shear stresses; in a Newtonian fluid, the viscous stress is proportional to the·rate of shear deformation.

Part II

APPARATUS & PROCEDURE: Glycerine is contained in the annular space between an outer circular cylinder and an inner concentric circular cylinder. The outer cylinder is stationary, while the inner cylinder may be rotated. The motion of the glycerine is made visible by drawing lines on the surface with a hypodermic needle filled with dyed glycerine.

SEQUENCES: (1) A straight radial line is drawn in the stationary glycerine. The inner cylinder is then rotated. The dyed glycerine immediately adjacent to the moving surface moves with the surface, showing that there is no relative motion between a liquid and an adjacent solid surface. (2) The entire straight line of 1. At first only the fluid near the inner moving cylinder moves. The motion is transmitted outward through the agency of viscous stresses, and as time proceeds more and more of the initially-straight radial line is brought into motion. This may be described as the growth with time of a laminar boundary layer near the moving surface. Alternatively, the vorticity may be thought of as diffused outward by viscous stresses. (3) A square particle of fluid is marked in the stationary glycerine, about midway between the inner and outer cylinders. The inner cylinder is rotated. The square is deformed into a parallelogram, indicating shear deformation. This shear deformation is related to the viscous stresses in the fluid, and accounts for the fact that a torque must be applied to the rotating cylinder in order to establish and maintain the motion.

QUESTIONS: (1) Do fluid particles undergo shear deformation in a non-viscous fluid? (2) In a viscous fluid undergoing shear deformation, does every pair of mutually-perpendicular lines exhibit a change in angle from the initially right angle? (3) If the inner cylinder, after being rotated in one direction, were rotated the other way by the same amount, would the marked lines return to their original configuration? (4) Form a criterion for the motion to be of "creeping" type, in terms of the two radii, the speed of the cylinder, the kinematic viscosity, and time. (5) If the fluid were so viscous that the motion were of "creeping" type, what would happen to a straight radial line?

CREDITS: Experiments designed and performed by A. H. Shapiro at Educational Services Inc. Edited by A. H. Shapiro, M.I.T., and W. Tannebring, E.S.I.

FM-4
Separated Flows — Part I

PURPOSE: To show some phenomena of boundary layer separation: (a) the difference between the separation of laminar and turbulent boundary layers on an unstreamlined body; (b) the effect of streamlining on the size of the separated wake; (c) the occurrence of separation at high Reynolds No., and its absence at low Reynolds No.

APPARATUS & PROCEDURE: (1) Sequences 1 to 5: A model is suspended by a sting in a vertical, open air jet, 8" diameter, at speeds up to 150 ft/sec. Titanium tetrachloride which reacts with water vapor in the air to form a dense fog is applied to the surface. (2) Sequences 6 and 7: A steel ball 3/4" in diameter, coated with dyed liquid, falls at terminal speed through a vertical tube of liquid.

SEQUENCES: (1) Air flows past a smooth sphere, 3" in diameter. Tetrachloride is placed at the nose. Initially the air speed is high, and the Reynolds No., such that the boundary layer is turbulent. The boundary-layer flow remains attached beyond the point of maximum thickness, and then the main flow separates because of backflow in the boundary layer. The wake is unsteady and eddying, and is narrower than the diameter of the ball. The air velocity, and with it the Reynolds No., are steadily reduced, until the boundary layer is laminar; the point of separation moves upstream of the point of maximum thickness, and the separated wake is considerably larger than the diameter of the sphere. At still lower speed, the point of separation moves further upstream and the wake becomes even wider. (2) Tetrachloride is placed on the side of the sphere slightly upstream of the shoulder. The boundary layer remains attached and thin up to about the maximum thickness of the sphere, and then there is a sharp and distinct separation. The separated boundary layer curves back into the wake, showing a recirculatory motion. (3) Tetrachloride is placed at the rear of the sphere near the sting. The smoke moves against the main flow until it reaches the point of separation, where it is entrained by the main flow. The separated wake has an eddying motion. (4) Tetrachloride is placed near the nose of a partially-streamlined body of revolution. The speed is low and the boundary layer laminar. The boundary layer remains attached almost to the rear of body. (5) Similar to 4, but higher speed. The boundary layer is turbulent. Separation occurs later than in 4, and the separated wake is narrower. (6) A ball coated with dyed water falls through water at a speed such that the boundary layer is turbulent. Boundary layer separation occurs, producing a broad, eddying wake, as in the smoke flow experiment. (7) A ball coated with dyed glycerine falls through glycerine at a Reynolds No. so low that the motion is of "creeping" type (inertial forces negligible compared with viscous forces). The dyed glycerine on the surface collects into a very thin stream tube at the rear, separation being virtually absent. Comparison of 6 and 7 shows that the cause of separation is the pressure field associated with a "slightly-viscous" flow at high Reynolds No.; specifically, boundary-layer backflow in regions of adverse pressure gradient. In creeping flow, the viscous region is too thick to be described as a boundary layer; furthermore, the surface pressure gradient is negative over the whole surface, and applies forces in the streamwise direction to the low-momentum fluid near the surface.

CREDITS: Experiments designed and performed by A. H. Shapiro at M.I.T. and at the studio of Educational Services Inc. Edited by A. H. Shapiro, M.I.T., and W. Tannebring, E.S.I.

parated Flows — Part II

PURPOSE: To show separated flows past rough bodies at low
& high Reynolds No., and that artificial introduction of turbu-
ce can suppress the separation of laminar boundary layers.

PARATUS & PROCEDURE: (1) Sequences 1, 2, and 3: A
del is suspended by a sting in a vertical open air jet, 8"
diameter, at speeds up to 150 ft/sec. Titanium tetrachloride,
ich reacts with water vapor in the air to form a dense fog,
applied to the surface. (2) Sequences 4, 5, 6 and 7: Water
ws from right to left in an open channel. A curve in the
nnel wall produces a constriction (a Venturi throat). The
id near the wall, outside the boundary layer, accelerates
to the throat, and then decelerates. By Bernoulli's equation
wall pressure decreases to the throat, then increases.
e flow is visible through small entrained air bubbles.

QUENCES: (1) A sphere 3" in diameter with grooves about
02" deep is in the air flow. A drop of tetrachloride is placed
the nose. The Reynolds No. is low, and the boundary layer
ninar, and separation occurs slightly upstream of the point
maximum thickness. The wake width is about the same as
e sphere diameter. (2) As in 1, except that the air speed is
gher, and the boundary layer turbulent. Separation occurs
wnstream of the point of maximum thickness of the sphere,
d the wake is considerably narrower than the sphere diam-
er. (3) As in 1, but the sphere has indentations to simulate
golf ball. At first the air speed is high, and the Reynolds
, such that the boundary layer is turbulent. Separation occurs
wnstream of the point of maximum thickness, and the wake is
rrower than the diameter of the ball. As the air speed de-
eases, the boundary layer becomes laminar, the point of
undary layer separation moves upstream of the point of
aximum thickness, and the wake broadens. At low Reynolds
., the laminar boundary layer is so thick that the surface
ughness is submerged and does not induce transition to
rbulence. (4) Low-speed flow in the water channel, with a
minar boundary layer. Separation occurs just downstream
the hump, very soon after the pressure gradient becomes
verse. In the separated region, the flow near the wall
oves upstream, until it approaches the point of separation,
here it is entrained by the main separated flow. (5) As in
but an obstruction is placed on the wall upstream of the
mp, making the boundary layer strongly turbulent. The
undary layer remains attached longer, and the region of
parated flow is markedly reduced by the momentum transfer
e to turbulence. (6) A close view of the laminar boundary
yer of 4. Note the directions of the motions in the sepa-
ated region. (7) A close view of the turbulent boundary layer
5. The extent of the separated region is markedly reduced
the turbulence, and there is little evidence of upstream
otion.

REDITS: Experiments designed and performed by A. H.
hapiro at M.I.T. and at the studio of Educational Services
c. Edited by A. H. Shapiro, M.I.T., and A. H. Pesetsky,
.S.I.

FM-6
Boundary Layer Formation

PURPOSE: To show how the boundary layer thickness next to
a flat plate increases with time when the flow is started impulsively.

APPARATUS & PROCEDURE: A long straight channel with
vertical side walls has a large reservoir at the upstream end
(to the right) and a gate at the downstream end (to the left).
It is filled with glycerine. The flow is made visible by drawing lines on the surface of the glycerine with a hypodermic
needle loaded with dyed glycerine.

SEQUENCES: (1) A line perpendicular to one of the vertical
walls is drawn on the motionless glycerine. This line identifies
a set of fluid particles of fixed identity. (2) The gate is raised
and the dyed line of particles moves downstream. The fluid
in contact with the wall does not move; it has no motion relative to the wall. The particles near the center of the channel
move fastest, and there is a gradation of velocity between the
wall and the center of the channel. (3) A close-up view of a
"U" formed by a line on the wall and two lines normal to the
wall. When the motion starts, the fluid at the wall does not
slip. (4) Repeat of 3. (5) A line is drawn normal to the wall
on the surface, and the flow is started. At a later time, there
is a region at some distance from the wall in which the speed
is uniform, and a region near the wall where the flow has
been decelerated by reason of viscous stresses. The latter
region, the viscous boundary layer, grows thicker as time
proceeds. (6) Sketch illustrating that the distance moved by
any dyed particle from its initial position is proportional to the
average velocity of that particle during the corresponding time
interval. (7) Sketch to show that the instantaneous shape of
the dyed line of particles is a rough indication of the velocity
distribution near the wall. The region of variable velocity
near the wall is the boundary layer. Outside the boundary
layer is the region of uniform velocity, where viscous action
has not yet diffused. (8) A superposition of shots of successive positions of the dyed lines of particles, showing that the
boundary layer region grows in thickness due to the diffusion
of vorticity by viscous stresses. (9) With low fluid speed and
high viscosity, the thickness of the boundary layer grows very
rapidly. (10) Repeat of 9. The boundary layer reaches to the
middle of the channel in a very short time. (11) The same flow
velocity as in 9 and 10, but with a lower viscosity. The thickness of the boundary layer increases less rapidly. (12) The
same reduced viscosity as in 11, but higher speed. The boundary layer thickness increases still less rapidly. Near the end
of the run it is only a small fraction of the thickness of the
channel, while in 9 it fills nearly the entire channel.

QUESTIONS: (1) Show by order-of-magnitude reasoning that the
boundary-layer thickness should grow in proportion to the
square root of the time and of the kinematic viscosity. (2)
Compare the flow of this experiment with the steady flow past
a semi-infinite flat plate. Are they comparable?

CREDITS: Experiments designed and performed by A. H.
Shapiro at the studio of Educational Services Inc. Edited
by A. H. Shapiro, M.I.T., and A. H. Pesetsky, E.S.I.

FM-7
Propagating Stall in Airfoil Cascade

PURPOSE: To show visual evidence of the phenomenon of propagating stall which occurs in axial-flow compressors ("rotating stall"), and which can be seen in air discharge louvers and venetian blinds.

APPARATUS: A stationary cascade of airfoils, representing a blade row of an axial-flow compressor, is mounted in a cascade wind tunnel. The air stream approaches the cascade in a horizontal direction, is turned approximately 90° by the cascade, and leaves in a generally vertical direction. The combination of solidity, turning angle, stagger angle and angle of incidence used are such that this particular cascade is stalled.

SEQUENCES: (1) Smoke streaklines from upstream pass through the cascade. The flow is viewed at normal camera speed. (2) The same action, seen in slow motion. (3) Smoke is emitted from downstream of the cascade, and reveals clearly regions of backflow.

DISCUSSION: Stall does not occur uniformly over the entire cascade. Instead, a wave-like propagation of a stall zone develops. This results from an instability. If a small group of blades is stalled, the local flow blockage thus produced diverts the approach flow around the stalled group. On one side of the stalled group the angle of incidence is thereby reduced, thus eliminating stall. On the other side of the stalled group the angle of incidence is increased, thus making for a worse stall. The stall region therefore moves toward the latter side, and thus the original stalled group tends to unstall. In an axial-flow compressor, the stall pattern rotates periodically around a blade row with respect to the blades. In a stationary airfoil cascade (or a louver, or a venetian blind), the stall pattern propagates with a characteristic speed along the length of the cascade and reflects from the ends.

CREDITS: Experiments and film by G. Sovran, General Motors Research Center; Consultant and Director, A. H. Shapiro, M.I.T.; Editor, J. Hirschfeld, E.S.I.; Art Director, R. P. Larkin, E.S.I.

FM-8
The Occurrence of Turbulence

PURPOSE: To show some turbulent flows, including examples in which laminar flows become turbulent with an increase of Reynolds number.

SEQUENCES: (1) Laminar and turbulent smoke jet. The smoke is titanium tetrachloride, issuing from a small tube. The Reynolds number is raised by raising the speed. (2) Laminar and turbulent water flow in a pipe, as seen from the appearance of the discharge, and ending with a still picture of the turbulent case. The glassy appearance of the laminar flow is succeeded by a fuzzy and unsteady look when the Reynolds number is increased and the flow becomes turbulent. (3) Turbulent mixing of cream in coffee, laminar mixing of white paint in molasses. Turbulence can be suppressed by decreasing Reynolds number through an increase of viscosity. (4) Laminar and turbulent wakes behind dye-coated spheres falling respectively through glycerine and water. The laminar momentum wake is of course not as thin as the dye wake; the molecular diffusivity of dye is much smaller than the molecular diffusivity of momentum, i.e., the kinematic viscosity. (5) Wrinkling of a bubble sheet in a turbulent boundary layer, viewed normal to wall. A fine wire is set normal to the flow and parallel to the wall near the middle of a fully turbulent water boundary layer. Hydrogen bubbles electrolyzed at the wire act as tracers. They follow the fluid motion fairly well because of their small size and correspondingly high drag relative to the fluid. (6) Turbulent wake of a supersonic projectile, shadowgraph picture. (7) Time lapse movies of clouds. These "speeded up" movies show turbulent agitation in cumulus clouds. (8) The Great Nebula in the constellation Orion. Many celestial objects show either actual turbulent motions or brightness distributions which seem to indicate the results of turbulent motions.

CREDITS: Experiments and film by S. Corrsin, S. J. Kline, A. H. Shapiro. Consultant, S. Corrsin, The Johns Hopkins University. Director, J. Churchill, E.S.I., and A. H. Shapiro, M.I.T. Editor, E. Carini, E.S.I.

rodynamic Heating as Shown by
mperature-Sensitive Paints

RPOSE: To show the patterns of aerodynamic heating
bodies in a high-speed flow.

PARATUS: The high stagnation temperature relative
a body in hypersonic flight is simulated by placing
dels in a supersonic jet coming from a high-tempera-
e reservoir. The models are sprayed with a tempera-
e-sensitive paint which assumes a series of different
rs at definite transition temperatures. In order of
nding temperature, the colors are: pale blue; medium
e; light purple-brown; deep purple-brown.

QUENCES: (1) A model being sprayed with paint. (2)
model in place at the nozzle discharge. (3) *Hypersonic
der at 15° Incidence*. The flow is started suddenly, as
all the experiments to follow. After the run is over, the
del is rotated on its supporting sting to make every
face visible. The heating is most intense and rapid at
nose, then along the leading edges of the delta wing,
n along the leading edges of the fins. (4) *Same Model
15° Incidence and 10° Yaw*. Similar to previous test.
the heating is asymmetrical. The advanced edge of
delta heats over a larger area. (5) *Sphere*. The heating
most intense at the nose and decreases downstream.
Ballistic Ogive at Small Incidence. The model has a
listic ogive nose, followed by a cylindrical section, and
ds in a flared conical "skirt". The heating is most in-
se at the nose. The lower (i.e. advanced) portion of the
rt heats up more rapidly than the cylindrical section.

EDITS: Experiments and film by the Office National
tudes et de Recherches Aérospatiales (O.N.E.R.A.),
ance; Consultant, A. H. Shapiro, M.I.T.; Producer, R.
rgman, E.D.C.; Editor, M. Chalufour, E.D.C.; Art Direc-
r, R. P. Larkin, E.D.C. Produced by the National Com-
ittee for Fluid Mechanics Films and Education De-
lopment Center, Inc. (formerly Educational Services
c.), with the support of the National Science Foundation.

Generation of Circulation
and Lift for an Airfoil

PURPOSE: To illustrate how the lift on an airfoil is associated
with the circulation around the airfoil produced by the vortex
"bound" in the airfoil; and to show how the establishment of the
bound vortex is related to the production of a starting vortex
owing to viscous action at the trailing edge.

APPARATUS & PROCEDURE: A carriage on rails tows the airfoil through a long tank filled with water. Aluminum particles
sprinkled on the water surface show the motion. When the
camera is mounted on the tank, it shows the flow relative to the
undisturbed water; when it is mounted on the moving carriage, it shows the flow relative to the model.

SEQUENCES: (1) The airfoil at angle of attack starts from
rest and is brought to a steady speed impulsively. The camera
moves with the airfoil, and shows the flow relative to the airfoil. A "starting" vortex is shed from the trailing edge because
of viscous boundary layer separation caused by adverse pressure gradients in the initial circulation-free, potential flow.
The starting vortex is carried downstream with the fluid in
which it is embedded, and a fictitious "bound" vortex of opposite sign is simultaneously created in the airfoil, as per
Kelvin's Theorems regarding the permanence of circulation
in a non-viscous, barotropic fluid not acted upon by nonconservative forces. By the Kutta-Joukowski Law, the lift
on the airfoil is associated with the circulation around the
airfoil. The circulation in turn is related to the bound vortex
according to Stokes' theorem (the circulation around a closed
curve equals the flux of vorticity through the bounded area).
(2) the same as 1, but observed with the camera fixed to the
tank. The starting vortex remains attached to the fluid in
which it is generated, as per Helmholtz's Vortex Law. (3)
Repeat of 1. (4) Repeat of 2. (5) The airfoil starts impulsively and generates a starting vortex. After a short distance of steady motion, it stops impulsively, and a stopping
vortex is shed. (6) Repeat of 5. The starting and stopping
vortices are of equal and opposite strength, and the circulation for a circuit enclosing the airfoil and both vortices is
zero, as per Kelvin's Theorem. The lift generated during
starting disappears during stopping. The vortex pair propels
itself downward, each vortex moving in the velocity field of
the other; this illustrates Helmholtz's Vortex Law that the
vorticity is, as it were, "frozen" in the fluid.

QUESTIONS: (1) Explain in detail the mechanism of establishment of the starting and bound vortices. (2) Is there any way
of producing lift in a totally non-viscous fluid at subsonic
speeds? At supersonic speeds? (3) Contrast the mechanisms
of lift generation at subsonic and supersonic speeds. (4) Can
lift be developed in a "creeping" flow? In a free-molecule
flow?

CREDITS: Experiments done under the direction of L. Prandtl
(Göttingen). Edited by A. H. Shapiro, M.I.T., and R. Bergman,
E.S.I.

FM-11
The Magnus Effect

PURPOSE: To illustrate how the lateral force produced on a rotating cylinder moving through a fluid is related to the circulation around the cylinder produced by the shedding of a starting vortex due to asymmetric viscous action.

APPARATUS & PROCEDURE: A carriage on rails tows the circular cylinder through a long tank filled with water. Aluminum particles sprinkled on the water surface show the motion. When the camera is mounted on the tank, it shows the motion relative to the undisturbed fluid; when it is mounted on the moving carriage, it shows the flow relative to the cylinder.

SEQUENCES: (1) The cylinder is impulsively rotated but not translated. Viscous action diffuses vorticity through the fluid. The boundary layer may be ascribed to a fictitious vortex "bound" in the cylinder. (2) The cylinder is impulsively rotated clockwise, and translated leftward at the same time. The upper surface moves with the fluid, and helps the boundary layer move against the adverse pressure gradient; the lower surface moves against the fluid, and strengthens the effect of the adverse pressure gradient. Because of this asymmetric viscous action, boundary layer separation occurs more easily on the lower side. A counterclockwise starting vortex is shed. By Kelvin's circulation theorem, a clockwise vortex is simultaneously "bound" to the cylinder. The circulation of the bound vortex is related to the lifting force on the cylinder. (3) Repeat of 2. (4) The translating and rotating cylinder stops translating but continues to rotate. The bound vortex remains in the cylinder, as in 1. The lift disappears since there is no translation relative to the fluid. (5) As in 4, but the rotating cylinder once again resumes translation. The bound vortex is carried along with the moving cylinder, and lift is re-established when the cylinder resumes translation. (6) The rotating, translating cylinder stops translating. After a pause, the rotation also stops. The cylinder then resumes translation. The vortex which had been bound in the cylinder remains in the fluid, and the cylinder which is now translating but not rotating has no circulation and no lift. (7) The cylinder first translates without rotation. Vortices are shed alternately, and the circulation oscillates between clockwise and counterclockwise. When the cylinder begins to rotate the boundary layer separation becomes asymmetric, and the circulation persists in one direction.

INTERPRETATION: The flow due to a rotating cylinder translating through a viscous fluid is very different from the potential flow of a non-viscous fluid. The potential flow purporting to represent the Magnus Effect is the superposition of a streaming flow with a doublet and vortex. In non-viscous flow, no mechanism exists for establishing the vortex with its circulation. With a real fluid, asymmetrical viscous action on the upper and lower surfaces causes the separation point to move off the axis of symmetry; this leads to a pressure difference between upper and lower surfaces, and net lift.

QUESTIONS: (1) Relate the experiments to Kelvin's circulation theorem applied to various circuits around the cylinder and the vortices. (2) Would a rotating cylinder translating through an extremely viscous fluid in "creeping" motion exhibit a Magnus Effect?

CREDITS: Experiments performed under the direction of L. Prandtl, (Göttingen). Edited by A. H. Shapiro, M.I.T., and R. Bergman, E.S.I.

FM-12
Flow Separation and Vortex Shedding

PURPOSE: To illustrate the development of separated f past bodies due to backflow in boundary layers facing adve pressure gradients.

APPARATUS & PROCEDURE: A carriage on rails tows model through a long tank filled with water. Aluminum p ticles sprinkled on the water surface show the motion. camera, mounted on the moving carriage, shows flow rela to the model.

SEQUENCES: (1) Development of the flow past a serie three ellipses with their major axes parallel to the flow. W thickness ratios of 0.5 and 0.15, the boundary layer ultima separates near the maximum-thickness position, and the se rated wake is approximately equal to the body thickness. first the separation occurs symmetrically on top and bott but, as the pattern develops, the separation becomes asy metrical, and vortices are shed alternately. With the flat pl (zero thickness ratio), a thin alternating wake of vortices produced from the upper and lower boundary layers. Development of the flow past the same ellipses, with th major axes normal to the flow. The separated wake is r wider than the maximum body thickness. At first, vorti are shed symmetrically at top and bottom, but as time p ceeds the vortex shedding becomes asymmetrical. (3) D velopment of the separated flow past a circular cylinder. T boundary layer separates at about the point of maximum thi ness. The vortex shedding starts symmetrically, then becom asymmetrical. The alternating lift force thus produced cause "singing" of telephone wires and severe oscillatio in tall chimneys and suspension bridges. (4) A close-up the rear side of the circular cylinder at the start of the fl The flow is at first attached, and is very close to the sy metrical potential flow of a non-viscous fluid. The bound layer on the rear side grows thicker through the diffusion vorticity, and becomes less able to move forward against adverse pressure gradient. When the low-momentum fl ultimately comes to rest, or "stalls," it is pushed backw by the pressure gradient. The backflow becomes stronger the boundary layer thickens, and ultimately results in sep ration of the main flow and a recirculating region which shed as a vortex. (5) Repeat of 4. (6) Detail of the devel ment with time of the flow past a body that is not well strea lined. As with the circular cylinder, the flow at first resembl a potential flow, but the combination of a thickening bounda layer and an adverse pressure gradient leads to backflow an region of separated flow. As the separated region grows, t point of boundary-layer separation gradually moves upstrea from near the tail to near the point of maximum thickne Similar occurrences near the trailing edge of an airfoil pr duce a shed starting vortex and a corresponding vortex "boun in the airfoil. (7) Repeat of 6. (8) Development of flow pa the edge of a flat plate set transversely to the stream. T events are similar to those in 4 for the circular cylinde During the period in which the attached potential flow prevai the adverse pressure gradient near the sharp edge is ve large, and separation seems to occur instantaneously. Actual it occurs over a finite interval of time.

CREDITS: Experiments performed under the direction of Prandtl, Göttingen. Edited by A. H. Shapiro, M.I.T., and Bergman, E.S.I.

JRPOSE: To illustrate Coriolis force, Kelvin's Circulation
eorem, and Helmholtz's Vortex Laws with the controversial
periment of the direction of swirl when water drains from a
rge vessel through a small hole.

PPARATUS & PROCEDURE: A circular tank, 6 ft. in diam-
er, with a 3/8" diameter hole at its center, is filled with
ter to a depth of six inches. After the water has stood for
hours, a valve at the exit of a long hose leading from the
ain hole is opened. A cross made of two horizontal pieces
wood, each 1" long, and pinned together with a fine vertical
re, is inserted above the drain. The wire protrudes through
e surface and the vanes are just below. A clock shows the
apsed time.

EQUENCES: (1) A nozzle fills the tank, and produces a
ockwise swirl. (2) Twenty-four hours later there is no
rceptible motion. Presumably, any residual motion would
a clockwise swirl. (3) The clock is started, and the valve
opened. The vorticity cross is inserted, and remains
ntered because it is stable in the sink-like flow toward the
rain. (4) At four minutes no rotation of the cross is observ-
le. (5) At six minutes, still no observable rotation. (6) At
ne minutes, a barely perceptible counterclockwise rotation.
) At sixteen minutes, a definite CCW rotation. (8) The globe
the northern hemisphere, showing the CCW Coriolis force
ssociated with the flow toward the drain. (9) At twenty-four
inutes the cross rotates about 30,000 times faster than the
arth at Boston.

TERPRETATIONS: (1) The CCW torque of the Coriolis
rces increases the CCW momentum as seen in the earth's
eference frame. (2) By Kelvin's Circulation Theorem, the
irculation in inertial space for a circular fluid line concentric
ith the vertical axis of the tank remains constant as time
asses. As the radius of this circle decreases, the tangential
elocity in inertial space must increase proportionately. Owing
the relation between velocities in the earth frame and in
nertial space, a circular fluid line with zero circulation in
he earth frame must acquire CCW circulation as time pro-
eeds. (3) A vertical fluid line on the axis starts with a vor-
icity corresponding to the local vector speed of rotation of
he earth. Due to the flow down the drain, this vortex line is
tretched. By Helmholtz's Vortex Law, the vorticity in in-
rtial space increases proportionately. Owing to the relation
etween velocities in the earth frame and in inertial space,
n earth observer sees the vorticity on the axis increase.

ADDITIONAL COMMENTS: (1) The everyday observations of
swirl in a sink or bathtub are not related to the earth's rotation. The earth's Coriolis force is so tiny that a special
apparatus is necessary to make it dominant over other effects,
.g., residual motions which persist for hours or days, asymmetrical viscous forces, buoyancy forces due to temperature
radients, non-uniform surface tension, and air currents.

QUESTIONS: (1) Compare the earth's Coriolis force with
gravity for a speed toward the drain of 0.01 ft/sec. (2) Com-
are the factor of 30,000 observed in the experiment with a
heoretical estimate.

CREDITS: Experiment designed and performed by A. H.
Shapiro, M.I.T., at Educational Services Inc. Edited by A.
H. Shapiro, M.I.T., and E. Carini, E.S.I.

FM-14A
Visualization of Vorticity
With Vorticity Meter – Part I

PURPOSE: To illustrate the concept of vorticity with a "vorticity meter" (continued on FM-14B).

APPARATUS & PROCEDURE: The vorticity meter is a hollow plastic cylindrical tube that floats upright in water. At
its upper end is fixed a horizontal arrow, to make rotation
of the tube visible. At its lower end is a pair of vertical
sheet-metal vanes rigidly fixed at right angles. The vanes
are driven by the fluid. Since the vanes are rigidly connected they cannot exhibit the shear deformation of the fluid,
and so the meter rotates, to a good approximation, with the
average angular velocity of two mutually-perpendicular fluid
lines. Using the definition of vorticity as the sum of the
angular velocities of two mutually-perpendicular fluid lines,
the meter rotates with an angular speed equal to approximately half the vertical vorticity of the lump of fluid in which
the vanes are immersed.

SEQUENCES: (1) The vorticity meter. (2) A "forced vortex"
in a tank of water on a turntable rotating like a solid body.
The vorticity meter rotates with the same angular velocity
as the solid-body rotation of the water. Every horizontal
line of water has the same angular velocity, hence the angular
velocity of the meter is half the fluid vorticity. (3) The
vorticity meter is put in the boundary layer flow of water
adjacent to the side wall of a long, straight open channel. The
streamlines in the boundary layer are nearly straight and
parallel, but the vorticity meter rotates, showing the presence
of vorticity in the boundary layer. This vorticity is related
to the velocity gradient in the boundary layer, or, alternatively,
to the gradient of total pressure normal to the streamline.
(4) A sketch of the plan view of a circular tank in which a
vortex-sink flow is formed. The water is introduced tangentially at the outer radius and spirals inward in a tight vortex
until it leaves by gravity through a small drain at the center.
(5) In the vortex-sink flow, the float moves in approximate
circles around the drainhole, but the arrow remains parallel
to itself; it undergoes translation, like a compass needle on
a rotating turntable. Hence the fluid has no vorticity. The
flow approximates a potential vortex, with concentric circular
streamlines and the velocity distribution $Vr=const$.

QUESTIONS: (1) Explain why the vorticity meter does not
indicate exactly the vorticity of the fluid. What factors enter
into the relationship between the rotation of the vorticity
meter and the fluid vorticity? (2) Using the theorem of
moment of momentum, derive the relation $V_t r=const.$ for the
vortex-sink flow, where r is radius and V_t is tangential
speed. What assumptions are necessary for this, and how
far might they be valid? (3) Consider a small fluid cross
made up of a piece tangential to a circle and a perpendicular
piece. Show that, in a potential vortex, these two lines have
equal and opposite angular velocities.

CREDITS: Experiments designed and performed by A. H.
Shapiro, M.I.T., at the Hydrodynamics Laboratory of M.I.T.
and at Educational Services Inc. Edited by A. H. Shapiro,
M.I.T., and E. Carini, E.S.I.

FM-14B
Visualization of Vorticity
With Vorticity Meter – Part II

PURPOSE: To illustrate the concept of vorticity by means of a "vorticity meter" (continuation of FM-14A).

SEQUENCES: (1) A continuation from Loop FM-14A of the vorticity meter in the vortex-sink flow. The float moves in a circulation path, but with nearly pure translation, hence vorticity is absent. (2) The float slowly circles in toward the drain in nearly pure translation, until it is trapped in the center of the vortex, where it rotates rapidly in the rotational (vortical) core of the vortex. In the actual fluid, because of viscosity, the singular point present in the theoretical free-vortex flow is spread out over a vortex core of intense, concentrated vorticity. At the center of the vortex, by reason of symmetry, two mutually-perpendicular fluid lines rotate in the same sense with the same angular velocity; at any other point they rotate with equal but opposite angular velocities. (3) Powder sprinkled on the surface exhibits the streamlines as logarithmic spirals. The spirals are very tight, and the streamlines are nearly circular. The tangential velocity increases toward the center, in accord with the law Vr=constant for a potential vortex. (4) View of the hyperboloidal free surface in the vortex-sink flow. The pressure on the free surface is constant. By Bernoulli's equation, the depression of the free surface is thus proportional to the square of the radius. As the flow velocity decreases, the depression decreases; as it increases, the depression increases.

QUESTIONS: (1) Show that in a real fluid the viscous stresses at the center of a potential vortex would be infinite. (2) Calculate the shapes of the streamlines in a two-dimensional vortex-sink flow. How might they differ from those of the experiments? (3) How would the flow near the floor of the spiral vortex tank compare with that near the surface? (4) Calculate the shape of the free surface in a non-viscous sink-vortex. (5) Show that all closed curves not enclosing the axis of the free vortex have zero circulation, while all that do enclose the axis have the same circulation. (6) For a sink-vortex make an order-of-magnitude estimate of the diameter of the vortex core, in terms of the circulation and the kinematic viscosity.

CREDITS: Experiments designed and performed by A. H. Shapiro, M.I.T., at the Hydrodynamics Laboratory of M.I.T. and at Educational Services Inc. Edited by A. H. Shapiro, M.I.T., and E. Carini, E.S.I.

FM-15
Incompressible Flow Through
Area Contractions and Expansions

PURPOSE: To show the effect of the rate of increase and decre of flow area on the incompressible flow of fluid in a duct.

APPARATUS: Water flows through a horizontal, two-dimensio open channel of varying cross-section. Hydrogen bubbles produ by electrolysis at wires in the water mark the flow. Success "time-lines" are produced by pulsing the current to the wir which are stretched normal to the flow direction.

SEQUENCES: (1) Step Contraction Causes Flow Separation: water flows through an abrupt step contraction in channel wi The flow separates from the projecting downstream corne Separated eddies appear also in the upstream corners. The acc eration of the flow through the contraction is shown by the incre in spacing of the time-lines. The downstream separation off corners creates a central jet which continues to contract, unt minimum area for flow ("vena contracta") is reached. Time-li from a wire downstream of the contraction show the circulatio fluid in the separation eddies and the almost uniform veloc profile in the central flow. (2) Gradual Contraction Avoids Se ration: When the same degree of contraction is made with a grad change in channel width, the entire flow accelerates smoothly thro the contraction without separation. (3) Step Expansion Causes F Separation: When there is an abrupt step increase in channel wid the fluid separates from the sharp corners and forms a jet issu into the expanded channel. There are recirculating eddies betwe the separated jet and the channel wall. Time-lines produced j downstream of the expansion show that the jet width is alm constant for some distance downstream of the step. They also sh the complexity of the flow in the corner eddies. Mixing at the boundaries produces turbulent eddies further downstream. The gradually expands to fill the channel. (4) Gradual Expansion Avo Flow Separation: A very gradual expansion is usually needed avoid separation. (5) If Expansion is Too Rapid, Separation M Occur: Although this expansion has the same wall contour as gradual contraction used in sequence (2), the flow separates fr both walls.

DISCUSSION: The phenomena in this film are explained by behavior of viscous boundary layers in favorable and adver pressure gradients. In favorable pressure gradients they do separate from the channel walls; in sufficiently adverse gradie they may separate, and thus disrupt the entire flow. These effe are especially pronounced where curvature of the wall is sha There is a strong adverse pressure gradient upstream of a conca sharp corner and downstream of a convex sharp corner. Visco stresses at the boundaries of the main flow drive the recirculati separation eddies.

QUESTION: Explain why there can be local adverse pressu gradients along the wall of a channel contraction, where the avera pressure across any cross section must decrease through t contraction as a consequence of the fluid acceleration.

CREDITS: Experiments designed and performed at E.S.I. P. W. Runstadler, Jr., Dartmouth College; Director, A. H. Shapir M.I.T.; Editor, J. Hirschfeld, E.S.I.

FM-16
Flow from a Reservoir to a Duct

PURPOSE: To compare the effects of a sharp entrance and a rounded entrance on the incompressible flow of a fluid from a reservoir to a duct.

APPARATUS AND PROCEDURE: Water flows from a reservoir into a two-dimensional, open channel. Hydrogen bubbles produced by electrolysis at wires in the water mark the flow. Successive "time-lines" are produced by pulsing the current to the wires, which are stretched normal to the flow direction.

SEQUENCES AND DISCUSSION: Flow in a channel or a duct often originates in a reservoir, a region where the fluid velocities are low compared to the average velocity in the duct. The behavior of the entrance flow in the duct immediately downstream of the reservoir depends upon the configuration of the entrance joining the duct to the reservoir. (1) Sharp Entrance Causes Flow Separation: Bubbles from a wire located in the reservoir show the pattern of the flow into a duct with a right-angle corner entrance. The entering flow separates from the corners, producing recirculating eddies just downstream of the corners on either side of the duct. A second bubble wire downstream of the entrance shows the backflow of fluid into the separation eddies. Viscous mixing at the boundaries of the flow produces turbulent eddies in the flow further downstream. (2) Bellmouth Entrance Avoids Separation: Even a slight rounding of the duct entrance corners is sufficient to avoid separation. (3) Flow Shown by Time Lines: Time lines of bubbles are produced at the reservoir wire for both the sharp and rounded entrance. An indication of the acceleration of the flow as it enters the duct is obtained by observing the increase in spacing between time lines. With the rounded entrance, a pulsed wire shows there is no back flow at the wall. These time lines also show the almost uniform velocity profile of the flow immediately after entering the duct and the development of the boundary layer along the duct wall.

INTERPRETATION: The phenomena in this film are explained by the behavior of viscous boundary layers in favorable and adverse pressure gradients. In favorable pressure gradients, they do not separate from the channel walls; in sufficiently adverse gradients they may separate, and thus disrupt the entire flow. These effects are especially pronounced where curvature of the wall is sharp. There is a strong adverse pressure gradient downstream of a convex sharp corner.

QUESTION: Explain why there can be local adverse pressure gradients along the wall of a channel contraction, where the average pressure across any cross section must decrease through the contraction as a consequence of the fluid acceleration.

CREDITS: Experiments designed and performed at E.S.I. by P. W. Runstadler, Jr., Dartmouth College; Director, A. H. Shapiro, M.I.T.; Editor, J. Hirschfeld, E.S.I.

FM-17
Flow Patterns in Venturis, Nozzles and Orifices

PURPOSE: To show the flow patterns of the incompressible flow of a fluid through a gradual or a sudden contraction and expansion in a duct.

APPARATUS: Water flows through contractions and expansions in a horizontal open channel of rectangular cross-section. Hydrogen bubbles produced by electrolysis at wires in the water mark the flow. Successive time lines are produced by pulsing the current to horizontal wires stretched normal to the flow direction.

SEQUENCES: (1) *Venturi — Gradual Contraction and Gradual Expansion:* The flow passes through a gradual contraction and a more gradual expansion with almost no flow separation from the passage walls. (A small amount of separation can be seen in the expansion portion of the passage, particularly on the wall near the top of the picture.) Time lines from a wire upstream of the contraction show the relative acceleration of the flow through the venturi. (2) *Nozzle—Gradual Contraction and Sudden Expansion:* Flow passes through a rounded contraction forming a nozzle in the channel. The flow leaves the nozzle as a jet. Time lines from a wire upstream of the contraction show the relative acceleration through the nozzle and regions of separated-recirculating flow in the corners upstream of the nozzle. Time lines from a wire downstream of the nozzle show the edges of the high-velocity jet, the uniform velocity across the jet, and large regions of backflow and recirculation downstream of the nozzle. (3) *Orifice—Sudden Contraction and Expansion:* Flow through an orifice plate installed in the channel produces large regions of separated flow upstream and downstream of the orifice. The separation upstream is in the regions of the concave corners where the plate joins the channel. The downstream separation off the corners of the opening creates a central jet which continues to contract until a minimum area for flow ("*vena contracta*") is reached. Time lines from a wire downstream of the orifice show the flow recirculation, turbulent eddies and mixing along the edges of the jet. The jet persists for some distance downstream of the orifice plate until entrainment at the jet boundaries gradually expands the jet to fill the channel.

DISCUSSION: The phenomena are explained by the behavior of a viscous boundary layer in favorable and adverse pressure gradients. In a favorable pressure gradient it remains attached to the channel wall. In a sufficiently large adverse gradient it separates from the wall; between the separation streamline and the wall is a recirculating flow, with flow near the wall moving upstream toward the separation point. The flow patterns observed in the open channel flow in this film are similar to those that occur in pipe flows where an orifice, nozzle, or venturi passage is used to meter the rate of flow. Nozzles and orifices which produce large regions of flow separation and turbulent mixing lead to an irreversible dissipation of mechanical energy in the flow by viscous action. A well designed venturi passage minimizes head losses.

CREDITS: Experiments designed and performed at E.S.I. by P. W. Runstadler, Jr., Dartmouth College; Director, A. H. Shapiro, M.I.T.; Editor, J. Hirschfeld, E.S.I. Produced by the National Committee for Fluid Mechanics Films and Educational Services Incorporated, with the support of the National Science Foundation.

FM-18
Secondary Flow in a Teacup

PURPOSE: To demonstrate the flow phenomena that cause tea leaves to collect in the center of the cup after rotary stirring.

APPARATUS: After a preliminary experiment with a cup of tea, more precise experiments are done in a laboratory apparatus. A large, transparent, cylindrical jar containing water is rotated on a turntable until the water comes into solid-body rotation. The turntable is then stopped abruptly; except for viscous layers on the bottom and side walls, the water continues to rotate. The motion is made visible by lines of hydrogen bubbles electrolyzed at and shed from a vertical wire placed about two-thirds of the distance from the axis of rotation to the wall. Two camera positions are used: one looks radially inward toward the axis of rotation, and shows circumferential components of velocity; the other is a "tangential" view, looking in a direction perpendicular to the radial line from the axis to the wire, and shows radial components of velocity.

SEQUENCES: (1), (2) A cup of tea is swirled with a spoon. The tea leaves, being more dense than water, sink to the bottom. From considerations of centrifugal force only, one would expect the leaves to move to the walls of the cup. Instead, they collect at the center. (3), (4) Sketches of the laboratory experiments. (5) Radial view of the experiment. One sees the main circumferential motion of the rotating flow and the viscous boundary layer adjacent to the stationary floor of the jar. (6) Sketch showing the "tangential" view. (7) Tangential view of the experiment. Away from the bottom the bubbles actually move in circles; they appear to move to the right as the fluid is displaced out of the plane tangential to the circular streamlines. Near the bottom, in the viscous layer, a strong radially-inward component of velocity is visible. (8) Repeat. (9) Radial view (left) and tangential view (right) displayed side by side. This juxtaposition shows that the layer of radial inflow coincides with the viscous boundary layer of the circumferential flow. (10) The teacup again. The radial inflow in the boundary layer on the bottom overcomes the effect of centrifugal force and carries the heavy tea leaves to the center.

DISCUSSION: Viscous boundary layers on a solid wall perpendicular to the axis of a rotating flow are known as Ekman layers. They have special properties due to the centrifugal pressure field and Coriolis forces. The pressure distribution in the thin boundary layer is established mainly by the rotating flow outside the boundary layer: the pressure increases radially outward from the axis. Since the boundary-layer fluid has a relatively low velocity, it is forced radially inward by the pressure field associated with the high-velocity main flow. Ekman layers sometimes have a spatially periodic velocity profile, as seen here especially in the radial view.

REFERENCES: "Boundary Layer Theory" Schlichting, H., p. 176 Fourth Edition, New York, McGraw-Hill, 1960.

CREDITS: Scenes from the film "Secondary Flow" by E. S. Taylor, M.I.T.; Consultants, A. H. Shapiro, M.I.T., and E. S. Taylor, M.I.T.; Producer, R. Bergman, E.D.C.; Editor, J. Hirschfeld, E.D.C.; Art Director, R. P. Larkin, E.D.C. Produced by the National Committee for Fluid Mechanics Films and Education Development Center, Inc. (formerly Educational Services Inc.), with the support of the National Science Foundation.

FM-19
Secondary Flow in a Bend

PURPOSE: To show the secondary flows induced when the streamlines of a flow are curved and there is a vorticity component perpendicular to the flow but in the plane of streamline curvature.

EQUIPMENT AND PROCEDURES: Water flows in an open, horizontal channel of rectangular cross-section. The channel has a 90°-bend in a horizontal plane, preceded by a long straight section. Graduated screens at the inlet to the straight section produce an approximately linear velocity profile in a vertical plane, with the vorticity increasing from the near-bottom of the channel (outside the boundary layer) to the free surface. The velocity vector of this shear flow is horizontal and perpendicular to the upstream velocity vector. The vorticity is nearly uniform over the cross-section. Hydrogen bubbles, produced upstream by electrolysis at fine platinum wires makes the motion of various fluid lines visible.

SEQUENCES: (1) Sketch of flow channel and initial velocity distribution. (2) Seen from the side, a vertical wire in the straight section releases a sheet of bubbles. The leading edge of this sheet shows the approximately-linear velocity profile (horizontal vorticity) produced by the inlet screens. (3) This initially vertical bubble sheet is seen from above as it passes through the bend. The sheet rotates: its upper edge moves toward the outer wall, and its lower edge toward the inner wall of the bend. (4) Viewed from above, a short horizontal line of fluid lying along the velocity vector moves around the bend. (5) Viewed from above, a short horizontal line of fluid perpendicular to the velocity vector moves around the bend. This second line is released at the same instant as the line seen in (4), and the two lines initially form a horizontal fluid cross. Since viscous effects in this part of the flow are nearly negligible, Helmholtz' vortex law requires that a particle of fluid, which has no vertical vorticity attached to it, remain free of vertical vorticity as it begins to go around the bend. Thus the two legs of the fluid cross rotate with equal and opposite angular velocities at the beginning of the bend. The transverse leg which carries the originally horizontal vorticity develops a "secondary" component along the streamline, which accounts for the streamwise component of swirl of the flow at the exit of the bend. (6) Looking upstream into the straight inlet, a fluid cross in a vertical plane normal to the flow moves without rotation, showing that there is no streamwise vorticity component upstream of the bend. (7) Looking downstream into the exit of the bend, the same type of fluid cross is seen to acquire a net angular velocity as it passes around the bend. (8) With the graduated screens removed, there is no horizontal vorticity in the straight channel except in the viscous boundary layer on the floor. Looking obliquely downstream into the bend, the secondary flow induced by this shear layer is seen to create a strong "passage vortex" whose core lies along the flow.

CREDITS: Scenes from the film "Secondary Flow" by E. S. Taylor, M.I.T.; Consultants, E. S. Taylor, M.I.T., and A. H. Shapiro, M.I.T.; Producer, R. Bergman, E.S.I.; Editor, J. Hirschfeld, E.S.I.; Art Director, R. P. Larkin, E.S.I. Produced by the National Committee for Fluid Mechanics Films and Educational Services Incorporated, with the support of the National Science Foundation.

FM-20
The Horseshoe Vortex

PURPOSE: To show the horseshoe vortex flow pattern produced when a fluid flowing along a surface, on which a boundary-layer has developed, encounters an obstacle mounted on the surface.

APPARATUS: Water flows in a rectangular open channel having a horizontal floor. A vertical circular cylinder sticks up from the floor. The flow is visualized by hydrogen bubbles electrolyzed at a fine horizontal wire which stretches across the channel, perpendicular to the flow direction. The wire is situated within the boundary-layer shear flow on the floor of the channel.

SEQUENCES: (1) The cylinder is placed in the channel. (2) Close-up of the cylinder, showing its optical reflection in the floor of the channel. (3) Explanatory sketch of the apparatus, viewed from the side. (4) Top view. A sheet of bubbles is generated at the wire. This sheet does not reach the cylinder, but rolls up ahead of the cylinder, and folds around the cylinder, forming a horseshoe pattern. (5) Oblique top view of same. (6) Side view of same. (7) Close-up side view, showing details of vortex as it rolls up ahead of cylinder.

DISCUSSION: Postulate initially a shear flow which remains in layers parallel to the floor. Such a flow would have a stagnation line along the nose of the cylinder, but the pressure along this line would decrease toward the floor owing to the velocity decrease in the boundary layer. This variation in pressure would produce curvature of the streamlines in the vertical plane — they could not remain in parallel layers, but would be forced to turn downward by the pressure gradient. Thus, a secondary flow is created. This secondary flow causes pressure at the cylinder nose/floor intersection to rise above the free-stream static value, and thus makes it impossible for the slow-moving boundary layer to enter the region close to the cylinder. The downward-moving fluid near the nose of the cylinder continues to the bottom and turns *upstream*, thus blocking the boundary layer and separating it from the bottom wall. Ahead of the cylinder the boundary layer is wound up into a vortex which then sweeps around the sides of the cylinder, forming the horseshoe pattern. We ignore viscosity in considering the behavior of vortex lines near the cylinder. The shear flow contains many horizontal vortex lines frozen to and convected with the associated fluid lines. Such a fluid line is arrested upstream of the cylinder and then folds around the cylinder in the shape of a hairpin or horseshoe. The vorticity frozen in the fluid line yields streamwise vorticity in the legs of the horseshoe. Stretching of the fluid line as it enfolds the cylinder intensifies the vorticity both near the vertex of the horseshoe and in the legs. Wind blowing past a tall building or tree, or water flowing past a bridge pier in a river, creates a secondary flow of this type — running down the face of the obstacle and upstream along the ground — together with a vortex that can produce scouring of the ground both upstream and under the long legs of the horseshoe.

CREDITS: Scenes from the film "Secondary Flow" by E. S. Taylor, M.I.T.; Consultants, E. S. Taylor, M.I.T., and A. H. Shapiro, M.I.T.; Producer, R. Bergman, E.S.I.; Editor, J. Hirschfeld, E.S.I.; Art Director, R. P. Larkin, E.S.I. Produced by the National Committee for Fluid Mechanics Films and Educational Services Incorporated, with the support of the National Science Foundation.

FM-21
Techniques of Visualization for Low Speed Flows, Part I

PURPOSE: To demonstrate various methods of visualizing low speed fluid flows.

SEQUENCES: (1) Aluminum powder on a water surface shows the flow past a cylinder started from rest. The technique is particularly useful in separated flows if the start-up process is shown since then the wake is made visible and the stability conditions can be examined. In using surface techniques a check on the two-dimensionality should be made to insure that the surface flow represents the whole flow field. (2) Licopodium powder is shaken onto the free surface of a sink vortex. This method requires extreme care to keep the free surface clean; a contaminated surface may give misleading results. Surface active agents are useful in providing good surface conditions. (3), (4) Dye injected from a hypodermic syringe into another sink-vortex flow shows first the primary logarithmic-spiral flow far from the bottom surface and then the approximately radially-inward flow in the boundary layer near the bottom surface (See FM-70). (5) Dye injection, if carefully regulated, can be used to show the flow very near solid boundaries. Here, dye flows from a vertical slit in the wall of a channel. After the flow is established by moving the wall into place, the dye flows gently from the slit and makes only minimal disturbances. While the title of this scene indicates turbulent flow, the pattern shown is turbulent only in "spots" where a wrinkled, rapidly oscillatory flow is observed. These turbulent spots move through a streaky laminar background indicating the final stage of laminar-turbulent transition (See FM-1) (7), (8), (9) Generation of very fine bubbles by electrolysis from small wires is particularly useful for visualizing discrete planes of flow and for making combined-time-streak markers. The method is more difficult than some others, but more powerful and versatile. A great variety of effects can be produced by controlling wire location, pulse on and off times, and by insulating sections of the wire. Deformation can be made visible using finite patches or bubbles. Even more significant, instantaneous velocity patterns can be obtained on lines or planes in the flow in unsteady motions. See REFERENCE for details of techniques and limitations.

REFERENCE: Schraub, F. A.; Kline, S.J.; Henry, J.; Runstadler, P. W.; Littell, A., "Use of Hydrogen Bubbles for Quantitative Determination of Time-Dependent Velocity Fields in Low-Speed Water Flows," Journal of Basic Engineering *87,* D (June 65).

CREDITS: Scenes from the films "Flow Visualization," S. J. Kline, Stanford University; "Vorticity" and "Pressure Fields and Fluid Acceleration," A. H. Shapiro, M.I.T.; "Secondary Flow," E. S. Taylor, M.I.T.; experiments and film by L. Prandtl, Gottingen. Consultants, B. A. Egan, E.S.I., and S. J. Kline, Stanford University; Producer, R. Bergman, E.S.I.; Editor, M. Chalufour, E.S.I. Produced by the National Committee for Fluid Mechanics Films and Educational Services Incorporated, with the support of the National Science Foundation.

FM-22
Techniques of Visualization for Low Speed Flows, Part II

PURPOSE: To demonstrate various techniques of visualizing fluid flows.

SEQUENCES: (1) Titanium tetrachloride is applied in liquid form to the nose of a sphere in a wind tunnel. Water vapor in the air reacts with the $TiCl_4^*$ to produce a dense fog which is visible by scattered light. We can discern the line of boundary-layer separation and the relatively stagnant wake behind the sphere. This exemplifies other visualization techniques which depend upon the rate of mass transfer of a flow. (2) In another wind tunnel the streamlines about an airfoil at small and large (stalled) angles of attack are visualized by "smoke" streamlines of kerosene fog droplets. These are produced by condensation of kerosene vapor. (3) Tufts or threads attached to an object give information about boundary-layer behavior. Intermittent or transitory stall is seen to be occurring on the upper surface of an airfoil (see also FM-39). (4) A matrix of tufts downstream of a lifting airfoil shows the wing-tip vortices and the induced downwash in the wake. (5) The cross flow perpendicular to the axis of a slender cylindrical body at a small angle of attack is shown. The camera looks upstream and a plane of light perpendicular to the body axis illuminates air bubbles in water; one sees the cross-flow components of velocity. (6) The flow in the annular gap of two concentric cylinders with the inner cylinder rotating is made visible by mixing a small amount of aluminum paint particles into kerosene. The bands are vortex rings caused by instabilities in the main vortex flow (see FM-32). Aluminum flakes are plate-like in shape and visualization is obtained by differences in reflectivity due to variations in orientation. They tend to line up approximately with the direction of the maximum rate of shear. (7) A shallow tank of water lit from below through a glass floor shows the wave patterns of a disturbance moving at speeds below and above the propagation speed of small waves. The light is refracted by the curved surfaces of the waves and forms patterns of light and dark on a horizontal screen above the water surface.

CREDITS: Scenes from the films "Doppler Effect", by J. Strickland, E.S.I.; "The Fluid Dynamics of Drag", by A. H. Shapiro, M.I.T.; "Boundary Layer Control", by D. C. Hazen, Princeton University; "Flow Instabilities" by E. Mollo-Christensen, M.I.T.; and Office National d'Études et de Recherches Aérospatiales (O.N.E.R.A.), France, Consultants, B. A. Egan, E.S.I., and S. J. Kline, Stanford University; Producer, R. Bergman, E.S.I.; Editor, M. Chalufour, E.S.I. Produced by the National Committee for Fluid Mechanics Films and Educational Services Incorporated, with the support of the National Science Foundation.

FM-23
Tollmien-Schlichting Waves

PURPOSE: To illustrate the presence of Tollmien-Schlicht waves in the process of transition from laminar to turbulent fl in the boundary layer on an airfoil.

APPARATUS AND PROCEDURE: All experiments were made i low-turbulence wind tunnel, two feet high by three feet wide, at Collins Aeronautical Research Laboratory. The air speed is a proximately 40 ft/sec. The wing is attached to one side wall of tunnel; the side wall thus represents a plane of symmetry. Smc flows steadily out of the wing through many fine holes spaced alc the leading edge. In sequences (1) through (3) the camera speed 1500 frames per second; thus the motion is slowed by a factor about 100. In sequence (4) the camera speed of 3300 frames p second slows the motion by a factor of about 200.

SEQUENCES: (1) The air flow past a straight wing at $4°$ inciden is viewed from the rear. Following a schematic of the experimen set-up, the experiment is repeated. At about one-fourth cho length aft of the leading edge, the individual smoke streaklin form into evenly spaced patches. These patches are in phase the different streaklines, and give the appearance of waves paral to the leading edge. (2) A similar flow observed from the tip of wing. Near the wing tip, the instabilities appear earlier and a1 plify to turbulence more quickly than in mid-span. (3) A sequent comparison of the flow over a straight wing and over a swept wi shows that the waves are parallel to the leading edge, thus cc firming that they are associated with the boundary layer behavic (4) A single smoke streakline released through a slot in the lead: edge of the straight airfoil is seen in a close-up view. The smc ribbon exhibits longitudinal wave-like disturbances. These ampl and then roll up into discrete vortices, which finally burst in irregular manner. Since the wing surface is highly polished, c sees both the smoke ribbon and its image in the surface of the wi1

INTERPRETATION: Tollmien-Schlichting waves are two-dime sional unstable disturbances of very small amplitude in the bounda layer. In certain experimental conditions, the exponential grov of these waves is the first stage in the transition from laminar turbulent flow. Although the evidence in the experiments sho here is not decisive, it is difficult to find any other explanation th that the observed periodic, wave-like disturbances were initiat by the Tollmien-Schlichting mechanism. At some stage in t amplification of the disturbances, however, the waves are no long of Tollmien-Schlichting type, because of non-linear and thre dimensional effects which then appear.

REFERENCE: Schlichting, H.: "Boundary Layer Theory", Ch. pp. 382-406, Fourth Edition, McGraw Hill, New York, 1960.

CREDITS: Experiments and film by A. M. Lippisch, Collins Rad Company; Consultant, R. W. Fox, Purdue University; Directc A. H. Shapiro, M.I.T.; Editor, E. Carini, E.S.I.; Art Directc R. P. Larkin, E.S.I.

FM-24
Wing Tip Vortex

PURPOSE: To illustrate the wing tip vortex on a wing of rectangular plan-form.

APPARATUS: (A) The first series of experiments is made in a low-turbulence wind tunnel, two feet high by three feet wide. The wing is attached to one side wall of the tunnel; the side wall thus represents a plane of symmetry. Smoke is employed as a means of flow visualization. (B) In the second series of experiments, a full wing is mounted centrally in a rectangular wind tunnel, by means of a narrow support from the floor. Downstream of the wing there is a grid of wires stretched across the tunnel; tufts of wool are attached at each intersection point. The camera looks upstream toward the tuft grid; the tufts show the pattern of cross flow velocities in the wake of the wing.

SEQUENCES: (1) End view of the wing at a small angle of incidence (hence small lift). Initially there is no air flow. Smoke issues in jets from holes spaced along the lower wing surface near the tip. The air flow begins; the smoke shows that air flows from below the wing, around the tip, to the upper surface and forms into a tight vortex which trails into the wake of the wing tip. (2) Rear view of the wing tip vortex of sequence (1). Because of the camera position, the tip vortex appears to break off the wing surface at a sharp upward angle, but this is illusory. The actual vortex direction is clearly seen in sequences (1) and (5). (3) Rear view of tip vortex from an airfoil at higher angle of incidence (hence greater lift). The vortex appears to be stronger. Additionally, smoke released at the leading edge shows some cross flow over the upper wing surface. (4) A sketch shows the positions of the tuft grid and camera behind the wing. (5) View of the tuft grid and wing as seen from downstream, with the wing initially at a low angle of incidence. Vortices are observed behind the wing tips. As the angle of incidence of the wing is increased, the vortices grow larger in extent. The tufts show that the vortex cores induce substantial cross-flow velocities at large distances from the cores themselves.

INTERPRETATION: The tip vortices are formed by air flowing from the high pressure region below the wing to the low pressure region above. This pressure difference increases as the lift of the wing increases; therefore the strength of the tip vortices also increases with lift. The tip vortices may be viewed as a continuation of the vortex "bound" in the wing; the bound vortex accounts for the circulation around the wing, and its lift.

CREDITS: Experiments and film by A. M. Lippisch, Collins Radio Company, and the National Advisory Committee for Aeronautics; Consultant, R. W. Fox, Purdue University; Director, A. H. Shapiro, M.I.T.; Editor, E. Carini, E.S.L; Art Director, R. P. Larkin, E.S.I.

FM-25
Low-Speed Jets: Stability and Mixing

PURPOSE: To illustrate the effect of Reynolds number on the instability of jets and on mixing between a jet and its surroundings.

SEQUENCES: (1) A jet of dyed liquid is discharged into a tank of the same liquid (clear). The viscosity of the liquid is controlled by mixing corn syrup and water. (1a) With a high viscosity, yielding a Reynolds number of about 200 based on initial jet diameter, the jet remains laminar. Since mixing occurs only by molecular diffusion, the jet shows very little decay, spreading, or mixing. (1b) With a lower viscosity, yielding a Reynolds number of about 3000, the jet is turbulent. Turbulent convection accelerates the mixing greatly. The spreading of the jet, signifying velocity decay coupled with entrainment of the surrounding liquid, is much larger than in the laminar case. (2a) An air jet of circular cross-section enters the atmosphere. The upper and lower edges of the jet are marked by smoke injection. The region of the jet boundary exhibits periodic oscillations of increasing amplitude. These are due to Kelvin-Helmholtz instability in the free shear layer at the jet boundary. (2b) As the Reynolds number is increased, the space-wise rate of amplification increases. The folding of the jet boundary develops equally-spaced vortex rings, which then break up as a result of a higher-order instability. (2c) When the flow is lit stroboscopically at the frequency of the initial instability, the initial vortex rings appear stationary. This indicates that the amplification of the instability is highly frequency-selective. Further downstream, the flow appears unsteady, due to the appearance of other frequencies in the higher-order instabilities. (3) A jet of dyed water is discharged into clear water. The successive instabilities ultimately produce fully-developed jet turbulence. The rate of spreading of the jet is dramatically greater after it has become turbulent. As the Reynolds number is increased by increasing the jet speed, the rapid change in broadening that characterizes the appearance of fully-developed turbulence occurs earlier. (4) Two jets of liquid are shown simultaneously, with Reynolds numbers of 20,000 and 400, respectively. The large-scale structure of the turbulence is seen to be nearly the same in both cases. Since it is the large-scale eddies that govern gross mixing and decay of a jet, the two jets have the same spreading angle. (5) An air jet marked by smoke is discharged into the atmosphere. Smoke injection points surrounding the jet mark streaklines which make visible the flow from the surroundings. The jet acts like a two-dimensional line sink of fluid: the surrounding air is drawn in at nearly right angles to the jet.

CREDITS: Scenes from the films "Flow Studies of Free Flight Delta Wing Models", A.M. Lippisch, Collins Radio Company; "Low Reynolds Number Flow", G.I. Taylor, Cambridge, England; "Turbulence", R.W. Stewart, University of British Columbia; "Übergang von Laminarer zu Turbulenter Strömung", R. Wille, Technische Universität, Berlin. Consultants, A.H. Shapiro, M.I.T., and R.W. Stewart; Producer, R. Bergman, E.D.C.; Editor, E. Carini, E.D.C. Produced by the National Committee for Fluid Mechanics Films and Education Development Center (formerly Educational Services Incorporated) with the support of the National Science Foundation.

FM-26
Tornadoes in Nature and the Laboratory

PURPOSE: To display a variety of natural tornadoes photographed by adventuresome cameramen; and to show the flow in a laboratory model of a tornado.

APPARATUS: The laboratory experiment employs a circular tank, open at the top. Water enters tangentially at the rim through a vertical slit extending over the entire depth of liquid. The water leaves in an upward direction, at the center of the tank, through a siphon tube whose opening is slightly below the surface. A spiral sink-vortex is formed. Dye streaklines show the flow.

SEQUENCES: (1) Natural tornadoes. (2) The laboratory experiment. Dye is injected near the "ground" of the tank, flows nearly radially inward to the center, then upwards to the discharge tube in a tight vortex. At some vertical position, the vortex core "explodes" to a large diameter. (3) Natural tornadoes.

DISCUSSION: The nearly-circular streamlines of the general spiral sink-vortex flow in the laboratory experiment establishes a pressure distribution such that the pressure increases with radius. In this primary sink-vortex flow, the centrifugal force of the fluid is just balanced by the pressure gradient. Near the "ground", however, the viscous boundary layer, which has only a small circumferential speed, is pushed inward in a nearly radial direction by the pressure gradient. Most of the flow into the discharge tube comes from this "secondary flow" of the viscous boundary layer on the floor of the tank. Because of the tendency to conserve moment of momentum, the fluid reaching the center forms a concentrated vortex core. The explosion of the vortex core is not well understood.

CREDITS: Experiment from SECONDARY FLOW; Consultant and Director, A. H. Shapiro, M.I.T.; Editor, R. Bergman, E.S.I.: Art Director, R. P. Larkin, E.S.I.

FM-27
Interaction of Oblique Shock with Flat-Plate Boundary Layer

PURPOSE: To show the behavior of a boundary layer when an oblique shock is incident on it; and to show how the shock-boundary layer interaction differs for laminar and turbulent boundary layers.

APPARATUS: The experiments are done with air in a supersonic wind tunnel of rectangular cross section. Two test Mach Numbers are used, 2.0 and 3.0. Mounted near the floor of the tunnel is a thin flat plate which extends from wall to wall and which is aligned with the flow. The boundary layer to be looked at forms on this plate. Near the top of the tunnel is a two-dimensional wedge which generates an oblique shock wave that impinges on the flat plate. For a fixed tunnel Mach Number and wedge angle, the position of impingement on the plate is virtually fixed. The Reynolds Number of the boundary layer at this position, based on distance from the leading edge, is altered by changing the pressure level in the tunnel. At the point of shock impingement, the length-Reynolds Number may be varied between 0.4 million, for which the boundary layer is laminar, and 2.1 million, for which it is turbulent. Density variations in the flow are made visible by the color schlieren technique.

SEQUENCES: (1) $Mach\ Number = 2$. The boundary layer is at first laminar at the point of shock incidence. The Reynolds Number is then increased, and the boundary layer becomes turbulent. A subsequent decrease in Reynolds Number returns the boundary layer to a laminar condition. (2) $Mach\ Number = 3$. The boundary layer is at first laminar. Then it becomes turbulent as a result of an increase in Reynolds Number. (3) $A\ repeat\ of\ sequence\ (2)$.

INTERPRETATION: The oblique shock wave tends to produce an abrupt pressure rise to which the boundary layer must adjust. The boundary layer separates from the surface, often in the form of a local separation bubble. This spreads the pressure rise along the wall over a considerable distance, thus "softening" the abruptness of the pressure rise. The separation and thickening of the boundary layer themselves generate an oblique shock-wave system upstream of the incident shock wave; this wave system then intersects with the incident shock wave itself. The reflection pattern of the incident shock wave is complicated by the fact that the low-speed zone under the separated boundary layer produces a reflection condition which is neither constant direction nor constant pressure. The reflection pattern is frequently a rarefaction wave which turns the flow toward the plate, followed by compression wavelets which straighten the flow to follow the wall. The phenomenon is usually unsteady, and the pattern near the impingement region oscillates at high frequency. These effects are much less pronounced with the turbulent boundary layer than with the laminar, because the greater momentum of the turbulent boundary layer better enables it to withstand an adverse pressure gradient.

CREDITS: Experiments and film by the National Physical Laboratory, England; Consultant, R. Fox, Purdue; Director, A. H. Shapiro, M.I.T.; Editor, E. Carini, E.S.I.; Art Director, P. Larkin, E.S.I.

M-28
ransonic Flow Past a Symmetric Airfoil

URPOSE: To show the pressure fields and wave patterns for a ubsonic airfoil in the transonic range of speeds.

APPARATUS: A symmetric airfoil of subsonic type, with a rounded ose, is mounted in a horizontal wind tunnel of variable Mach umber. The flow is visualized by color Schlieren with light stops rranged vertically. Dark blue signifies a density increase (compression) from left to right; orange signifies a density decrease rarefaction) from left to right. Thus the color patterns indicate, a a qualitative way, the pressure distributions in regions outside e boundary layers.

EQUENCES: (1) The airfoil is at small incidence, and the tunnel peed is increased from a very low subsonic Mach Number (esentially incompressible flow) to a slightly supersonic value. At ery low Mach Number, both the upper and lower surfaces show ompression near the nose, followed by rarefaction regions, folowed by recompressions to the tail. As the speed increases, the olors intensify, corresponding to increases in the pressure and ensity gradients. At higher subsonic Mach Numbers, shock waves ppear at the end of the rarefaction zone on the upper (suction) urface, thus indicating that a portion of the flow in this zone is upersonic. With further increases of Mach Number, the shocks trengthen, move downstream as the size of the supersonic zone acreases, and develop into a "lambda" form owing to interaction f the shock with the boundary layer. Ultimately a similar shock ystem appears on the lower surface, and both shock systems move ownstream until they become attached to the tail. The free stream a still subsonic, while the flow over the airfoil is mainly superonic. As the free stream becomes supersonic, a detached, nearlyormal, bow shock appears from far upstream and comes closer to ne airfoil as the Mach Number increases; the flow downstream of ne shock, however, is slightly subsonic, and the flow past the airbil is virtually the same as for a slightly subsonic free-stream ow. (2) Flow with steady free-stream Mach Number of 1.4. The enerally oblique bow shock wave is slightly detached because of ne round nose. From the nose to the tail the pressure falls, and aint orange-colored Mach waves spring from the upper and lower urfaces. Oblique shock waves at the tail allow the upper and ower flows to rejoin with equal pressures and directions. Reflecions of the bow waves from the tunnel walls are also visible. 3) With fixed Mach Number of 0.8, the incidence is varied by ±2°. arge changes in flow pattern ensue, showing that at high subsonic peeds the lift and pitching moment are very sensitive to incidence. 4) Similarly, with fixed Mach Number of 1.4, the incidence is varied y ±2°. Relatively small changes in flow pattern occur, indicating esser sensitivity of lift and moment at supersonic speeds beyond he transonic range.

CREDITS: Experiments and film by National Physical Laboratory, ngland; Consultant, R. Fox, Purdue University; Director, A. H. hapiro, M.I.T; Editor, E. Carini, E.S.I.; Art Director, R. P. arkin, E.S.I.

FM-29
Supersonic Zones on Airfoils in Subsonic Flow

PURPOSE: To show the development of supersonic regions on an airfoil at high subsonic flight Mach Numbers; and the associated phenomena of shock waves and shock-boundary layer interactions.

APPARATUS: A symmetric airfoil of subsonic type is mounted in a subsonic wind tunnel capable of producing test-section Mach Numbers from very low values (essentially incompressible) to nearly unity (transonic). The airfoil support protrudes into the picture from the left. Density variations in the flow are made visible by the schlieren technique; the knife edge is aligned parallel to the flow, thus showing density gradients transverse to the flow.

SEQUENCES: (1) Small Incidence, Variable Mach Number. The airfoil is mounted at a small, constant angle of attack. Initially the free-stream Mach Number is so low that the flow is everywhere subsonic, and no waves are present. As the Mach Number increases, a supersonic zone appears in the flow adjacent to the upper surface; it is terminated by a shock wave which is nearly normal at the airfoil and which extends upward to the limit of the supersonic zone. At first the extent of the supersonic zone and the strength of the terminal shock wave are small, but both grow in size as the Mach Number increases. At the same time the terminal shock wave moves downstream. As the shock becomes stronger it produces a substantial boundary layer separation and a complicated shock-boundary layer interaction. Furthermore, as the Mach Number increases, a similar supersonic zone and terminal shock system ultimately appear on the lower surface. Finally, when the Mach Number is nearly unity, virtually the entire upper and lower surfaces are supersonic, and the two terminal shocks have become trailing-edge oblique shocks associated with the finite angle of the trailing edge. A seemingly intense oblique wave (exaggerated by the knife-edge orientation) on the upper surface, at about 15% aft of the leading edge, is due to a blowing slot at that position, which however, is not used in this film. (2) Repeat of sequence (1). (3) Constant Mach No. ($M_\infty = 0.75$), Variable Incidence. At zero incidence, the flow is entirely subsonic; no waves are seen. As the incidence increases, a supersonic zone terminated by a shock appears on the upper surface, close to the leading edge. The size of the supersonic zone and the strength of the shock increase as the incidence grows larger. Moreover, the shock produces a boundary layer separation which ultimately becomes so great that the flow separates at the leading edge and eliminates altogether the supersonic zone and shock pattern. The airfoil is "stalled." The apparently great thickness of the shock is the result of unsteadiness, the shock position oscillating back and forth at high frequency. When the incidence is decreased, the reverse sequence of events take place, but with great unsteadiness and probably some hysteresis. (4) Constant Mach Number ($M_\infty = 0.80$), Variable Incidence. At zero incidence, there are a small supersonic zone and terminal shock wave on the upper surface. Otherwise, the events are similar to those in sequence (3).

CREDITS: Experiments and film by the National Physical Laboratory, England; Consultant, R. Fox, Purdue; Director, A. H. Shapiro, M.I.T.; Editor, E. Carini, E.S.I.; Art Director, P. Larkin, E.S.I.

FM-30
Shock and Boundary Layer Interaction on Transonic Airfoil

PURPOSE: To show that the flow pattern near an airfoil at transonic flight speeds is greatly influenced by the interaction between the shock wave terminating the supersonic zone and the boundary layer on the airfoil surface.

APPARATUS: A symmetrical airfoil of subsonic type is mounted in a subsonic wind tunnel. The free-stream speed in the test section is in the high subsonic range. The flow is visualized by the schlieren method. With the schlieren knife edge perpendicular to the flow, density gradients along the flow are shown; with it set parallel to the flow, transverse gradients are shown.

SEQUENCES: (1) Mach No. = 0.82; incidence = 0°; Schlieren Knife Edge Perpendicular to Flow. The boundary layer is laminar on both top and bottom surfaces. Adjacent to each surface is a zone of supersonic flow terminated by a shock wave which (but for boundary-layer separation) is normal near the airfoil. The terminal shock produces a severe boundary-layer separation. An oblique shock is generated near the point of separation, which is considerably upstream of the terminal shock which is responsible for the separation. This oblique shock is reflected from the nearly-normal shock terminating the supersonic zone, thus producing a triple shock intersection in the form of a typical "lambda" configuration. (2) Mach No. = 0.82; Incidence = 0°; Schlieren Knife Edge Perpendicular to Flow. Everything is the same as in sequence (1) except that the boundary layers are made turbulent by the "tripping" action of small jets emitted about 15% aft of the leading edge. The extent of the boundary layer separation is greatly reduced, and there is much less upstream influence of the terminal nearly-normal shock; the latter extends virtually to the surface. Details of the shock-boundary layer interaction near the surface are obscured by high-frequency unsteadiness in this region which makes the shock appear much broader than it really is. (3) Mach No. = 0.82; Incidence = 0°; Schlieren Knife Edge Perpendicular to Flow. Again, the circumstances are as in sequences (1) and (2). Now, however, the boundary layer on the lower surface remains laminar, while on the upper surface it is alternately laminar and turbulent. (4) Mach No. = 0.81; Incidence = 2°; Schlieren Knife Edge Perpendicular to Flow. The boundary layer on the lower surface is turbulent. On the upper surface it is alternated between laminar and turbulent by the triggering jet. Changes in the shock configuration on the upper surface produce shifts in the strength and position of the shock on the lower surface (see below). (5) Schlieren Knife Edge Parallel to Flow. On the lower surface the boundary layer is turbulent. On the upper surface it is alternated between laminar and turbulent. As in sequence (4), a change in the shock configuration on the upper surface shifts the position of the shock on the lower surface. The reason for this is that the flows on both the upper and lower surfaces are subsonic between the shocks and the trailing edge. Hence changes in the pressure field on one side of the airfoil are communicated around the trailing edge to the other side.

CREDITS: Experiments and film by National Physical Laboratory, England; Consultant, R. Fox, Purdue University; Director, A. H. Shapiro, M.I.T.; Editor, E. Carini, E.S.I.; Art Director, R. P. Larkin, E.S.I.

FM-31
Instabilities in Circular Couette Flow

PURPOSE: To illustrate two successive instabilities which c occur in the annular gap between concentric cylinders when t inner cylinder is rotated.

APPARATUS: The mean diameter of the glass cylinders is abc 4 inches, and the height is about 7 inches. The radial gap is abo 1/4 inch. The working fluid is a silicone oil with a kinemat viscosity of about 0.1 cm²/sec. The inner cylinder is rotated speeds near 40 rpm, giving Reynolds numbers of about 150 bas on gap width. Aluminum flakes in the oil (about one part of pigme in 2000 parts of oil by weight) make the flow visible.

SEQUENCES: (1) The apparatus is assembled. (2) A jar of oil co taining aluminum flakes is agitated to demonstrate the flow-visua ization method. (3) The operating condition is defined by plotti rotation speed of the outer cylinder as abscissa and rotation spe of the inner cylinder as ordinate. The theoretical Taylor instabili boundary is shown. Below this line, the flow is stable. The fir operating point is on the vertical axis. (4) With the outer cylinde at rest, the inner cylinder speed increases slightly and crosses t Taylor boundary. Alternating light and dark rings appear in th fluid, showing the secondary flow resulting from Taylor instabilit This secondary flow is a series of vortex rings or cells of alter nating sense, stacked up along the axis of rotation like washe on a bolt. (5) A sketch of the vortex cell pattern. Each cell very nearly square in cross section. (6) Operation as in sequenc (4), but lighted from inside the inner cylinder. (7) A graph as sequence (3), showing the experimentally determined position of second instability boundary lying above the first one. The new op erating point is also on the vertical axis, at a higher speed than sequence (4). (8) The inner cylinder speed increases slightly ar crosses the second boundary. Tangential waves appear, supe imposed on the Taylor cells seen in earlier sequences. The wave are in perfect phase along the cylinders as they travel circum ferentially around the working space. When the outer cylinder at rest, the angular velocity of the wave pattern is about 1/2 1/3 of the angular velocity of the inner cylinder. (9) Close-u view of the waves. (10) Distant view as in sequence (8), but lighte from inside. (11) Close-up view, lighted from inside.

INTERPRETATION: The primary flow is purely rotational, and th streamlines are circles around the axis of rotation. The first in stability, discovered by G. I. Taylor in 1923, is a classical examp of instability to small disturbances, and is well understood. results from the centrifugal force field established by the curve streamlines, and leads to the observed steady pattern of vortex cell The second instability, which occurs at higher speeds and leads t a non-steady doubly-periodic flow, was first accurately describe by Schultz-Grunow and Hein in 1956. This second instability i not yet understood.

REFERENCES: (1) S. Chandrasekhar, "Hydrodynamic and Hydro magnetic Stability", Oxford, 1961. (2) D. Coles, Transition i Circular Couette Flow, J. Fluid Mech. (in press, 1964).

CREDITS: Experiments and film by D. Coles, C.I.T.; Director A. H. Shapiro, M.I.T.; Editor, E. Carini, E.S.I.; Art Director R. P. Larkin, E.S.I.

xamples of Turbulent Flow
etween Concentric Rotating Cylinders

URPOSE: To illustrate several fully turbulent flows which can ccur for different combinations of cylinder speeds.

PPARATUS: The mean diameter of the annular working space is bout 4 inches, and the height is about 7 inches. The radial gap is bout 1/4 inch. The working fluid is a silicone oil with a kinematic scosity of 0.01 cm^2/sec (about the same as water). The cylinders re rotated at speeds which give Reynolds numbers in the range ,000 to 20,000, based on radius and surface velocity. A small nount of aluminum paint pigment (about one part of pigment in 00 parts of oil by weight) is added to the oil to make the flow sible. The apparatus is usually lighted from inside the inner /linder during operation.

EQUENCES: (1) The apparatus is assembled. (2) A jar of oil con- ining aluminum flakes is agitated to demonstrate the flow visual- ation method. (3) The cylinders rotate in opposite directions at early equal speeds. The flow is fully turbulent, with a relatively parse scale. (4) Close-up view of the flow in sequence (4). (5) A ow similar to the one in sequence (3), but with the speed of both /linders approximately doubled. The scale of the turbulence is oticeably reduced. (6) The inner cylinder rotates at the same speed s in sequence (5), but the outer cylinder is at rest. The flow is rst lighted from outside. The strong axial periodicity is related the phenomenon of Taylor instability (see FM-31). (7) The same ow, but lighted from inside.

EFERENCES: (1) S. I. Pai, Turbulent Flow Between Rotating ylinders, NACA TN 892, 1943. (2) D. Coles, Transition in Circular ouette Flow, J. Fluid Mech. (in press, 1964).

CREDITS: Experiments and film by D. Coles, C.I.T.; Director, . H. Shapiro, M.I.T.; Editor, E. Carini, E.S.I.; Art Director, . P. Larkin, E.S.I.

FM-33
Stagnation Pressure

PURPOSE: To illustrate: (1) the idea of a stagnation point; (2) the measurement of stagnation pressure by means of a pitot probe; and (3) the determination of velocity from measurements of stagnation pressure and static pressure.

THEORY: For inviscid, incompressible, steady flow without body forces, Bernoulli's integral states that the sum of the static pressure (p) and the dynamic pressure ($\rho V^2/2$) remains constant on each streamline. If there is a point of zero velocity (i.e., a stagnation point) on some streamline, the static pressure is a maximum there. It is called the stagnation pressure.

EXPERIMENTS: (1) Visualization of Motion Near Stagnation Point on a Blunt Body. Hydrogen bubbles electrolyzed at a wire in a water flow are swept off by the stream, and light reflected from the bub- bles makes the motion visible. In successive sequences, streaklines and patches of hydrogen bubbles are emitted on the stagnation streamline approaching a two-dimensional body with a round nose. The patches decelerate as they approach the body. The wide streak- line, which is also a streamtube in this steady flow, widens as it nears the body, thus also illustrating the decrease of velocity. The stagnation point itself is a singular point, so there is no finite region where one can observe zero velocity. The stagnation streamline, likewise, is a singular curve. To the side of the stagnation stream- line, the fluid first decelerates, then accelerates as it passes around the nose. (2) Water Flow in a Two-Dimensional Contraction. A closed horizontal duct of rectangular cross-section has a built-in contraction. At the upstream and downstream sections of the con- traction are installed static pressure manometers, and manometers connected to pitot tubes facing upstream. The nose of a pitot tube, where the pressure is measured, is a stagnation point. The stagna- tion pressures shown by the manometers attached to the pitot tubes are virtually equal, because the viscous forces are relatively small during the acceleration through the short contraction. However, the observed static pressure is less downstream of the contraction than upstream. From Bernoulli's integral, this is required by the ve- locity increase associated with the reduction of cross-sectional area. Consequently, the dynamic pressure (i.e. the difference between stagnation pressure and static pressure) increases through the contraction. The combination of pitot and static pressure meas- urements illustrated here is often used to measure velocity, be- cause the dynamic pressure is proportional to the square of the velocity.

CREDITS: Scenes from the film "Pressure Fields and Fluid Ac- celeration", by A. H. Shapiro, M.I.T.; Editor, R. Bergman, E.S.I.; Art Director, R. P. Larkin, E.S.I.; Director, A. H. Shapiro, M.I.T.

FM-34
The Coanda Effect

PURPOSE: To illustrate experimentally the attachment of a free jet to a convex surface (the "Coanda effect"), and to exhibit the pressure difference associated with the curvature of the jet.

THEORY: When a free jet flows around a convex surface, the pressure on the outside of the jet is that of the ambient atmosphere. Because of the streamline curvature, a normal pressure gradient is present which makes the pressure at the interface of the jet and the convex surface less than atmospheric. The net pressure force on an element of the jet is thus toward the convex surface, and this accounts for the net change of momentum flux of the jet as it bends around the surface. The net pressure force thereby induced on the body is toward the jet, because of the reduced pressure at the jet interface.

EXPERIMENTS: (1) Bending of a Rectangular Water Jet Near a Finger. A rectangular water jet discharges vertically downward from a nozzle. A finger is pushed horizontally into the jet. The jet bends toward the finger; that is, the jet "attaches" to the finger and bends around the convex surface of the finger. (a) Rectangular Water Jet Flowing Around a Circular Cylinder. In a similar experiment, the rectangular water jet passes near a horizontal, circular cylinder which hangs freely on pendulum supports. When the cylinder is swung into contact with the jet, it remains there against the force of gravity, owing to the Coanda effect. The jet bends around the cylinder. A pressure tap in the cylinder in the region of contact with the jet is connected to a manometer. The latter shows that the pressure at the cylinder surface is sub-atmospheric. This sub-atmospheric pressure at the jet-cylinder interface accounts for the force on the cylinder that holds the pendulum off the vertical. When the water is turned off, the pendulum swings back to the vertical. (3) A Heavy Sphere is Supported by a Jet Inclined Obliquely to the Vertical. An air jet about 3/16-inch in diameter issues at high speed from a nozzle. It is directed generally upward, but at about 30° from the vertical. A billiard ball held to the lower side of the jet is "sucked" into the jet, and can be supported completely by the jet. The equilibrium of the sphere is determined by the vectorial balance between the weight (vertically downward), the Coanda force (approximately normal to the jet), and the drag (approximately along the jet). (4) Application of Coanda Effect to an Aircraft Wing Flap. This experiment is performed on a free-surface water table marked by floating particles. A jet of water issues from the rear of an airfoil section. Also at the rear of the airfoil section, below the jet, is a movable high-lift flap. When the flap is not near the jet, the jet moves in a straight line. As the flap is moved up into the jet, the latter is suddenly "caught" by the flap, and flows in an attached manner around the convex surface of the flap. The jet thus energizes the boundary layer on the flap, and prevents flow separation. This is a practical illustration of how the Coanda effect may be used to guide a jet over a convex surface.

CREDITS: Scenes from the films "Pressure Fields and Fluid Acceleration", by A. H. Shapiro, and "Boundary Layer Control", by D. Hazen; Editor, R. Bergman, E.S.I.; Art Director, R. P. Larkin, E.S.I.; Director, A. H. Shapiro, M.I.T.

FM-35
Radial Flow Between Parallel Discs

PURPOSE: To illustrate by experiment and explain with Bernoulli integral why the radial flow between two parallel discs tends force the discs together. A parlor trick, involving a wooden sp and a sheet of paper, is based on this experiment.

APPARATUS AND PROCEDURE: A plastic disc is attached pependicularly to a tube such that air passing through the tube issu at the center of the disc. The disc is placed above and parallel to second disc such that the air from the tube flows radially outwa in the gap between the discs. In the first experiment the lower di is light and free to move upward. In the second experiment, t lower disc is stationary and contains three spacers which establi the plate separation; moreover, the lower disc is instrumented wi pressure taps connected to manometer tubes which show how th static pressure varies with radial position. The manometer tube rise vertically from a common reservoir; as the pressure at static tap falls, the corresponding manometer level rises, a vice versa.

SEQUENCES: (1) The upper disc is above the loose lower dis When the air is turned on, the lower disc is lifted up to a positi close to the upper disc. When the air is shut off, the lower di falls to its original position. (2) The upper disc is placed on to of the instrumented lower disc. The air flow is turned on. At th axis, the pressure is greater than atmospheric. Over most of th disc, however, the pressure is less than atmospheric, and rise to atmospheric at the outer edge. (3) Reprise of sequence (1).

INTERPRETATION: The center of the lower disc is a stagnatio region, and is at a pressure greater than atmospheric by the amount of the dynamic pressure of the jet. Beyond a radial distance of a few gap spacings from the center, the velocity is radia and the flow approximates a two-dimensional source flow. The cross-sectional area increases with radius; by continuity, the velocity decreases as the radius increases. Both viscous force and pressure forces are present, but the latter dominate. Hence the fluid deceleration is balanced mainly by the pressure gradient and, according to Bernoulli's integral, the pressure increases wit radius. Since the flow escapes subsonically to the atmosphere the pressure at the outer edge of the discs is atmospheric. Over most of the disc area, therefore, the pressure is less than atmospheric. Thus the two discs are forced toward each other because of the sub-atmospheric pressure between them.

QUESTION: What establishes the plate separation when the lower disc is loose?

CREDITS: Experiments from PRESSURE FIELDS AND FLUID ACCELERATION: Consultant and Director, A. H. Shapiro, M.I.T.; Editor, R. Bergman, E.S.I.; Art Director, R. P. Larkin, E.S.I.

enturi Passage

PURPOSE: To show the incompressible flows and pressure dis- butions in venturi passages. Further, to illustrate use of a nturi as a jet pump. Further, to illustrate cavitation at the throat a venturi.

ART I (PRESSURE DISTRIBUTION AND FLOW). (1) Schematic horizontal test section used for showing pressure distribution. ater flows horizontally through a two-dimensional venturi having e curved wall and one straight wall. Pressure taps in the raight wall lead to a bank of manometer tubes open at the top d inclined at 45° to the horizontal. (2) Apparatus is viewed from ove. One sees the plan view of the venturi and a projected view the manometer bank, the tubes being aligned with their corre- onding pressure taps. The venturi has a small divergence angle. the flow begins, the pressures at points along the venturi change; e ultimate steady-state pressure distribution is then shown. The essure falls in the contraction, reaches a minimum somewhat wnstream of the throat, and rises again in the divergent section t does not recover its initial value owing to frictional losses. A similar experiment but with an excessively large divergence gle. The pressure distribution is roughly similar, but the pres- ure recovery from throat to exit is only about half its previous lue. (4) Hydrogen bubble visualization of the flow in a diffuser small divergence angle. The flow fills the passage. (5) Similar sualization with very large divergence angle. There is backflow ong the walls, and the flow from the throat emerges as a jet.

ART II (VENTURI USED AS A JET PUMP). (1) Schematic sketch f a venturi tube supplied with air at a pressure above atmospheric, nd discharging to atmospheric pressure at the exit of the diffuser. he venturi throat is connected to a tube dipping vertically down- ard into an open beaker of dyed water. (2) Before the air flow tarts, the liquid levels are the same in the vertical tube and eaker. When the air is turned on, the water rises in the tube, howing that the pressure at the venturi throat is sub-atmospheric; uis comes about because there is a large pressure drop associated ith the air acceleration in the contraction, followed by a rise to tmospheric pressure because of deceleration in the diffusing ection. As the driving air pressure is increased, the throat pres- ure continues to decrease, until the water enters the throat and atomized by the air stream. Venturis are used as jet pumps in ectors, injectors, aspirators, atomizers.

ART III (CAVITATION). (1) A horizontal two-dimensional venturi assage, carrying water, discharges to the atmosphere. As the riving pressure and the flow increase, a vacuum gage shows the ressure at the throat decreasing below atmospheric. When the acuum reaches nearly 30" Hg, cavitation zones (seen as dark egions) appear near the throat. These are regions of vapor aused by boiling of the water when the pressure goes below the apor pressure. Collapse on the walls of the vapor bubbles in the iffuser can cause great mechanical damage.

CREDITS: Scenes from the film "Pressure Fields and Fluid Ac- eleration", by A. H. Shapiro, M.I.T.; Editor, J. Hirschfeld, E.S.I.; rt Director, R. P. Larkin, E.S.I.; Director, A. H. Shapiro, M.I.T.

FM-37
Streamline Curvature and Normal Pressure Gradient

PURPOSE: To illustrate by experiments that in a steady inviscid flow the pressure gradient normal to the streamline is related to the streamline curvature.

THEORY: Euler's equation for steady, inviscid motion without body forces, in the direction n normal to the streamline and in the plane of principal curvature, is $\partial p/\partial n = \rho V^2/R$. Here p is pressure, n is measured outward from the center of curvature, ρ is mass density, V is velocity, and R is the principal radius of curvature. Thus, with curved streamlines, p always increases in the direction outward from the center of curvature.

EXPERIMENTS: (1) Flow in a Bend. Water flows around a hori- zontal bend in a closed duct of rectangular cross-section. A group of three static pressure manometers is placed at the straight sec- tion upstream of the bend, a second group is placed at the center of the bend, and a third group is placed at the straight section down- stream of the bend. Each group of three manometers is spaced laterally across the duct. In addition, a single manometer tube placed near the upstream group of three shows the stagnation pres- sure in the supply tank. The difference between this stagnation pressure and the static pressures in the tubes approaching the bend is a measure of the average velocity pressure of the flow. Up- stream of the bend, the pressure is seen to be uniform across the duct; there is no streamline curvature to induce a normal pressure gradient. Downstream of the bend, the pressure is almost uniform across the duct; the slight variation observed is due to a compli- cated secondary flow created by boundary layer effects in the bend. At the center of the bend, where the streamlines are curved, the pressure is seen to vary normal to the streamlines, and is higher on the outside. (2) Solid-Body Rotation. A closed circular cylinder containing water is rotated about its axis, which is vertical, until solid-body rotation is attained. Static pressure manometer tubes in the cover of the cylinder show how static pressure varies with radius. As the water in the cylinder accelerates, a radial pressure gradient develops because of the circular streamlines. The tan- gential velocity in the steady state varies directly with radius, while the pressure increases according to the square of the radius, giving a parabolic shape to the pressure distribution. (3) Parallel Flow with Non-Uniform Speed. A rectangular channel is divided into two halves by a splitter plate aligned with the flow. One side has a large flow resistance installed, with the result that when the two streams rejoin downstream of the splitter, they flow side by side but have different speeds. Static pressure manometers show that the two streams, after rejoining, have the same pressure. How- ever, pitot tubes show a difference in stagnation pressure and, hence, of velocity. Although the velocities are unequal, the static pressures are equal, because there is no streamline curvature to induce a pressure gradient normal to the streamlines. Similarly, in a viscous boundary layer on a wall, just as in this example of a free boundary layer, the pressure is nearly constant across the layer, even though the velocity is reduced to zero at the wall.

CREDITS: Scenes from the film "Pressure Fields and Fluid Ac- celeration" by A. H. Shapiro, M.I.T.; Editor, R. Bergman, E.S.I.; Art Director, R. P. Larkin, E.S.I.; Director, A. H. Shapiro, M.I.T.

FM-38
Streamwise Pressure Gradient in Inviscid Flow

PURPOSE: The object is to illustrate by experiment the one-dimensional (or stream filament) approximation to the dynamics of a steady incompressible flow in circumstances where viscosity is relatively unimportant. Changes of pressure and velocity in the streamwise direction only are considered. Attention is focused (a) on the connection between the duct geometry and the average velocity; (b) on the dynamical relation between the pressure gradient and the acceleration.

THEORY: Streamwise variations in velocity are governed by the one-dimensional equation of continuity, which is, for an incompressible flow, $Q = AV$; this shows that the velocity increases when the cross-sectional area decreases, and vice versa. Once the velocity variation is determined by the geometry of the duct, the pressure variation is determined by Euler's streamwise equation of steady inviscid motion without body forces, $\partial p/\partial s = -\rho V\, \partial V/\partial s$; this shows that a streamwise pressure decrease is associated with a streamwise velocity increase, and vice versa. Thus, an area decrease yields a pressure decrease, and vice versa.

EXPERIMENTS: (1) Two-dimensional Contraction. Successive patches of hydrogen bubbles, electrolysed at a wire, mark water flowing in an open, horizontal duct. The bubble patches show that the water accelerates as it flows through the contraction. (2) Two-dimensional Contraction. In the same apparatus, hydrogen bubbles emitted from ten wires spaced laterally across the duct generate ten streaklines in the contraction; since the flow is steady, these are also streamlines. The spacing between streamlines generally decreases from upstream to downstream, thus indicating a general velocity increase along each streamline. Near the curved wall, the streamline spacing increases at two locations, thus indicating local velocity decreases resulting from the flow being two-dimensional rather than strictly one-dimensional. (3) Two-dimensional Symmetrical Contraction. Water flows in a horizontal, closed duct. Manometer tubes show the static pressure upstream and downstream of the contraction. The static pressure decreases, corresponding dynamically to the velocity increase. (4) Flow Withdrawal from Manifold. Water flows through a pipe of constant area. At each of four cross-sections spaced along the duct, a series of holes around the circumference lead to a collecting ring from which water may be bled out through a valve. The design of the withdrawal system is such as to minimize head losses resulting from viscosity, flow separation, and eddies. Manometer tubes displaying the static pressure are arranged upstream and downstream of each bleed ring. (a) Bleed valves closed. With no water withdrawal, the pressure is nearly constant along the length of the tube, the slight pressure drop being due to friction. (b) Bleed valves opened. When water is withdrawn, the pressure rises along the length of the tube. This results dynamically from the streamwise deceleration caused by a reduction of volume flow through a tube of constant area.

CREDITS: Scenes from the film "Pressure Fields and Fluid Acceleration", by A. H. Shapiro, M.I.T.; Editor, R. Bergman, E.S.I.; Art Director, R. P. Larkin, E.S.I.; Director, A. H. Shapiro, M.I.T.

FM-39
Interpretation of Flow Using Wall Tufts

PURPOSE: To demonstrate some uses of tufts (of wool or similar materials) attached to a surface for interpretation of flow patterns. The tufts can indicate both the type of flow and the streamlines near the surface.

APPARATUS: Two experimental pieces of apparatus are used: (i) An airfoil with tufts attached to the upper surface in a low-speed wind tunnel. (ii) A water channel with wool tufts attached to a flat side surface; tufts were made neutrally buoyant.

FILM SEQUENCES: (1) Introductory view of airfoil with tufts in wind tunnel. (2) Sketch of tuft attached to surface. The tuft is very light and has a large surface area. If the tuft is free to pivot about the point of attachment, the drag forces applied by a moving fluid tend to align it with the instantaneous direction of flow. The unattached end of each tuft in the water channel is marked with a white band. (3) Side view of tufts attached to vertical wall of water channel with flow from left to right. (4) In laminar flow the tufts are virtually motionless. (5) In turbulent flow the tufts wiggle over a small range of angles about a mean value. (6) Turbulent flow over the upper surface of the airfoil. (7) Intermittent formation and subsequent washing away of reversed flow (transitory stall) on the upper surface of the airfoil. The tufts flop back and forth, completely reversing direction. (8, 9) Sketch and tuft visualization of steady two-dimensional separation along a flat surface. The separation was produced by placing a step about 2 boundary layer thicknesses high on the wall of the channel downstream of the field of view. Separation is indicated by the fact that the tufts on either side of a separation-streamline point toward each other. (10, 11) Sketch and tuft visualization of steady, two-dimensional reattachment along a flat surface. The reattachment of the flow occurs downstream of essentially the same obstruction used in (9); the obstruction is now upstream of the field of view. Tufts on either side of a dividing reattachment streamline point away from each other. Dye emitted through holes on opposite sides of the line of reattachment provides further visualization of the directions of flow.

REFERENCES: (1) D. L. Cochran, S. J. Kline, NACA TN 4309 (1958). (2) E. C. Maskell, Report AERO 2565 Bri. RAE (1956).

CREDITS: Experiments and film by T. B. Morrow, J. R. MacDonald and S. J. Kline, Stanford University and excerpts from the film "Boundary Layer Control", D. C. Hazen, Princeton University. Consultant, A. H. Shapiro, M.I.T.; Producer, R. Bergman, E.D.C.; Editor, W. Tannebring, E.D.C.; Art Director, R. P. Larkin, E.D.C. Produced by the National Committee for Fluid Mechanics Films and Education Development Center, Inc. with the support of the National Science Foundation.

FM-43
Buoyancy-Induced Waves in Rotating Fluid

PURPOSE: To show "jet-stream" waves analogous to large-scale atmospheric flows resulting from coupled buoyancy and Coriolis forces in a rotating fluid.

APPARTUS: Water is contained in the circular annulus of a cylindrical open vessel which has a central cylindrical core. The entire vessel is rotated (counterclockwise as seen from above) with steady angular speed about the vertical axis of symmetry. A heater ring around the vessel maintains the outer wall of the annulus at high temperature. Cold water circulated through the central core maintains the inner wall of the annulus at low temperature. The density differences thus created produce (with the vessel not rotating) a motion which is radially inward near the free water surface, downward near the inner annulus wall, radially outward near the bottom of the vessel, and upward near the outer annulus wall. Rotation forces the development of velocity components in the tangential direction which may be either axially symmetrical or disturbed by waves. The motion is observed from above using a rotating camera aligned with the axis. The water is marked either by powder on the surface or by colored inks. Except as noted, the camera rotates with the vessel; thus one sees the motion relative to the rotating fluid. The whole apparatus roughly simulates the atmospheric conditions in the Northern Hemisphere, with heating near the equator and cooling near the North Pole.

SEQUENCES: (1) Sketch of apparatus. (2) With Rossby No. (average jet speed divided by rim speed of cylinder) = 0.25, a Rossby-wave pattern with 3 lobes propagates slowly toward the east, i.e. counterclockwise. A "jet stream" is visible at the top surface. (3) With higher rotational speed, giving a lower Rossby No. = 0.20, the wave pattern has 4 lobes ($n = 4$). (4) With the mode $n = 4$, the flow details are observed with the camera rotating at a speed to make the wave appear stationary. Red ink near the surface shows the high-speed jet. Blue ink, after it sinks, shows velocity patterns corresponding to a sea level map. (5), (6), (7), (8) Depending upon annulus width, mean diameter, rotative speed, and temperature difference, Rossby wave modes with $n = 4$, $n = 5$, $n = 6$ and $n = 10$ are shown using internal powder tracer, which shows the upper-level flow patterns.

DISCUSSION: The type of instability leading to the final wave state is called "baroclinic instability". There is a strong kinetic constraint against axially symmetric flows due to the conservation of circulation for horizontal circular fluid rings. This constraint is much weaker for wave modes with horizontal side-by-side, inward and outward currents. The disturbed density field gives buoyancy induced circulations in vertical planes in the tangential direction that draw on gravitational potential energy to amplify or support the wave motion.

REFERENCES: (1) Fultz, D., et al (1959): *Meteor. Monogr.*, 4, No. 21. (2) Kuo, H. L., (1957): *J. Meteor.*, 14, 553-558. (3) Brindley, J. (1960): *Phil. Trans. Roy. Soc. London* (A), 253, 1-25. CREDITS: Scenes from a film by D. Fultz and R. Kaylor, University of Chicago, and from the film "Planetary Circulation of the Atmosphere" (a production of the American Meteorological Society) by R. Hide, M.I.T.; Consultant, A. H. Shapiro, M.I.T.; Editor, J. Hirschfeld, E.D.C.; Art Director, R. P. Larkin, E.D.C. Produced by the National Committee for Fluid Mechanics Films and Education Development Center (formerly Educational Services Incorporated), with the support of the National Science Foundation.

FM-44
Elastoid-Inertia Oscillations in a Rotating Fluid

PURPOSE: To illustrate some of the possible standing oscillations in a perturbed rotating fluid.

APPARATUS AND PROCEDURE: Water (depth 8.2 cm) in a right circular cylinder (radius 4.2 cm) rotating about its vertical axis of symmetry is perturbed by oscillating a small circular disk up and down on the axis. Ink is released from a hypodermic in the tube carrying the disk.

SEQUENCES: (1) Cylinder not rotating, ink released from edge of disk. Note vortex ring shedding from disk and eventual irregular spreading of ink throughout cylinder. Considerable time elapses during fadeout in mid-sequence. (2) Fundamental $(n_\theta, n_r, n_z) = (0,1,1)$ axisymmetric elastoid-inertia mode with rotation period $\tau = 2.52$ sec and oscillation period $\tau_{os} = 1.32\tau$, observed at 1/2 natural speed. n_θ is the number of wavelengths of any perturbation quantity in azimuth, n_r the number of half-wavelengths radially (unchanged in remaining sequences), and n_z is the number of half-wavelengths between bottom and top. Ink was released from an axis point and, in contrast to sequence (1), remains confined to a limited cylindrical column except for gradual spread due to viscous secondary flows. Note out-of-phase radial contraction of ink column above and below disk, corresponding axial extension or compression, and spin acceleration or deceleration visible in ink details below disk. This is a manifestation of the fundamental kinetic mechanism for the oscillations which depends on the material conservation of vortex tube strength in the Helmholtz vortex theorems. (3) Harmonic axisymmetric mode $(0,1,2)$ with nodal plane at mid-depth with $\tau = 3.94$ sec and $\tau_{os} = 0.79\tau$, observed at 1/2 natural speed. The generator disk is at the upper cell anti-node. The outer canopy of ink reached its position due to secondary flow. Note out-of-phase radial displacements between bottom and top half in oscillations from barrel to hour-glass shape. Jagged pattern at surface of ink column is residual from vortex rings shed by disk. (4) Next axisymmetric mode $(0,1,3)$ with 2 nodal planes, $\tau = 3.92$ sec, and $\tau_{os} = 0.64\tau$, observed at 2/3 natural speed. As n_z increases, τ_{os} decreases toward a limit $1/2\tau$ (the half-pendulum day) for nodal surfaces spaced at distances small compared to the radius. For the modes (not shown) where n_r increases, τ_{os} increases (contrary to the usual behavior of higher harmonics), and is ultimately proportionally to n_r. (5) Oblique view of asymmetric mode $(5,1,3)$ with $\tau = 3.91$ sec and $\tau_{os} = 0.65\tau$, observed at 2/3 natural speed. The pentagonal shape of the ink column is produced by spontaneous excitation of 5 azimuthal wavelengths (no change in generator disk). Asymmetric modes are progressive and occur in rotationally split pairs which propagate in opposite directions relative to the cylinder. This mode has a positive angular phase speed (i.e., rotates faster than the cylinder).

REFERENCES: D. Fultz (1959): J. Meteor., 16, 199; S. Chandrasekhar (1961): Hydrodynamic and Hydromagnetic Stability, Oxford, p. 272 ff. See these for original Kelvin and Bjerknes references.

CREDITS: Experiments and film by D. Fultz, Univ. of Chicago; Director, A. H. Shapiro, M.I.T.; Editor, J. Hirschfeld, E.S.I.; Art Director, R. P. Larkin, E.S.I.

FM-45
Velocities Near an Airfoil

PURPOSE: To illustrate the relation between velocities and lift on an airfoil.

APPARATUS: Two apparatuses are used. The first is a conventional low-speed wind tunnel equipped with smoke visualization; view is from side through tunnel window. The second is an open-surface water channel; view is from above; visualization employs sheets of hydrogen bubbles made by electrolysis at a fine wire (ref. 1). The bubble motions show streaklines, and the bubble sheets are interrupted to provide widely-spaced timelines.

SEQUENCES: (1) Flow pattern at approximately zero lift (zero circulation). Pattern is shown by streaklines of smoke in the wind tunnel. (2) A similar picture is shown in the second apparatus for much lower Reynolds number. The timelines rejoin except for the wake region at the trailing edge; this shows that the average speed is approximately the same above and below the airfoil. (3) Two flow patterns at positive lift around the airfoil are shown using smoke in the air tunnel; the angle of incidence increases in two steps showing first an unstalled and then a stalled flow pattern. (4) The pattern at positive lift with unstalled flow is shown in the water apparatus. The timeline over the top surface outruns that beneath the airfoil; thus the flow has higher average speed over the top surface than beneath. (5) The airfoil is shown with negative lift in the water apparatus. The timeline beneath the airfoil now outruns that above it.

QUESTIONS: (1) How can the bubble timeline pictures be interpreted to indicate the lifting force on the airfoil? (2) How is it possible in a potential flow for part of a timeline to go faster than another part? (3) A brochure provided to passengers by one commercial airline states that aerodynamic lift comes about because fluid particles above and below the wing must traverse the length of the wing in the same time, and, because of camber, the speed above the wing is therefore less than that below; do you agree?

REFERENCES: 1. "Use of Hydrogen Bubbles for Quantitative Determination of Time-Dependent Velocity Fields in Low-Speed Water Flows", Schraub, Kline, Henry, Runstadler and Littell, TASME, Series D, Vol. 87, June 1965.

CREDITS: Scenes from the films "Flow Visualization" by S. J. Kline, Stanford University and "Boundary Layer Control" by D. C. Hazen, Princeton University; Consultants, S. J. Kline, Stanford University and A. H. Shapiro, M.I.T.; Director, J. Churchill, E.S.I.; Editor, R. Bergman, E.S.I.; Art Director, R. P. Larkin, E.S.I.

FM-46
Current-Induced Instabilities of a Mercury Jet

PURPOSE: To exhibit two forms of instability in a current-carrying jet in an applied axial magnetic field.

APPARATUS AND PROCEDURE: A liquid jet of mercury, 4 mm in diameter, falls vertically from a copper vessel to a copper plate 14 cm below. A 6-volt battery supplies a current of up to 400 amps which flows along the mercury jet. Solenoids above and below can apply a vertical (axial) magnetic field, variable up to 8000 gauss. The jet behavior is photographed at 3300 frames/sec; the motion appears slowed down by a factor of 200. The pendulum and falling droplets at the left of the picture provide a time scale.

SEQUENCES: (1) With no applied magnetic field ($\bar{B}=0$), and with currents of 200 amps and 300 amps, a "sausage" instability occurs. (2) With a weak applied magnetic field ($\bar{B}=300$ gauss), and with currents of 100 amps and 300 amps, a "corkscrew" instability occurs. (3) With a strong applied magnetic field ($\bar{B}=8000$ gauss), and with a current of 300 amps, the corkscrew instability is so strong that the jet breaks up, thus interrupting the current until a new section of jet reaches the copper plate.

DISCUSSION: (a) No Applied Field. The current in the jet itself creates an azimuthal $\bar{B}$ field which varies as $1/r$. If the jet diameter locally is perturbed to a smaller diameter, the local values of $\bar{J}$ and $\bar{B}$ both increase; the radial $\bar{J} \times \bar{B}$ force is accordingly increased, thus squeezing the jet further. Liquid is pushed away from the contracted section, and forms a swelling above and below, which augments the instability. Periodic sausage-like bulges are produced, much like those due to surface tension in Rayleigh instability. (b) With Applied Vertical (Axial) Field. Suppose the jet is locally perturbed by a kink in a vertical plane. The vertical $\bar{B}$ field of the solenoids then acts with the radial components of $\bar{J}$ to create azimuthal $\bar{J} \times \bar{B}$ forces. The latter forces are of opposite direction for the upper and lower parts of the kink; hence the upper and lower parts move azimuthally in opposite directions, giving the jet a corkscrew character. The azimuthal component of $\bar{J}$ thus generated interacts with the vertical $\bar{B}$ to create a radial $\bar{J} \times \bar{B}$ force which makes the helix accelerate outwards.

REFERENCE: Dattner, A.: "Current-induced instabilities of a mercury jet", Arkiv f. Fysik, 21, 7, pp. 71-80, 1962).

CREDITS: Experiments and film by A. Dattner, Royal Institute of Technology, Stockholm; Consultant and Director, A. H. Shapiro, M.I.T.; Editor, E. Carini, E.S.I.; Art Director, R. P. Larkin, E.S.I.

FM-47
Pathlines, Streaklines, Streamlines and Timelines in Steady Flow

PURPOSE: (1) To define and illustrate the concepts of pathline, streakline and timeline. (2) To show the relationships between pathlines, streaklines and streamlines in steady flow. (3) To illustrate several ways of making velocity patterns visible with streaklines, pathlines, timelines and combined-time-streak markers.

APPARATUS: One-sided contraction in an open surface water channel. Views are from above; black masks are used so that only the flow inside the channel is seen. Flow is from left to right at about 1/3 foot per second in entry. Visualization is based on hydrogen bubbles generated by electrolysis at wires of various shape (see ref. 1).

SEQUENCES: (1) A pathline is illustrated by the motion of a single particle; the pathline is marked by an optically superposed line. Note that a pathline is defined as the locus of points through which a single particle passes; the superposed line makes all the points visible at once. (2) Comparison of the motion of several elements with the previously marked pathline shows that successive pathlines coincide in steady flow. (3) A streakline is illustrated by continuous generation of marking bubbles at a fixed location; a streakline is defined as the locus of all particles passing through a fixed point in space. Comparison with the pathline established in (1) shows that the streakline generated by particles from the same initial point coincides with the pathline. (4) Multiple streaklines are shown, and it is noted they show the streamline pattern in steady flow. (5) Timelines are illustrated by pulsing a bubble generation wire normal to the flow; the definition of a timeline is the instantaneous location of a line of fluid particles marked at a given earlier time. (6) Combined-time streak markers are illustrated.

QUESTIONS: (1) Can the velocity pattern of a flow be found from streaklines alone? Does your answer apply to pathlines and timelines also? (2) In a steady flow, how many different ways can you devise to find the velocity pattern from a set of combined-time-streak markers? (3) A solid line of dye is introduced at one point in a flow field through a small tube. The flow is understood to be in oscillatory motion superposed on a steady through flow at the injection point. The dye is observed and does take on an oscillatory motion in one plane. It is proposed to use the spacing of the dye oscillations divided by the mean speed of the flow to obtain the period of oscillation of the flow. Will the result obtained be reliable?

REFERENCES: (1) "Use of Hydrogen Bubbles for Quantitative Determination of Time-Dependent Velocity Fields in Low-Speed Water Flows", Schraub, Kline, Henry, Runstadler and Littell, ASME, Series D, Vol. 87, June 1965.

CREDITS: Scenes from the film "Flow Visualization" by S. J. Kline, Stanford University; Consultants, S. J. Kline, Stanford University and A. H. Shapiro, M.I.T.; Director, J. Churchill, E.S.I.; Editor, R. Bergman, E.S.I.; Art Director, R. P. Larkin, E.S.I.

FM-48
Pathlines, Streaklines and Streamlines in Unsteady Flow

PURPOSE: To illustrate the differences between streaklines, streamlines and pathlines in unsteady flow. (Refer to loop FM-47 for steady flow.)

APPARATUS: An oscillating flat plate creates a periodic flow in an open surface water channel. Flow visualization is by the hydrogen bubble method (ref. 1). View is from above open surface. Flow velocities are roughly one half foot per second.

SEQUENCES: (1) Flow field about the oscillating plate is made visible using a "grainy" sheet of bubbles. (2) A pathline is shown by releasing a single bubble patch from an initial point; the pathline is marked by an optically superposed line. (3) A second pathline, passing through the same initial point at a different time in the cycle, is similarly marked. (4) The two distinct pathlines are compared. (5) The continuous succession of time-dependent streaklines, all originating from the same initial point as the pathlines of (2) and (3) are produced by releasing bubbles continuously from that point; every instantaneous streakline is seen to differ from each of the pathlines marked previously. (6) A coarse grid of small combined-time-streak markers is introduced. Two still frames slightly separated in time are superposed optically, and the direction field of the flow is used to construct the instantaneous streamline field from the superposed pictures. (7) The particular instantaneous streamline through the initial point of the streaklines and pathlines of (2) and (3) is marked (note: It would be different at every instant); it is compared first to the two pathlines and then to the streaklines and is seen to differ from all of these. (8) A number of streaklines are shown continuously; they move as a function of time. The title states that motion of the streakline normal to itself is a telltale which signifies unsteady motion and indicates caution is necessary in interpreting streakline patterns; in particular, they cannot usually be interpreted as streamlines.

QUESTIONS: (1) Can any streakline or pathline coincide with a streamline in unsteady flow? Can the complete velocity field in unsteady flow be found from streaklines or pathlines alone? If not, what is needed? (2) As a single element approaches the plate in the pathline sequence, very strong deformation of the element is seen to occur; can this strong deformation be a potential flow? (3) Compare the sequences in this loop with those of loop FM-47, "Pathlines, Streaklines, Streamlines and Timelines in Steady Flow". What are the essential differences and similarities?

REFERENCES: (1) "Use of Hydrogen Bubbles for Quantitative Determination of Time-Dependent Velocity Fields in Low-Speed Water Flows", Schraub, Kline, Henry, Runstadler and Littell, TASME, Series D, Vol. 87, June 1965.

CREDITS: Scenes from the film "Flow Visualization" by S. J. Kline, Stanford University; Consultants, S. J. Kline, Stanford University and A. H. Shapiro, M.I.T.; Director, J. Churchill, E.S.I.; Editor, R. Bergman, E.S.I.; Art Director, R. P. Larkin, E.S.I.

FM-49
Flow Regimes in Subsonic Diffusers

PURPOSE: To illustrate the major patterns of flow in subsonic diffusers.

APPARATUS: Horizontal, open-surface water table with adjustable vertical walls. The view is looking down on the open table with flow from left to right. Inlet conditions are the same in all sequences. The mean speed is roughly one-half foot per second at the inlet. Flow is made visible by hydrogen bubbles from electrolysis at a wire (refs. 1 and 2). Black masks are used so that only the channel is seen except at the far right where the flow leaves the diffuser and enters a plenum.

SEQUENCES: (1) Flow in the passage with zero divergence angle between the walls; the flow is unstalled and the boundary layer is laminar. (2) The vertical walls are pivoted outward around the throat section to a total included angle of about 7°; the flow remains essentially unstalled, but the boundary layer is now turbulent. (3) A further increase in divergence angle between the walls changes the flow pattern to transitory stall. The flow is grossly unsteady; large areas of stall develop and then wash out. This process will repeat indefinitely; no steady flow pattern exists with geometries in this region. (4) Further increase in divergence angle causes a change to a fully-developed stall; it is a relatively steady, asymmetric flow. During this sequence the positions of the throughflow and of the recirculating eddy are interchanged by temporary insertion of a vane deflector. Once this is done the pattern remains reversed, illustrating the bi-stable nature of this type of flow. (5) An increase in divergence angle to very large values causes separation from both walls, creating a jet flow pattern. (6) One means of stall control, short vanes, is illustrated.

DISCUSSION: Only one kind of diffuser and fixed values of many parameters are displayed, but the patterns illustrate the important features of flow in almost all geometries of subsonic diffusing ducts for a very wide range of the controlling parameters. The four flow patterns observed (unstalled, transitory stall, fully-developed stall, jet flow) are determined primarily by angle and area ratio; the flow pattern is only very weakly dependent on other parameters. More complete data maps of flow regimes and discussion of controlling parameters are in refs. 3 and 5. An extensive discussion of flow separation in ducts is in ref. 4. Ref. 1 illustrates the use of flow pattern correlations for engineering design and for formulation of mathematical models.

QUESTIONS: Can you visualize use of the bi-stable property of fully stalled flows as a switching element? Can you think of an application where the large oscillations of transitory stall might be desirable?

REFERENCES:
1. "Flow Visualization " 1/2 hour, 16 mm, sound film by NCFMF; Educational Services Inc., Watertown, Mass.
2. Use of Hydrogen Bubbles for Quantitative Determination of Time-Dependent Velocity Fields in Low-Speed Water Flows, Schraub, Kline, Henry, Runstadler and Littell, TASME, Series D, Vol. 87, 1965.
3. Optimum Design of Straight Walled Diffusers, Kline, Abbott and Fox, TASME, Series D, Vol. 81, 1959.
4. On the Nature of Stall, S. J. Kline, TASME, Vol. 81, Series D, 1959.
5. Flow Regimes in Curved Subsonic Diffusers, R. W. Fox and S. J. Kline, TASME, Series D, Vol. 84, 1962.

CREDITS: Scenes from the film "Flow Visualization" by S. J. Kline, Stanford University; Consultants, S. J. Kline, Stanford University and A. H. Shapiro, M.I.T.; Director, J. Churchill, E.S.I.; Editor, R. Bergman, E.S.I.; Art Director, R. P. Larkin, E.S.I.

FM-50
Flow Over an Upstream Facing Step

PURPOSE: To show the flow pattern, separation region, and trapped vortex when a high-Reynolds-number flow along a straight wall meets (1) an upstream-facing two-dimensional step, and (2) a two-dimensional jet issuing perpendicularly to the flow from a slit nozzle in the straight wall.

APPARATUS AND PROCEDURE: Water flows horizontally on the upper side of a flat plate. The flow is obstructed either by an upstream-facing step or by a two-dimensional jet of water issuing vertically from the plate. The flow is visualized either (a) by light scattered from fine air bubbles dispersed in the water, or (b) by injection of colored dyes.

SEQUENCES: (Experiments with Step): (1) Flow over step is visualized by air bubbles in the water. (2) Explanatory schematic of same, showing: (a) the main streamlines; (b) the line of separation between the external flow passing over the step and the trapped recirculating flow near the concave corner of the step; (c) the position of laminar boundary layer detachment; (d) the main vortex trapped in the concave corner; (e) the position of reattachment of the main flow on the vertical face of the step; (f) longitudinal velocity profiles showing, as the flow approaches the step, reverse flow near the wall. The separation of the main flow and the recirculating eddy near the plate are the results of boundary-layer separation and back flow in an adverse pressure gradient associated with the tendency for a concave corner to be a stagnation point in a potential flow. The consequent flow pattern is produced by a coupling between the inviscid external flow and the low-momentum, viscous boundary layer flow. (3) Three scenes of similar experiments in close-up, with dye injection to show additional details. To the left of the main trapped vortex (rotating clockwise) may be seen a smaller trapped vortex (rotating anticlockwise).

SEQUENCES: (Experiments with Perpendicular Jet): (1) At first the water flows undisturbed along the flat plate, with the flow visualized by air bubbles. The jet is then turned on. It is bent downstream by the momentum of the main flow, and acts as a surface of separation which causes the main flow to pass, as it were, over a rounded step. However, it does not produce exactly the same effects as a solid step of the same shape because the jet itself entrains much of the boundary layer flow approaching it along the plate. The character of the total flow pattern is created by the interaction of the boundary layer flow with the main inviscid flow and by the entrainment in the jet. Upstream of the jet is a separation bubble containing a trapped vortex. Separation occurs because of the adverse pressure gradient produced by the turning of the main flow as it approaches the jet. The separation region is closed, on the other hand, by suction of the boundary layer into the jet itself. (2) A close-up view of same, with dye injection to show details of the separated region.

CREDITS: Experiments and film by Office National d'Études et de Recherches Aérospatiales (O.N.E.R.A.), France; Consultants, R. Fox, Purdue University, and A. H. Shapiro, M.I.T.; Director, R. Bergman, E.S.I.; Editor, E. Carini, E.S.I.; Art Director, R. P. Larkin, E.S.I. Produced by the National Committee for Fluid Mechanics Films and Educational Services Incorporated, with the support of the National Science Foundation.

FM-51
Simple Supersonic Inlet
(Axially Symmetric Geometry)

PURPOSE: To show the choked and unchoked operating regimes of a simple duct inlet at supersonic speeds.

APPARATUS & PROCEDURE: The wind tunnel has a horizontal test section with Mach Number 1.8. An air intake is simulated by a tube aligned with the flow, with a diffusing section at its forward end, and an adjustable nozzle at its rear end. The latter permits variations in internal resistance, such as might be produced by a propulsion engine with a variable propulsion nozzle. Schlieren photography, with the knife edge perpendicular to the flow (i.e., vertical) is used for visualization.

SEQUENCES: (1) Schematic drawing of wind tunnel, model, variable exit nozzle, and viewing windows. (2) Close-up of same, showing the portion of the flow field visible in the windows. (3) As the tunnel flow is turned on, the "starting" shock of the tunnel passes downstream over the model. The tunnel is said to be "started", and produces the operating Mach Number of 1.8. The outlet nozzle area in this run is large enough (or, alternatively, the internal resistance is low enough) for the central part of the bow shock on the model to be "swallowed"; 100% of the approach flow corresponding to the "capture area" of the inlet is actually captured by the inlet and passes through the model. This is the maximum amount of flow that can enter during supersonic flight. The exterior shock wave system is attached to the inlet lip. (4) Repeat of the established-flow portion of the foregoing sequence. (5) The internal resistance is now gradually increased by reducing the area of the exit nozzle. This ultimately reduces the amount of flow that can enter the inlet below that corresponding to the maximum. A detached bow shock wave then forms. This provides the necessary mechanism by which the flow in the "capture area" can partially spill over the lip of the inlet (note that the region downstream of the nearly-normal central portion of the bow shock is now subsonic). As the internal resistance continues to increase, the bow shock moves further upstream, thus permitting greater spillage and a lesser percentage of maximum flow actually captured. (6) The internal resistance now decreases, and the bow shock moves closer to the inlet plane until, when it becomes a normal shock in the inlet plane, it is said to be "swallowed"; at this condition the maximum possible inlet flow is once again captured.

CREDITS: Experiments and film by Office National d'Études et de Recherches Aérospatiales (O.N.E.R.A.), France; Consultant and Director, A. H. Shapiro, M.I.T.; Editor, E. Carini, E.S.I.; Art Director, R. P. Larkin, E.S.I.

FM-52
Supersonic Conical-Spike Inlet

PURPOSE: To show different regimes of operation, including "buzz", of a conical-spike inlet at supersonic speeds.

APPARATUS AND PROCEDURE: The wind tunnel has a horizontal test section with Mach Number 1.8. Installed in the test section is a tube aligned with the flow, with an annular conical-spike diffuser at its forward end, and an adjustable nozzle at its rearward end. The diffuser simulates an air intake, while the nozzle permits changes in internal resistance, such as might be produced by a propulsion engine having a variable fuel rate or variable propulsion nozzle. The wave system is made visible by the schlieren technique, with the knife edge perpendicular to the flow.

SEQUENCES: (1) Schematic drawing of wind tunnel, model, variable exit nozzle, and viewing windows. (2) Close-up of same, showing the portion of the flow field visible in the windows. (3) and (4) "Starting" of tunnel. The flow approaching the model is at first subsonic. As the driving pressure ratio is slowly increased, a normal shock wave in the wind tunnel nozzle gradually moves downstream. When this "starting" shock wave has passed into the test section and gone beyond the model, the tunnel is "started", and produces a Mach Number of 1.8 in the flow approaching the model. In this run the outlet-nozzle area is not large enough for the maximum flow to enter the inlet. Downstream of the conical shock from the spike can be seen another shock just upstream of the annular intake; this "detached" lip shock has subsonic flow behind it which allows some of the flow to spill around the inlet lip. (5) This run begins with a large enough outlet-nozzle area to permit capture by the inlet of the maximum flow; there is no lip shock. As the run proceeds, the outlet-nozzle area is gradually reduced. A lip shock forms and detaches, thereby allowing some of the flow to spill around the lip, and thus reducing the amount of flow captured. When the lip shock has moved out almost to the tip of the spike, a high frequency oscillation (commonly called "buzz") appears very suddenly. The lip shock oscillates rapidly back and forth with a large displacement, and produces an unsteady, more or less periodic flow, with large velocity and pressure amplitudes within the inlet. The "buzz" results from an instability produced by a combination of inertia in the approach flow and mass storage in the interior volume of the inlet. (6) Starting with the buzz condition, the outlet-nozzle area is gradually increased. The buzz disappears and the lip shock moves closer to the annulus lip.

CREDITS: Experiments and film by Office National d'Études et de Recherches Aérospatiales (O.N.E.R.A.), France; Consultants, R. W. Fox, Purdue University, and A. H. Shapiro, M.I.T.; Director, R. Bergman, E.S.I.; Editor, E. Carini, E.S.I.; Art Director, R. P. Larkin, E.S.I. Produced by the National Committee for Fluid Mechanics Films and Educational Services Incorporated, with the support of the National Science Foundation.

FM-54
Leading-Edge Separation Bubble in Two-Dimensional Flow

PURPOSE: To show boundary-layer separation and re-attachment near the nose of two-dimensional bodies of the type which produces a "separation bubble" in which there is a trapped zone of recirculating flow.

EXPERIMENTS: I. *Air flows past a thick plate with a semicircular nose.* Smoke streaklines show the streamline pattern. The main flow separates about where the circular nose joins the plate, but later re-attaches to the plate. A "separation bubble" is therefore formed, containing a region of recirculating, trapped fluid. II. *Air flows past a thin, symmetrical airfoil at zero incidence.* Smoke streak-lines show the flow. The pressure distribution on the air-foil is such that no separation bubble is present. III. *Water flows past a thin, symmetrical airfoil which has a jet flap.* Air bubbles dispersed in the water show the fluid motion. A "jet flap" at the trailing edge (a two-dimensional jet issuing downwards) produces circulation and lift. The re-sulting pressure distribution is such that a leading-edge separation bubble is present on the upper surface. Dye injected from a hole in the upper surface shows details of the flow in the separation bubble. Dye from a hole in the lower surface shows that there is no separation on the lower surface. IV. *Effect of drooping the leading edge.* Same experiment as in III, except that the leading edge is drooped. The effect of the droop is to eliminate the separa-tion bubble. This is associated with the change in pressure distribution. V. *Effect of drooping the leading edge.* Same experiment as in IV, except in a wind tunnel, with smoke streaklines.

DISCUSSION: The phenomenon of the leading-edge separa-tion bubble is associated with flows in which there is a sharp adverse pressure gradient a short distance aft of the forward stagnation point, where the boundary layer is still thin. Separation and backflow in the boundary layer initiate the separation bubble. The re-attachment results from a combination of two effects: (a) entrainment of the rela-tively small amount of "dead" fluid in the separation region by the high-speed external flow; and (b) a steep adverse pressure gradient followed by a less steep adverse pressure gradient, or even a favorable pressure gradient. The separation bubble itself alters the pressure distribu-tion, reducing adverse gradients. The separation bubble can be eliminated entirely through use of a profile shape which avoids adverse pressure gradients strong enough to produce boundary-layer separation.

CREDITS: Experiments and film from Office National d'Études et de Recherches Aérospatiales (O.N.E.R.A.), France, and from "Boundary Layer Control", by D. C. Hazen, Princeton University; Consultants, R. Fox, Purdue University, and A. H. Shapiro, M.I.T.; Director, R. Berg-man, E.S.I.; Editor, M. Chalufour, E.S.I.; Art Director, R. P. Larkin, E.S.I. Produced by the National Committee for Fluid Mechanics Films and Educational Services In-corporated, with the support of the National Science Foundation.

FM-55
Bow Waves in Hypersonic Flow

PURPOSE: To illustrate bow shock waves for several body shapes at very high Mach Numbers.

APPARATUS: The models are placed in supersonic free jets. Provision is made for changing the attitude of the model. The flow is visualized by schlieren photography.

EXPERIMENTS: I. *A 75°-delta glider re-entry model at $M = 7$.* The schlieren knife-edge is parallel to the free stream. A detached bow shock wave is formed. As the in-cidence increases, the shock on the lower (high-pressure) side becomes stronger and comes closer to the surface; while that on the upper (low-pressure) side becomes weak-er and moves away from the surface. On the lower side the shock is nearly straight, while on the upper side it curves and becomes weaker because of rarefaction waves ema-nating from the upper surface. At high incidence the shock wave comes so close to the lower surface that it merges with the viscous boundary layer on the lower surface. II. *Flat 75° delta wing at $M = 10$.* Schlieren knife-edge is parallel to the free stream. At an incidence of about 30°, the lower shock wave is straight and attached; it is very close to the wing, merging with the viscous boundary layer. The upper shock wave is very weak and curved and is present only because of the slight thickness of the wing. The pressure on the upper surface is far below free-stream pressure. At an incidence of about 45°, the lower shock is still attached and is merged with the viscous boundary layer; the upper shock is extremely weak and is followed by a rarefaction zone which is not visible. At an incidence of about 80°, the shock system is detached and curved. On the lower side, which is now facing nearly forward, the shock is very close to the wing. The fluid downstream of the shock wave escapes around the wing edges by flowing nearly perpendicular to the free stream in the thin region behind the bow shock wave and the wing. At an incidence of 90°, there is a nearly normal bow shock wave, close to the wing. It is asymmetrical because of the delta shape. III. *Model of "Project Mercury" re-entry space capsule.* Color schlieren is used, shock waves showing in orange. The bow shock is very close to the blunt nose. IV. *A different model of a re-entry space capsule.* Shock waves show as orange.

CREDITS: Experiments and film by Office National d'Études et de Recherches Aérospatiales (O.N.E.R.A.), France, and the National Aeronautics and Space Adminis-tration (N.A.S.A.); Consultant, A. H. Shapiro, M.I.T.; Di-rector, R. Bergman, E.S.I.; Editor, M. Chalufour, E.S.I.; Art Director, R. P. Larkin, E.S.I. Produced by the National Committee for Fluid Mechanics Films and Educational Services Incorporated, with the support of the National Science Foundation.

FM-56
Three-Dimensional Boundary-Layer Separation

PURPOSE: To show the nature of boundary-layer separation when it occurs in a three-dimensional flow.

APPARATUS: Two model shapes are tested in a water tunnel. Colored dyes are emitted through holes in the surface of each model. When the dye is injected at a high rate, it penetrates into the main flow. As the dye rate is reduced, the dye streaklines tend toward coincidence with the undisturbed streamlines in the boundary layer.

SEQUENCES: (1) *Ellipsoid of Revolution at Zero Incidence.* A curved "separation line" is observed on the body surface near the aft end, where there are large adverse pressure gradients. The boundary-layer streamlines curve laterally as they approach the separation line, leave the vicinity of the body surface, and flow near the "separation surface" in the fluid that springs from the separation line. Underneath the separation surface is a zone of recirculating flow. (2) *Ellipsoid at 6° Incidence.* View of the lee side of the same ellipsoid set at 6° incidence. The boundary-layer flow, separation line, separation surface, and recirculating region are qualitatively as before, except that the separation line is further upstream. (3) *Delta Wing at Incidence.* View of lee side. Separation occurs in the region of adverse pressure gradient near the trailing edge, with features generally similar to those for the ellipsoid.

DISCUSSION: Boundary-layer separation in a three-dimensional flow is fundamentally different from separation in a two-dimensional flow. (a) *Two-Dimensional.* There are no lateral pressure gradients and the longitudinal pressure gradient does not vary transversely. No lateral velocities are induced anywhere and all the boundary-layer events remain two-dimensional. The separation surface dividing the main flow from the recirculating flow between it and the body surface is cylindrical. The separation line, where the separation surface intersects the body surface, is perpendicular to the plane of the flow, and each point on it is a special form of stagnation point. (b) *Three-Dimensional.* On the ellipsoid and delta wings lateral pressure gradients do exist and the longitudinal pressure gradient varies with lateral position. The separation surface is doubly-curved and intersects the body at a curved separation line. Except at one point on the axis of symmetry, the separation line is not a stagnation region. Instead, the boundary-layer streamlines curve laterally as they approach the separation line, and the boundary-layer flow sweeps in a lateral direction as the boundary layer lifts from the body surface and moves along the separation surface. The same is true for the upstream-moving boundary layer in the recirculating region. The boundary-layer fluid is displaced sideways from the one stagnation point on the separation line.

CREDITS: Experiments and film by Office National d'Études et de Recherches Aérospatiales (O.N.E.R.A.), France; Consultant, A. H. Shapiro, M.I.T.; Producer, R. Bergman, E.S.I.; Editor, M. Chalufour, E.S.I.; Art Director, R. P. Larkin, E.S.I. Produced by the National Committee for Fluid Mechanics Films and Educational Services Incorporated, with the support of the National Science Foundation.

FM-57
Effect of Axial Jet on Afterbody Separation

PURPOSE: To show how a jet (e.g., from a rocket or turbojet engine) issuing from the rear of a body with a blunt base eliminates "base separation" of the main flow streaming past the body.

APPARATUS: Two different experiments are shown. In each one, a body of revolution with a more or less blunt base is in a water tunnel, with the external water flow streaming past the body. Each body has a nozzle built into its rear end, and provision is made to feed water under pressure into the interior of the body so that it issues as a jet through the nozzle, directed downstream.

SEQUENCES: *Experiment I.* The body is a cylinder with a flat base. The flow is visualized by light scattered from small air bubbles in the main stream; a thin vertical sheet of light is used, illuminating the flow in a diametrical plane. At first no jet issues from the rear, and there is a cylindrical separated wake with about the same diameter as the body itself; that is, the main flow separates at the sharp corner of the base. When the jet is turned on, the separated wake becomes very small compared with its original size. (The symbol Vj on the film indicates that jet velocity is increasing.) *Experiment II.* The body is an ellipsoid of revolution. The flow in the boundary layer and the wake is made visible by colored dyes issuing from holes in the body near the aft end. At first there is no jet; the rear of the body is sufficiently blunt to produce boundary-layer separation and a large separated wake. When the jet is turned on, the boundary-layer separation and the separated wake seem to be completely eliminated. In interpreting these pictures, one must recognize that the boundary layer and wake are correctly visualized only when the dye injection rate is reduced to a very low value; at a high dye rate, the dye momentum projects it as a small jet into the main flow, giving a false impression.

INTERPRETATION. The jet, through viscous and turbulent action, entrains the surrounding flow. The jet therefore acts as a line sink into which the surrounding flow is sucked. This entrainment draws in the boundary layer on the rearward part of the body and so alters the pressure distribution on the rearward part as to reduce or entirely eliminate the boundary-layer separation and separated wake. Without the jet there is a large "base drag" owing to the relatively low pressure on the aft portion of the body; with the jet, this base drag is much reduced.

CREDITS: Experiments and film by Office National d'Études et de Recherches Aérospatiales (O.N.E.R.A.), France; Consultants, R. W. Fox, Purdue University, and A. H. Shapiro, M.I.T.; Director, R. Bergman, E.S.I.; Editor, M. Chalufour, E.S.I.; Art Director, R. P. Larkin, E.S.I. Produced by the National Committee for Fluid Mechanics Films and Educational Services Incorporated, with the support of the National Science Foundation.

FM-58
Effect of Jet Blowing Over Airfoil Flap

PURPOSE: To show how a trailing-edge flap on an airfoil deflected at a large angle can be made more effective by energizing the boundary layer on the upper surface with a jet.

APPARATUS: Most of the experiments are in a water tunnel. In each case the trailing-edge flap is deflected downward at a large angle. The jet fluid under pressure escapes from a slot in the upper surface of the wing. It is directed to flow over the upper surface of the flap in a rearward direction. The flow is made visible either by dye emitted from a hole in the upper surface of the wing or by numerous small air bubbles entrained in the water.

SEQUENCES: (1) Sketches showing the experimental arrangement. (2) Dye injection from the upper surface of the wing, seen in an oblique view. At first there is no jet, and the flow over the upper surface separates. The flow does not follow the flap because of boundary-layer separation in the adverse pressure gradient downstream of the convex corner. Hence the flap is not effective in increasing lift. When the jet blowing begins, the boundary layer is energized, and the separation is suppressed. The flow follows the flap, and high lift is developed. (3) Side view of wing profile, with air bubbles in the water. The blowing jet is first off, then on, then off again. When on, it is extremely effective in eliminating boundary-layer separation. As a result, the large circulation (and with it, the large lift) predicted by potential theory is achieved. With very intensive blowing, the lift associated with the potential flow past the wing with deflected flap is actually surpassed; the jet flowing downward with high momentum acts like an additional flap, producing what is known as "super-circulation". (4) Test on an airplane in a hangar, with airflow over the airfoil produced by a propeller. A long view shows the flap being deflected. A medium view followed by a close-up shows the smoke being injected and following the flap at a large deflection angle. (5) A similar test on an airplane with a double flap, with air injection into the boundary layer at each of the two deflection locations. Smoke is injected through the upper surface of the airfoil and also at the second slot. At the end of the run, the blowing ceases, and severe separation is observed.

DISCUSSION: Boundary-layer energization by means of a blowing jet makes high-lift flaps effective. This is a form of boundary-layer control for the purpose of avoiding separation in adverse pressure gradients. The energized boundary-layer fluid is able to move through the strong positive pressure gradient, and follow the upper surface of the deflected flap, thus enabling the latter to produce high circulation and high lift.

CREDITS: Experiments and film by Office National d'Études et de Recherches Aérospatiales (O.N.E.R.A.), France; Consultant A. H. Shapiro, M.I.T.; Producer, R. Bergman, E.S.I.; Editor, E. Carini, E.S.I.; Art Director, R. P. Larkin, E.S.I. Produced by the National Committee for Fluid Mechanics Films and Educational Services Incorporated, with the support of the National Science Foundation.

FM-59
Leading-Edge Vortices on Delta Wing in Subsonic Flow

PURPOSE: To show the vortex system generated by the sharp leading edges of a thin delta wing at an angle of incidence in a subsonic flow.

(Note: FM-60, "Breakdown of Leading Edge Vortices on Delta Wing in Subsonic Flow", is a sequel to FM-59, showing rapid "bursting", or increase of core diameter of the vortex systems, at high incidence).

APPARATUS AND PROCEDURE: A thin delta wing having sharp leading edges is mounted in a water tunnel. It is maintained at zero yaw angle, but the angle of incidence is changed. The flow is visualized by two different methods: (a) injection of colored dyes through small holes; (b) light scattering from small air bubbles dispersed in the water.

SEQUENCE: (1) *Wing is at fixed incidence.* The "upper" surface ("leeward" surface; "suction" surface) is viewed in an oblique plan view. Dye is injected through five holes in each leading edge. Each dye streakline shows a steady-flow streamline. The several streaklines from each leading edge roll up into a vortex whose core is approximately at zero incidence but which is yawed at approximately the sweep angle of the leading edge. (2) *Plan view of upper surface,* with wing at fixed incidence. A succession of five scenes is shown, each showing one of the five pairs of corresponding dye streaklines. Each streakline is approximately helical in shape, indicating a vortex flow with a superimposed longitudinal velocity (similar to the trailing tip vortices of an unswept wing of rectangular planform and finite aspect ratio). (3) *Plan view of upper surface,* with five dye streaklines issuing from each leading edge. The angle of incidence increases from zero. At zero incidence each streakline is straight and parallel to the main flow. As the incidence increases the streaklines develop a helical structure and also yaw laterally so as to be approximately along the leading edges. (4) Same action as preceding, but viewed in oblique plan view. (5) *Plan view of upper surface,* with dye additionally injected from six points on the upper surface near the middle of the planform, and from six points on the upper surface near the trailing edge. These dye streaklines show the direction of boundary-layer flow. As the incidence increases the boundary-layer flow is diverted laterally more and more away from the central axis of symmetry, showing that the surface pressure is a maximum along the center axis of the wing and falls off to the edges. (6) *The wing is at a small fixed incidence,* and the water in the tunnel is aerated with many fine bubbles. A thin sheet of light is passed through the tunnel. The plane of light is perpendicular to the free-stream flow, and the camera looks upstream, along the plane of the wing. Hence one sees only the components of flow velocity transverse to the wing. This scene shows a cross-sectional view of the vortices seen in the earlier scenes. The observed transverse flow is similar to the flow past a transverse flat plate with sharp edges: on the lee side, boundary layer separation produces a vortex behind each sharp edge. This experiment shows that for thin bodies at small incidence, the flow past the body may be treated as the sum of two flows: that due to the component of free-stream velocity parallel to the plane of the body, and that due to the component normal to the plane of the body.

CREDITS: Experiments and film by Office National d'Études et de Recherches Aérospatiales (O.N.E.R.A.), France; Consultants, R. W. Fox, Purdue University, and A. H. Shapiro, M.I.T.; Director, R. Bergman, E.S.I.; Editor, M. Chalufour, E.S.I.; Art Director, R. P. Larkin, E.S.I. Produced by the National Committee for Fluid Mechanics Films and Educational Services Incorporated, with the support of the National Science Foundation.

FM-60
Breakdown of Leading-Edge Vortices on Delta Wing in Subsonic Flow

PURPOSE: To illustrate that the leading-edge vortices from a delta wing in subsonic flow can "burst", i.e. can undergo a rapid increase in diameter of the vortex cores. (Note: FM-59, LEADING-EDGE VORTICES ON DELTA WING IN SUBSONIC FLOW shows leading-edge vortices without breakdown.)

APPARATUS AND PROCEDURE: A delta wing having sharp leading edges is placed at quite high angle of incidence (compared with the experiments in FM-59) in a water tunnel. The camera is placed so as to see the "leeward" surface ("upper" surface; "suction" surface) in plan view. The flow is visualized by injection of colored dye at symmetrical positions on the two leading edges just behind the apex of the wing.

SEQUENCES: (1) Zero Yaw Angle. The vortices produced at the apex by the sharp leading edges are at first sharply defined with a narrow core. At some position there is a rapid transition from the sharp narrow cores to turbulent cores of larger diameter. As seen in FM-59, this "bursting" does not occur at low angles of incidence, i.e. with weaker vortex strengths. The reasons for the bursting are not understood at this writing. Vortex bursting is also observed in sink-vortex flows, as in tornadoes (see FM-26, TORNADOES IN NATURE AND THE LABORATORY; and FM-70, "THE SINK VORTEX"). It is thought by some that the bursting is similar to a hydraulic jump, with the gravity field of the latter analogous to the centrifugal force field in the vortex; this view is consistent with the observation that vortex bursting is associated with high fluid speeds in the direction of the vortex core. (2) and (3) are closer views of same. (4) Suction is applied near the trailing edge on right-hand side, at the position of the superimposed arrow. This suppresses the bursting of the right-hand vortex. (5) When the wing is yawed (at constant incidence), the vortex off the leading side bursts earlier, and the vortex on the rearward side persists longer before bursting. (6) Repeat of (5).

CREDITS: Experiments and film by Office National d'Études et de Recherches Aérospatiales (O.N.E.R.A.), France; Consultants, R. W. Fox, Purdue University, and A. H. Shapiro, M.I.T.; Director, R. Bergman, E.S.I.; Editor, M. Chalufour, E.S.I.; Art Director, R. P. Larkin, E.S.I. Produced by the National Committee for Fluid Mechanics Films and Educational Services Incorporated, with the support of the National Science Foundation.

FM-61
Ablation of Ice Models in a Water Tunnel

PURPOSE: To show the relative rates of ablation at different positions on several nose shapes. The ablation of re-entry bodies due to the high temperatures generated by high-speed motion through the atmosphere is approximately modeled by the melting of ice models in a stream of water.

APPARATUS AND PROCEDURE: Models built of ice are placed in a water tunnel. The ice melts (ablates) and the melted layers flow downstream in the boundary layer. In each sequence the original shape of the model, before the water flow begins, is shown by a superimposed dashed curve. This provides a reference for observing the relative rates of ablation at different positions.

SEQUENCES: (1) *Cone with spherical nose.* The ablation is most rapid at the stagnation point. The spherical nose ablates much more rapidly than the cone. The forward part of the cone ablates more rapidly than the rearward part. (2) *Circular cylinder with "Rankine nose."* The ablation is most rapid at the stagnation point. The nose ablates more rapidly than the cylinder. On the cylinder itself the rate of ablation decreases in the downstream direction. (3) *Cylinder with blunt nose.* The ablation is most rapid near the slightly rounded shoulder of the blunt nose, and the blunt nose becomes less blunt. The ablation of the nose is, however, more uniform than in the two previous experiments. On the cylinder itself, the ablation rate decreases with downstream distance. (4) Same experiment, showing later stages of ablation, with the model virtually disappearing. Repeat of same. (5) *Cylinder with blunt nose, at angle of incidence.* When placed at an angle of incidence, the nose ablates into a shape which is still nearly axi-symmetric. Compared with the experiment at zero incidence, there is an increase in the ablation rate of the cylinder relative to that of the nose. Nevertheless, the nose ablates more rapidly than the cylinder.

DISCUSSION: The melting rate of the ice is governed by heat transfer in the thermal boundary layer, and is affected by the shielding action of the cold water released along the surface by the melting itself. The combination of the growth of the thermal boundary layer and of the continuous introduction of cold melted water into this layer explains why the melting is usually most rapid at the stagnation point and decreases as one goes downstream along the surface. In an ablating vehicle entering a planetary atmosphere, one should note that additional phenomena affect the ablation rates: (a) chemical reactions between the air and the ablating material, with associated energy release; (b) part of the ablated material may change from the liquid to the gaseous phase; (c) energy is deposited in the viscous boundary layer by viscous dissipation ("aerodynamic heating"); (d) the inviscid external flow is colder than the body surface; (e) the flow patterns are different because of compressibility and shock waves. The state of the boundary layer — whether laminar or turbulent — is also very important to the ablation rate.

CREDITS: Experiments and film by Office National d'Études et de Recherches Aérospatiales (O.N.E.R.A.), France; Consultants, R. Fox, Purdue University, and A. H. Shapiro, M.I.T.: Director, R. Bergman, E.S.I.; Editor, E. Carini, E.S.I.; Art Director, R. P. Larkin, E.S.I. Produced by the National Committee for Fluid Mechanics Films and Educational Services Incorporated, with the support of the National Science Foundation.

FM-62
Interactions Between Oblique Shocks and Expansion Waves

PURPOSE: To show the interactions between waves moving in the same direction in a supersonic flow.

APPARATUS: A symmetric airfoil with a rounded nose is tested in a subsonic wind tunnel. At high subsonic tunnel speeds, supersonic flow is established over a large part of the airfoil. The airfoil is at zero incidence except for a trailing-edge flap which is deflected slightly upward. Lateral density gradients in the flow are made visible by a schlieren optical system whose knife edge is parallel to the flow (horizontal).

SEQUENCES: (1) Test-section Mach number increases from a low value to 0.9. At first the flow around the airfoil is entirely subsonic; the only density gradients visible are near the stagnation region at the nose, in the boundary layers, and in the wake. As the tunnel speed increases, local supersonic zones occur on the top and bottom surfaces; they can be identified by the shock waves at the downstream ends of these zones. These terminating shock waves move downstream and increase in lateral extent as the tunnel speed increases. At $M_\infty = 0.9$ the flow over the rear half of the airfoil is supersonic; waves originate on both the upper and lower surfaces at the sharp junction with the flap and at the trailing edge. (2) Attention is called to the waves from the upper surface. The oblique shock from the flap junction intersects with the oblique shock from the trailing edge, resulting in a continuing shock of increased strength. (3) Attention is called to the waves from the lower surface. The centered rarefaction wave (Prandtl-Meyer expansion fan) from the flap junction intersects with the oblique shock from the trailing edge. The interaction causes mutual cancellation and weakening of both waves. (4) Repeat of sequence (1), $M_\infty \cong 0$ up to $M_\infty = 0.9$.

DISCUSSION: Waves of the same family in a supersonic flow may intersect and thus interact. Two adjacent oblique shocks always tend to intersect. A rarefaction wave intersects an oblique shock ahead of it and also an oblique shock behind it. It is only adjacent rarefaction waves that do not intersect. The component of flow speed normal to a rarefaction wave is exactly sonic both ahead of and behind the wave. For an oblique compression shock, the normal component of flow speed is supersonic ahead of the shock, and subsonic behind. Stated differently, the front of a rarefaction wave propagates normal to itself with exactly the sonic speed in the gas ahead; similarly for the back of a rarefaction wave. An oblique shock, on the other hand, propagates normal to itself with a speed greater than the speed of sound in the gas ahead, and with a speed less than the speed of sound in the gas behind. Thus an oblique shock overtakes a rarefaction wave ahead of it, and is overtaken by a rarefaction wave behind it. Further, an oblique shock overtakes an oblique shock ahead of it.

CREDITS: Experiments and film by the National Physical Laboratory, England; Consultant, A. H. Shapiro, M.I.T.; Producer, R. Bergman, E.S.I.; Editor, E. Carini, E.S.I.; Art Director, R. P. Larkin, E.S.I. Produced by the National Committee for Fluid Mechanics Films and Educational Services Incorporated, with the support of the National Science Foundation.

FM-63
Slot Blowing to Suppress Shock-Induced Separation

PURPOSE: To show how energization of the boundary-layer by blowing from a slot can reduce separation on an airfoil resulting from shock-boundary layer interactions at transonic speeds.

APPARATUS: An airfoil at fixed angle of attack is mounted in a subsonic wind tunnel capable of producing test-section Mach Numbers as high as 0.95. On the upper surface of the airfoil, about 15% aft of the leading edge, a slot is arranged so that an energizing jet of high-speed air can be blown into the boundary layer, parallel to the surface. The flow is visualized by a schlieren system; the knife edge is parallel to the flow, thus showing transverse density gradients.

SEQUENCES: (1) A preliminary scene and sketch shows the arrangement of the blowing slot, and an arrow identifies its position. (2) Experiment with No Blowing. There is no jet. The Mach Number is increased continuously from 0.6 to 0.95. At first the flow is entirely subsonic. As the Mach Number increases, a supersonic zone appears on the upper surface, terminated by a nearly-normal shock wave. The shock wave moves aft with rising Mach Number. Interaction of the shock with the boundary layer causes a severe boundary-layer separation over the whole aft end of the airfoil, and also produces a wave system of "lambda" type originating at the airfoil surface. The oblique wave originating from the disturbance at the blowing slot is clearly visible in this and the next sequence. It is actually weaker than the nearly-normal terminal shock, but appears stronger because of the knife-edge orientation and because it is relatively steady. (3) Experiment with Blowing from Slot. Again the Mach Number increases continuously from 0.6 to 0.95, but the blowing jet is in operation. The shock-boundary layer interaction is much reduced, and there is relatively slight boundary layer separation. Of particular interest are the intersections of the oblique waves originating at the slot and at the jet with the nearly-normal shock wave which terminates the supersonic zone. Vortex sheets spring from the intersection points. (4) Comparison of "No Blowing" with "Blowing". The portions of sequences (2) and (3) for which the free-stream Mach Number is 0.95 are compared directly, to show more clearly the effect of slot blowing.

CREDITS: Experiments and film by the National Physical Laboratory, England; Consultant, R. Fox, Purdue University; Director, A. H. Shapiro, M.I.T.; Editor, E. Carini, E.S.I.; Art Director, R. P. Larkin, E.S.I.

FM-65
Wide-Angle Diffuser with Suction

PURPOSE: To show how boundary-layer suction can eliminate separation owing to an adverse pressure gradient. Further, to show how a wide-angle diffuser can be made to operate effectively with the help of boundary-layer suction.

APPARATUS AND PROCEDURE: Water flows through a two-dimensional diffuser having a wide angle of divergence. Aluminum powder on the free surface makes the flow visible. The diffuser has four locations at which suction may be applied, two on each wall.

SEQUENCES: (1) With no suction, the flow separates from both walls and issues from the venturi throat as a jet. (2) Suction is applied at the upper wall. The flow attaches to the upper wall, and remains detached from the lower wall. (3) With suction at both walls, there is no separation; the diffuser runs full. When the suction is turned off at the upper wall, the flow separates there, and the jet follows the lower wall.

DISCUSSION: Separation normally occurs in diffusers with large divergence angles because the low-momentum fluid in the boundary layer cannot proceed against the adverse pressure gradients which would prevail in the absence of separation. The separation itself reduces greatly the pressure gradient of the main flow, thus defeating the purpose of the diffuser. When the boundary layer is sucked off at a wall, the cause of separation disappears, and the diffuser operates effectively.

CREDITS: Experiments and film by L. Prandtl, (Göttingen); Consultant and Director, A. H. Shapiro, M.I.T.; Editor, E. Carini, E.S.I.

FM-66
Thin Bodies of Revolution at Incidence

PURPOSE: To show the character of the flow past a thin body of revolution at a small angle of incidence to the free stream.

APPARATUS AND PROCEDURE: Each model is set at a fixed angle of incidence in a water tunnel. The flow near and along the model is visualized by the injection of colored dyes from holes in the model. The component of the flow perpendicular to the model axis (i.e. the transverse, or cross flow) is visualized by dispersing air bubbles in the stream and by looking along the body axis to see the bubbles in a transverse plane as they are illuminated by a sheet of light whose plane is perpendicular to the model axis.

EXPERIMENTS: (A) *CONE OF SMALL INCLUDED ANGLE AT INCIDENCE*. (1) *Side view*. The dyed streaklines near the model surface are approximately helical in shape, and extend downstream. (2) *View of leeward surface from above*. The dyed streaklines are seen to wind around two vortex cores which are on the lee (upper) side and whose axes are approximately along the surface of the model. (3) *Transverse flow as viewed by scattering from light sheet*. The bubble motions in the transverse plane show plainly the separated cross flow and the two vortex cores on the lee side of the cone. These vortices resemble closely the shed vortices on the lee side of a circular cylinder whose axis is normal to the free stream. (B) *OGIVE-CYLINDER AT INCIDENCE*. (1) *View of leeward surface from above*. The dyed streaklines near the model surface show two vortex cores on the lee side, whose axes are approximately parallel to the model axis. (2) *Transverse flow as viewed by scattering from light sheet*. As with the cone, the transverse flow is separated on the lee side and two vortices appear in the separated wake.

DISCUSSION: The flow past a thin body of revolution at small incidence may be approximately resolved into the sum of two flows: (i) The flow associated with the component of approach velocity that is parallel to the model axis; this parallel component of flow corresponds to flow past the model at zero incidence. (ii) The flow associated with the component of approach velocity that is perpendicular to the model axis; this cross flow, like its two-dimensional approximation, involves flow separation and the formation of two vortices in the wake. The combination of the cross flow with the parallel flow produces a separated zone on the lee side of the inclined body, with vortices whose cores are nearly parallel to the model surface. The streamlines around the vortices are somewhat helical in shape; the longitudinal component of velocity is large compared with the swirling component of velocity.

CREDITS: Experiments and film by Office National d'Études et de Recherches Aérospatiales (O.N.E.R.A.), France; Consultant, A. H. Shapiro, M.I.T.; Director, R. Bergman, E.S.I.; Editor, M. Chalufour, E.S.I.; Art Director, R. P. Larkin, E.S.I. Produced by the National Committee for Fluid Mechanics Films and Educational Services Incorporated, with the support of the National Science Foundation.

FM-67
Flow Through Right-Angle Bends

PURPOSE: To show the development and establishment of steady-state flow in right-angle bends of several geometries.

APPARATUS AND PROCEDURE: Water flows through a horizontal, two-dimensional right-angle bend. Aluminum flakes on the free surface show the fluid motion. In each sequence the flow begins from rest, so that the development of separated regions may be observed.

SEQUENCES AND DISCUSSION: (1) Medium-Radius Elbow: Just after the flow starts, the motion is very close to an inviscid potential flow. As a result of streamline curvature, the static pressure in the middle of the bend rises from a low value at the inner wall to a high value at the outer wall. Along the inner wall, the pressure decreases from its upstream value to a minimum value near the mid-point of the bend, and then rises again to its downstream value. Along the outer wall, the pressure increases to a maximum near the mid-point of the bend, then falls to its downstream value. Owing to the adverse pressure gradients thus created on the outer and inner walls, the viscous boundary layers "stall", and regions of separated flow develop. On the outer wall the stall begins upstream of the mid-point of the bend; on the inner wall the stall begins downstream of the mid-point. Dissipation of energy in the recirculating regions is one of the important sources of head loss in a bend. (2) Elbow with Sharp Outer Corner: The stall develops more rapidly at the outer wall, and the separated, recirculating region there is much larger than when a radius is present. Head losses are greater in sharp elbows than in gradual bends. (3) 90°Turn Through a Plenum: At first a double vortex is formed at the inlet. Gradually a separated flow develops; a curved jet flows from inlet to outlet, with recirculating eddies in the spaces above and below the jet.

CREDITS: Experiments and film by L. Prandtl, Göttingen; Consultant and Director, A. H. Shapiro, M.I.T.; Editor, E. Carini, E.S.I.; Art Director, R. P. Larkin, E.S.I.

FM-68
Flow Through Ported Chambers

PURPOSE: To show the development and establishment of steady-state flow through a circular chamber having four pipe-like ports arranged at 90° intervals. The model simulates a manifold arrangement. The role of viscosity and the complex character of the flow are made evident.

APPARATUS AND PROCEDURE: The model is two-dimensional and horizontal. Aluminum flakes on the free surface of the water make the flow visible. In each sequence the flow begins from rest. When the appropriate valves are opened, flow from the inlet port enters the chamber and leaves from one or more exit ports.

SEQUENCES: There are three sequences: (1) Flow enters at 9 o'clock; leaves at 3 o'clock. (2) Flow enters at 9 o'clock; leaves at 12 o'clock and 6 o'clock. (3) Flow enters at 9 o'clock; leaves at 12 o'clock and 3 o'clock.

DISCUSSION: In each case, the motion in the first instants of starting is a potential flow filling the circular chamber. Shortly, however, the effects of viscosity coupled with the adverse pressure gradient at the junction of the inlet port with the chamber cause reverse flow on the chamber wall. The ensuing separation of the inlet flow, and its formation as a jet, generates two vortices at the inlet lip. The flow then changes rapidly in time, and ultimately characterized by unsteadiness, complexity, eddy generation and shedding, and jets and jet mixing. It is evident that large head losses occur.

CREDITS: Experiments and film by L. Prandtl, (Göttingen); Consultant and Director, A. H. Shapiro, M.I.T.; Editor, E. Carini, E.S.I.

URPOSE: To show the development and establishment of the
uid motion in a two-dimensional model of a tee-fitting, with dif-
rent combinations of flows in the two outlet branches.

PPARATUS AND PROCEDURE: Water enters a straight-through
g of a horizontal, two-dimensional tee-fitting. Valves are posi-
oned in the other straight-through leg and in the side branch.
luminum flakes on the free surface show the fluid motion. In each
se the flow starts from rest, so that the development of the
eady-state flow may be seen.

EQUENCES AND DISCUSSION: (1) Straight-Through Flow, with
ide Branch Closed: The throughflow, by means of viscous stresses,
rives a recirculating eddy in the closed-off side branch. (2)
0°-Turn into Side Branch, with no Straight-Through Flow: Be-
ause of the adverse pressure gradient on the downstream side of
e convex corner, the flow entering the side branch separates at
e corner. Back flow of the viscous layer produces a recircu-
ating eddy which constricts the area available for the main flow.
hus a relatively high speed jet is formed. The increased kinetic
nergy of this jet in the end augments the loss in head through the
itting. In the closed-off straight-through branch, viscous stresses
pplied by the main flow drive a low-speed recirculating eddy.
3) All Passages Open. The flow divides between the side branch
nd the straight-through branch. Because of boundary-layer
eparation, relatively high-speed jets are formed in both branches.
he stalled region in the straight-through passage is brought about
y boundary layer separation in an adverse pressure gradient; a
igher local pressure occurs at the separation location owing to
urvature of streamlines into the side branch.

CREDITS: Experiments and film by L. Prandtl, Göttingen; Consultant and Director, A. H. Shapiro, M.I.T.; Editor, E. Carini,
E.S.I.; Art Director, R. P. Larkin, E.S.I.

FM-70
The Sink Vortex

PURPOSE: To show experimentally several aspects of a quasi-
two-dimensional sink-vortex water flow in a vessel having a free
surface and a bottom, namely: (a) the starting vortex; (b) the
streamlines and velocity distribution at the surface; (c) the vorticity
about midway between the free surface and the bottom; (d) the
secondary flow near the bottom.

APPARATUS: In all the experiments, water enters tangentially into
a circular tank and flows out by gravity through a drain hole centrally located in the bottom of the tank. The tank has no cover, so
that the free surface of the water is at atmospheric pressure. Except for the viscous boundary layer on the bottom, the flow is approximately the combination of a potential vortex and a radially-
inward sink flow.

EXPERIMENTS: (1) Starting Vortex. When the flow begins, a
starting vortex is shed from the sharp lip where the tangential
entry meets the circular wall. The starting vortex is convected
to the center by the flow. It stabilizes itself at the center, thus
establishing the steady-state flow. (2) Velocity Distribution at
Free Surface. A powder sprinkled on the surface shows that the
streamlines are tightly wound spirals (for a two-dimensional,
potential sink-vortex, they would be logarithmic spirals). A view
from above shows that the tangential velocity increases as the
radius decreases, in agreement with the distribution for a potential
vortex, $Vr = constant$. A perspective view shows the depression of
the free surface near the center, associated (by Bernoulli's integral) with the high velocity in that region. (3) Vorticity. The
"vorticity meter" is a cylindrical shaft with a pair of crossed vanes
at its lower end, and a horizontal arrow at its upper end. It floats
in a vertical position, with the vanes immersed in the water and
with the arrow above the water surface. The crossed vanes are
driven by the vertical component of vorticity (or angular velocity).
The rate of turning of the arrow is a good approximation to the
vertical vorticity. (a) Away From the Center. When the vorticity
meter is placed at any radius away from the center, it moves in
pure translation, showing that the flow is generally free of vorticity. (b) At the Center. When the vorticity meter reaches the
center, it suddenly begins to rotate. The center of the vortex is a
"singular" region of highly concentrated vorticity which accounts
for the circulation $\oint \overline{V} \cdot d\overline{r}$ of the entire flow. (4) Boundary Layer
on Bottom. Dye injected near the bottom runs nearly radially to
the drain hole. Most of the discharge in fact comes from the
viscous layer on the floor. The radial pressure gradient associated with the centrifugal acceleration of the main flow drives
the viscous layer, which has a relatively low tangential velocity,
radially inward.

CREDITS: Experiments and film by L. Prandtl, Göttingen, A. H.
Shapiro, M.I.T., and E. S. Taylor, M.I.T.; Editor, J. Hirschfeld,
E.S.I.; Art Director, R. P. Larkin, E.S.I.; Director, A. H. Shapiro,
M.I.T.

FM-71
Flow Near Tip of Lifting Wing

PURPOSE: To show the flows on the upper and lower surfaces near the tip of a lifting wing, and the relationship between these flows and the development of the trailing tip vortex.

APPARATUS: A rectangular half-wing with a symmetric airfoil cross-section is mounted on one wall of a water tunnel, and simulates a wing in free flight with an aspect ratio of about 2.5. The wing-tip is far from the other wall of the tunnel. Colored dyes are injected from holes at several positions on the upper and lower surfaces of the wing, and the resulting streaklines show the streamlines near these surfaces. *Experiment I*. The airfoil is at zero incidence, but lift can be generated by means of a "jet flap" (a two-dimensional jet blowing downward) issuing from the trailing edge. *Experiment II*. A similar experiment, but lift is generated by setting the wing at an angle of incidence.

SEQUENCES: (1) *Plan view of wing lower surface*, showing dye injection under conditions of zero lift. The dye streaklines are virtually parallel to the free-stream flow. (2) *End view of wing* showing the airfoil in profile. Initially the jet flap is not on, and there is no lift. When the jet flap is turned on, lift, and hence circulation, are produced. Dye streaklines from the lower ("pressure") surface flow around the wing tip to the upper ("suction") surface. This flow around the wing tip, from a region of relatively high pressure to a region of relatively low pressure, is one way of explaining the vortices which trail back from the tip. (3) *Plan view of wing upper surface*. The jet flap produces a pressure difference between the lower and upper surfaces, causing a flow around the wing tip. This flow, in turn, causes the streamlines on the upper surface to turn *away* from the wing tip. The trailing tip vortex is strongly evident. (4) *Plan view of wing lower surface*. The jet flap again produces a flow around the wing tip from the lower surface to the upper surface. The streamlines on the lower surface turn toward the wing tip. (5) *Side view of wing at incidence (no jet flap)*. The dye streaklines are turned on one at a time, and, by their helical shape, exhibit the trailing vortices. (6) Close-up view of same. The trailing vortex persists with a narrow core as it moves downstream.

DISCUSSION: The flow patterns on the upper and lower surfaces, and the trailing tip vortex, may be explained in two different ways: (a) By consideration of the pressure field, as discussed above. (b) Since vortex lines cannot end, the bound vortex in the airfoil, which accounts for the circulation and lift, trails downstream from the tip. The bound vortex, the trailing vortices, and the "starting" vortex (shed from the airfoil when the lift was first developed) form closed vortex lines in the shape of approximately rectangular loops.

CREDITS: Experiments and film by Office National d'Études et de Recherches Aérospatiales (O.N.E.R.A.), France; Consultant, A. H. Shapiro, M.I.T.; Director, R. Bergman, E.S.I.; Editor, E. Carini, E.S.I.; Art Director, R. P. Larkin, E.S.I. Produced by the National Committee for Fluid Mechanics Films and Educational Services Incorporated, with the support of the National Science Foundation.

FM-72
Examples of Surface Tension

PURPOSE: To illustrate the physical reality of surface tension showing several phenomena which clearly appear to be affected tensile forces of the liquid surfaces.

EXPERIMENTS: (1) Water is seen held up in a dripping fauc until the drop becomes large and falls. The water holds itself against gravity until its weight becomes too much for the tensi of the water surface to support. (2) A soap film on a wire fram pulls on the thread used to close the open end of the frame. T thread is pulled into circular arcs, showing the film tension to l uniform in all directions. Moving the thread to enlarge the fil requires work; it takes energy to pull molecules up into the sui face. (3) A jet of water impinging on an axially symmetric conca surface flows radially outward as a sheet which breaks up in drops. At times, because of surface tension, the sheet is pulle back sharply upon itself. (4) Water drops on a waterproof cloth not wet the cloth. Their curved surfaces keep them from enterii small holes. (5) A limp loop of thread on water snaps into a circl when the water inside is touched by soap. The soap molecules lowe the surface tension of the water within the loop. (6) Water pull itself up between two wetted glass plates, highest where the plate touch. The plates form a wedge, and the water rises within to for a hyperbolic curve. (7) A soap bubble ruptures, seen slowed dow 400 times. The edge can be seen advancing on the otherwise un disturbed film. (8) A milk drop falling on a thin layer of wate sends up a coronet. In slow motion one sees a succession of shape as inertia forces and surface tension forces compete.

UNDERSTANDING SURFACE TENSION: Liquid surfaces act as they are in tension. This tension can be measured. For pure water the tension is a little over 70 dynes per centimeter. Other liquids except for a few such as liquid metals, have lower surface tensions.

This tension of surfaces can be understood in terms of forces between molecules. Molecules repel each other when close, attrac each other when farther away, and at some in-between distance neither repel nor attract. Molecules in a surface are, on the average, separated from each other by more than the neutral intermolecular distance. They therefore attract each other, causing the tension force that is observed.

Surface tension phenomena can equally well be understood in terms of surface energies. Just as it takes energy to evaporate molecules from the closely packed liquid state into the widely separated gaseous state, it takes energy, about half as much, to move molecules from the interior of a liquid up into surface positions, where they have close neighbors on only one side. This energy of surface molecules (free energy in the thermodynamic sense) is equivalent to surface tension, and either concept can be used to explain surface phenomena. Since fluid motions are perhaps more easily imagined in terms of forces than in terms of energies, the concept of surface tension is widely used in fluid mechanics.

CREDITS: Scenes from the film, "Surface Tension in Fluid Mechanics", by Lloyd Trefethen, Tufts University; Director, Jack Churchill, E.S.I.; Editor, Richard Bergman, E.S.I.

urface Tension and Contact Angles

URPOSE: To show how the shape of a liquid surface is influenced
 the angles the liquid assumes at its edges. These angles, called
contact angles", provide one of the three boundary conditions
eded to express mathematically the effects of surface tension in
uid mechanics.

XPERIMENTS: (1) Soap bubbles blown between two glass plates
ow many intersections of the soap films. Three soap films meet-
g at an intersection pull equally, and therefore must meet sym-
etrically, always at 120° angles. Four soap films meeting at an
tersection are in unstable equilibrium; any disturbance unbalances
e forces in such a way as to produce two three-film intersections.
) A drop of water is placed on a bar of wax. The water does not
et the wax; that is, the total water angle where the water, air, and
ax meet is greater than 90°. (3) Mercury drops on a glass
imilarly do not wet the solid. (4) A water drop on a bar of soap
lopts a liquid angle of less than 90°, and is therefore said to wet
e soap. (5) Water drops on a waterproof cloth do not penetrate
e cloth. A liquid that wets a solid will pull itself into holes in that
olid; when it doesn't wet the solid, the liquid surface will curve in
 direction that resists the liquid's entering holes. Note: The word
wet" is sometimes defined differently by chemists, who are perhaps
nore interested in whether a liquid will of itself spread over a flat
urface than in whether it will enter a porous surface. Chemists
ay that a liquid "wets" only when it will spread completely over a
lat solid surface, that is, when the liquid takes a zero contact angle.

ISCUSSION: Contact angles are usually determined experimental-
y, not predicted. Contact angles are usually explained by an
quation, called "Young's Equation":

$$\sigma_{solid/gas} = \sigma_{solid/liq} + \sigma_{gas/liq} \cos(\text{contact angle})$$

which is a statement that the three surface tensions pulling on the
ntersection of a gas, a liquid, and a solid should balance along a
lat surface. Since it is not readily apparent that solid surfaces exert
 pull (although the concept is consistent with experiments), the
dea of surface energies rather than tensions is often used for
olids. The above equation is still used, the σ's standing for the
ree energies of the interfaces. The equation then amounts to a
statement that any perturbation of a stable intersection increases
the total free energy of the surfaces.

Young's Equation is not very useful in practice. The surface energies of real solids are often not known. The energy of a surface
over which liquid is advancing is often different from that of the
same surface if the liquid has retreated. (If it were not for such a
hysteresis effect, water drops could not stand still on windowpanes.)
A thermodynamically correct statement would be that the contact
angle will be that angle which leads to the minimum available
energy of the system, a principle even more difficult to apply than
Young's Equation. Hence, contact angles are usually measured.

CREDITS: Scenes from the film, "Surface Tension in Fluid
Mechanics", by Lloyd Trefethen, Tufts University; Director Jack
Churchill, E.S.I.; Editor, Richard Bergman, E.S.I.

FM-74
Formation of Bubbles

PURPOSE: To illustrate how bubbles are usually formed as
pinched-off enlargements of bubbles which grow from vapor pockets
which remain in cracks on the surface of solids.

EXPERIMENTS: (1) Water is shown boiling slowly at the bottom
of a heated beaker. The beaker had been rather well cleaned so
that bubble columns arise from only a few sites on the glass surface. Each nucleation site is seen to be a continuous source of
bubbles. (2) A small glass tube, pinched near one end to form an
artificial cavity, is lowered into the superheated layer of water
on the beaker bottom. Bubbles emerge continuously from the
cavity. When the tube is moved up to cooler water, the vapor in
the cavity condenses and water enters the tube, but not all the way
into the tapered end. The gas cavity never completely disappears;
when it is lowered back into the superheated layer, it again acts
as a nucleation site and can generate any number of new bubbles.
(3) Bubbles are shown being generated from two nucleation sites
on the side of a glass of beer. Although supersaturated with carbon
dioxide, the bulk liquid provides no opportunity for the carbon
dioxide molecules to move from their positions within the liquid
into vapor spaces. It is only at the wall that vapor spaces exist and
provide liquid-gas interfaces across which the carbon dioxide
molecules can pass from solution into the gaseous state. (4) The
roughened end of a glass rod is introduced into the beer, and causes
many bubbles. The rough glass surface contains numerous vapor
pockets, which enable the carbon dioxide molecules to come out of
solution more rapidly.

DISCUSSION OF BUBBLE GENERATION: Until recently, people
have found it difficult to understand how bubbles can be generated in
boiling water and other bubbling systems. New bubbles, to start
from a zero size, would require an infinite internal pressure, because surface tension causes the pressure within a bubble to be
higher than outside by the amount $2\sigma/r$, which is infinite if r, the
radius, is zero. Even an assumption that bubbles start from
statistical voids, of sizes near the intermolecular distance, required
that initial vapor pressures be thousands of pounds per square inch.
About 1955, people began to adopt the idea illustrated in these
pictures, that new bubbles are usually just pinched-off enlargements
of old bubbles remaining in cracks on solid surfaces. Subsequent
research has supported this now generally accepted idea. New
bubbles can be generated in a liquid where there were none before,
in instruments called bubble chambers. There, a high energy
particle collides with atomic particles in the liquid. These collisions
cause local regions of such high momentary energy that voids are
created, voids large enough to grow into bubbles which mark the
path of the particle. While in principle it should be possible to heat
a large volume of water to such high temperatures that voids would
form and enlarge into bubbles, this has not yet been achieved
experimentally. The required superheat would be orders of magnitude larger than those encountered in normal boiling.

CREDITS: Scenes from the film, "Surface Tension in Fluid
Mechanics", by Lloyd Trefethen, Tufts University; Director, Jack
Churchill, E.S.I.; Editor, Richard Bergman, E.S.I.

FM-75
Surface Tension and Curved Surfaces

PURPOSE: To show that the pressure on the concave side of a curved liquid surface is greater than on the convex side. This pressure difference provides one of the three boundary conditions needed to express mathematically the effects of surface tension in fluid mechanics.

EXPERIMENTS: (1) Smoke is used to blow a soap bubble at the end of a tube. The surface tensions of both the inner and the outer surfaces of the soap film compress the gas inside. When the other end of the tube is uncovered, the smoke is pushed out through the tube by this higher pressure inside. (2) The water in a dripping faucet holds itself up between drips. The curved water surface causes a pressure rise over atmospheric pressure which is just enough to equal the higher hydrostatic pressure within the water. The surface curvature is therefore sharper at the bottom of the drop where the hydrostatic pressure in the water is highest. (3) Water and alcohol are shown dripping from two burettes of the same size. The drip rate is the same, but the alcohol drops are smaller than the water drops. This difference in size is caused by the lower surface tension of alcohol. (4) The lower ends of two identical glass tubes are dipped into separate dishes of water and alcohol. Both wet their tubes, but the water rises higher than alcohol. The larger weight of water is held up by the larger pressure difference the water surface achieves across its curved miniscus. (5) A stream of water from a faucet breaks up into drops. In slow motion, slowed down 200 times, small irregularities are seen to amplify as they move with the jet, and the jet pinches itself off into a series of drops.

DISCUSSION: If an interface is curved, surface tension causes a pressure difference between the fluids on each side of the interface. The pressure difference is in the direction to resist a reduction in surface area; that is, the pressure is always higher on the concave side. A force balance across surfaces shows this pressure difference to be:

$$\Delta P = \sigma \left(\frac{1}{R_1} + \frac{1}{R_2} \right)$$

where σ is the surface tension, in units of force per unit length, and R_1 and R_2 are the two radii necessary to express the curvature of a surface. The pressure inside a cylindrical jet of liquid is therefore σ/r higher than outside. Irregularities in a liquid jet therefore amplify themselves; the higher pressure in each smaller region forces liquid away, into the adjacent, bigger regions. Inside a spherical bubble, where $R_1 = R_2 = r$, the pressure is $2\sigma/r$, and inside a soap bubble, which has both an inner and an outer surface, the pressure is $4\sigma/r$.

Liquid will be sucked into a hole or a tube if it wets the wall material. In a small tube, the liquid surface will be spherical, with $R_1 = R_2 = r/\cos\theta$, where θ is the liquid contact angle. If the pressure difference, $2\sigma\cos\theta/r$, is equated to the hydrostatic head, ρgh, the height of the capillary rise can be calculated to be $h = 2\sigma\cos\theta/\rho gr$. Liquid will rise only if θ is less than 90° (i.e., if it wets the tube).

CREDITS: Scenes from the film, "Surface Tension in Fluid Mechanics", by Lloyd Trefethen, Tufts University; Director, Jack Churchill, E.S.I.; Editor, Richard Bergman, E.S.I.

FM-76
Breakup of Liquid into Drops

PURPOSE: To show that the pressures caused by curved surfac[e] lead to the breakup of jets and sheets of liquids.

EXPERIMENTS AND DISCUSSION: A high speed camera was use[d] in order to slow the motion, by a factor of about 200 in the fir[st] experiment, and by a factor of 400 in the others. (1) Water emerg[ing] ing vertically downwards from a faucet as a smooth cylinder be[comes] comes pinched at intervals as it moves down, and still farth[er] down changes itself into drops. All liquid cylinders are unstabl[e] since any irregularity causes a pressure difference that amplifi[es] the irregularity. These pictures show how complex the process [is]; capillary waves run up and down the pinched-off liquid segment[s]; there are small as well as large drops; and the final drops osci[l]late about a spherical shape as kinetic energy of the distorti[ng] liquid alternates with the potential energy of an elongated, large[r] surface area. The breakup of liquids into drops always involves [a] reduction of the total surface area. (2) A jet of water impinging o[n] an axillary symmetric concave surface becomes a radial shee[t]. The edge of the sheet, pulled by surface tension forces, accumulate[s] into rolls which become drops. Irregularities in the production o[f] drops are probably related to irregularities in the impinging je[t]. There is little understanding as yet of the details of how sheet[s] break up. (The most revealing work to date is that of G. I. Taylo[r], reported in the <u>Proceedings of the Royal Society</u> of 1959 and 1960[.] (3) A soap bubble is touched by a pencil that has been dipped i[n] alcohol. A hole appears and rapidly grows larger, consuming th[e] bubble. The alcohol ruptures the liquid sheet by reducing th[e] surface tension of a small portion of the surface. The higher sur[-]face tension of surrounding surface stretches the weaker regio[n] until its thickness reduces to about that of the distance betwee[n] molecules, and a hole appears. Its edge is pulled by surfac[e] tension at a velocity determined by the rate at which the momentu[m] flux of liquid into the edge equals the pull of the surface tension[s]. (4) A drop of milk hits a layer of water. The depth of the water i[s] about half the radius of the drop. A sequence of events occurs[.] The impacting drop spreads sideways, meets the stationary shee[t] of water, and sends up a cylindrical sheet of liquid, the base o[f] which increases in radius at a rate approximately half that of th[e] original drop velocity. The sheet projected upwards has a leading edge which is pulled back by the surface tension of the two sides[,] causing a rolled up edge. This is unstable, and its tendency to become drops results in a coronet-like shape of surprising regularity. This coronet is pulled down, by surface tension, and par[t] of its liquid moves inwards, focusing on the center where it moves upwards as a liquid spike. This is unstable, and its top liquid pinches off to form a drop. Because of its upward momentum, the drop continues to move up, but the remainder of the spike is pulled down, by surface tension.

CREDITS: Scenes from the film, "Surface Tension in Fluid Mechanics," by Lloyd Trefethen, Tufts University; Director, Jack Churchill, E.S.I.; Editor, Richard Bergman, E.S.I.

Motions Caused by Composition Gradients Along Liquid Surfaces

PURPOSE: To show that liquids move when there are surface tension gradients along their surfaces. Such gradients must be balanced by shear forces in the bounding fluids. Fluids in shear must be moving fluids. These shear forces provide one of the free boundary conditions needed to express mathematically the effects of surface tension in fluid mechanics. In these experiments, the surface tension gradients are caused by variations of kinds of molecules along surfaces.

EXPERIMENTS: (1) A limp loop of thread on water snaps into a circle when the water inside is touched by soap. The soap molecules lower the surface tension within the loop. (2) Bits of camphor scraped onto water move around rapidly. Camphor molecules lower the surface tension. Each particle dissolves unevenly, and the higher pull on the side where it dissolves the least pulls the particle along the surface. (3) A bit of wood with camphor at one end moves like a boat. The greater surface tension on the bow pulls the boat ahead. (4) Ether vapor from a swab held over a thin water layer causes the water to move away and leave a bare spot. A drop of alcohol also causes a bare spot. In both cases, the contaminating molecules lower the surface tension locally. The surface is pulled towards the surrounding regions of higher surface tension, dragging along the liquid underneath. (5) A water-alcohol mixture sloshed around in a glass leads to "wine tears", drops which move up and down the sides of the glass. Evaporation of alcohol lowers the alcoholic content of the liquid film on the glass and increases its surface tension. The surface is therefore continuously pulled from the bulk liquid up the side of the glass, pumping up liquid which accumulates to form the wine tears.

DISCUSSION: When the surface tension is unbalanced around an unrestrained floating solid, the solid must move in the direction of the unbalanced force. The loop of thread becoming a circle and the motion of camphor particles on water are examples of such unbalanced forces.

Variation in surface tension along a surface will also cause liquid surfaces to move. A force balance on a unit area of interface between a liquid, A, and another liquid such as a gas, B, indicates that the surface must move, in the direction of higher surface tension pull, at such a velocity that the imbalance in surface tension is equalled by the shear forces, τ_A and τ_A which the moving surface exerts on the bounding fluids. This leads to a boundary condition that must be met by free liquid surfaces:

$$\text{Grad } \sigma = \tau_A + \tau_A$$

These shear forces produce boundary layers in the fluids adjacent to the moving surface. The moving surface must in this way drag along its bounding fluids.

Since the types of molecules in surfaces affect the surface tension, concentration gradients along free liquid surfaces will cause motion of the liquid. Similar motions are also caused by electrical and temperature gradients along surfaces.

CREDITS: Scenes from the film, "Surface Tension in Fluid Mechanics", by Lloyd Trefethen, Tufts University; Director, Jack Churchill, E.S.I.; Editor, Richard Bergman, E.S.I.

Motions Caused by Electrical and Chemical Effects on Liquid Surfaces

PURPOSE: To illustrate by the two experiments known as the "beating heart" and the "mercury amoeba" how electrical charges and chemical effects at interfaces can cause motions of liquids.

THE "BEATING HEART" EXPERIMENT: A pool of mercury in a dish of dilute sulfuric acid with a bit of potassium dichromate added moves suddenly when touched by an iron nail. With just the right position of the nail, the mercury sets itself into sustained oscillation.

The explanation of these motions lies in the effect that electrical charges have in reducing surface tension. In the beating heart experiment, the mercury and the nail are two different metals in an electrolyte, and therefore form a battery. The mercury surface becomes electrically charged, and its surface tension is reduced by the mutually self-repellent forces of these charges along the surface. When the nail touches the mercury, the battery is short circuited and the mercury is no longer charged. Its surface tension therefore increases, and the large flat drop pulls itself up a bit. When the motion of tightening separates the mercury from the nail, the mercury again acquires a charge, slumps, touches the nail, and continues alternately touching and drawing away from the nail. The slight energy input of each cycle builds up oscillations of appreciable magnitude, oscillations which can continue for an hour or more, until the chemical source of the electrical energy is exhausted.

THE "MERCURY AMOEBA" EXPERIMENT: A pool of mercury is placed in dilute nitric acid, and a crystal of potassium dichromate is added. The mercury moves towards the crystal, at times so rapidly that it fragments itself into several drops. The motion of the mercury in this experiment is caused by changes in the surface tension along the mercury-acid interface. The surface tension becomes lower near the potassium dichromate crystal. The mercury surface therefore moves along itself, away from the crystal, and the viscous drag of this surface on the bounding acid picks up a boundary layer of acid which is jetted away from the far side. Because of this jet-thrust action, the mercury can swim towards the crystal. The jetting effect is appreciable, and it can be seen that the surface of the mercury drop acts as an effective agitator, forcing convection and diffusion of the potassium dichromate throughout the liquid.

How the crystal reduces the tension of the nearby mercury-acid interface does not appear to be well known. Presumably it is not a temperature effect, but is instead either an induced electrical effect or a molecular-concentration effect at the interface, or both.

CREDITS: Scenes from the film, "Surface Tension in Fluid Mechanics", by Lloyd Trefethen, Tufts University; Director, Jack Churchill, E.S.I.; Editor, Richard Bergman, E.S.I.

FM-79
Motions Caused by Temperature Gradients Along Liquid Surfaces

PURPOSE: To show that liquids move when there are temperature gradients along their free surfaces.

EXPERIMENTS: (1) A thin metal plate is covered by a layer of silicone oil. When a hot soldering iron touches the underside of the plate, a bare spot appears in the liquid above. As the iron is moved around under the plate, the bare spot moves along with it. (2) An ice cube instead of a soldering iron causes the liquid film above to hump up over the cold region. Such results with hot and cold regions reflect the change in surface tension caused by temperature. Increasing the temperature always lowers surface tension, because it becomes easier to pull molecules up into the surface. Above the soldering iron, the hot liquid has a lower surface tension than the cold liquid around it, so the surface is pulled to the cold regions. In moving, the surface drags the underlying liquid away from the hot region, leaving a bare space. Above a cold spot, the higher surface tension pulls nearby surface towards the cold spot, and liquid accumulates as a hump wherever the ice cube has cooled the plate. (3) An air bubble in a silicone oil in a horizontal glass tube moves itself towards the hot tip of a soldering iron held against the glass to one side of the bubble. Two bubbles swimming together in this way coalesce quickly on meeting. Such bubbles swim whenever one side is hotter than the other. The higher surface tension of the cold side continually pulls surface around from the hot end. This moving surface, through viscosity, moves liquid away from the hot side and deposits it on the cold side. In this way the bubble moves itself continually in the hot direction. When two bubbles are close and the liquid between is hot, this moving-surface effect stretches and ruptures the liquid film. (4) An electrically-heated horizontal wire in a mixture of acetone and about 2% water generates bubbles of acetone and water vapor at nucleation sites along the wire. At low heating rates, the bubbles, instead of rising off the wire at the nucleation sites, swim along the wire, jetting hot liquid away from their cold ends. Because of the mixture of vapors in these bubbles, they can have appreciable temperature differences around their surfaces (bubbles in a pure liquid would not). The differences in temperature cause surface-tension gradients that move surface to the cold end, and jet the accompanying hot liquid off the cold end. As a bubble grows at its nucleation site, its jet forms a ring vortex above the bubble. Once perturbed, the bubble moves along the wire, still jetting back hot liquid. These jets show periodic variations reflecting the alternating current in the wire, and the variations in these jets move at speeds that suggest that the bubble surfaces are moving quite fast, about 1,000 bubble diameters a second. Once bubbles start moving, the wire is colder where they have been and hotter where they are going, so they just keep going to where it is hotter.

CREDITS: Scenes from the film, "Surface Tension in Fluid Mechanics," by Lloyd Trefethen, Tufts University; Director, Jack Churchill, E.S.I.; Editor, Richard Bergman, E.S.I.

FM-80
Hele-Shaw Analog to Potential Flows
Part I: Sources and sinks in uniform flow

PURPOSE: To demonstrate the superposition of source, sink, and streaming flows.

APPARATUS AND PROCEDURE: Water flows in the thin (0.020 in.) gap between parallel transparent plates (6 in. x 20 in.). The camera looks normal to the plates; the side gaskets which establish the plate separation are visible along the top and bottom of the field of view. The water can be made to enter or leave the flow gap uniformly from the left or right ends, as well as from any of five small holes (sources and sinks) which pierce one plate and are located on a line midway between the side gaskets. The left-most hole is surrounded by eight dye-injection holes. Dye holes (out of view) mark the uniform flow when it enters from the left.

DISCUSSION: Because of very small speed and gap width, the Reynolds Number is low enough to make viscous forces dominant and inertial forces negligible. Hence the velocity averaged across the gap is proportional to the gradient of the pressure, and the pressure obeys Laplace's equation. (See references.) Thus the Hele-Shaw flows are formally analogous to two-dimensional fields in (i) irrotational flow of an incompressible liquid, (ii) electrostatics, (iii) magnetostatics, and (iv) steady heat flow in a medium of uniform conductivity; and the pressure in the Hele-Shaw experiment is the potential, corresponding respectively to (i) the velocity potential, (ii) the electrostatic potential, (iii) the vector magnetic potential, and (iv) the temperature.

SEQUENCES: (1) *Source Near Walls*. Only one hole is open, and both ends are open equally to the drain. A source flow, marked by dye, is suddenly started. Initially the dye front is circular, but as it enlarges, the influence of the horizontal walls distorts it. (2) Dye is emitted from holes around the source, thus showing the streaklines (and streamlines) of a source flow centered between two walls. (2) *Rankine Half-Body*. Initially there is a steady, uniform flow from left to right, with straight, parallel dye streamlines. Flow from one source is added; the separation streamline forms a semi-infinite Rankline "half-body". (4) The streamline dividing the uniform flow from the source flow is replaced by a solid boundary; the external streamlines are unaltered, illustrating the principle that any streamline may be replaced by a solid wall. (5) Uniform flow plus source flow. The relative strength of the source flow is varied, and the size of the Rankine body changes. (6) *Rankine Oval*. Initially, a uniform flow plus source flow, making a Rankine half-body. A sink is added downstream of the source, modifying the shape of the half-body. When the sink strength is adjusted to equal the source strength, a streaming flow past a closed "Rankine oval" is produced. (7) Uniform flow, plus a source flow, plus an equal sink flow from the hole closest to the source. The distance between the source and sink is small compared with the size of the Rankine oval, and a doublet is approximated. The Rankine oval is nearly circular. (8) Uniform flow, plus source flow plus two sinks equalling together the source flow strength. This produces a pear-shaped closed streamline, and illustrates that potential flow past a desired shape may be simulated by superposition of distributed sources and sinks of which the algebraic sum is zero.

REFERENCES: (1) Hele-Shaw, H. S., Trans. Inst. Nav. Arch. Vol. XL. 1898. (2) Schlichting, H., Boundary Layer Theory, McGraw-Hill, 1960.

CREDITS: Experiments and script designed by C. Conn; Experimental technician, G. Fardy; Director, A. Pesetsky; Cinematography by A. Morochnik; Editor, W. Hansard. Produced by the National Committee for Fluid Mechanics Films and Educational Services Incorporated, with the support of the National Science Foundation.

FM-81
Hele-Shaw Analog to Potential Flows
Part II: Sources and sinks

PURPOSE: To demonstrate the superposition of source, sink, and streaming flows.

APPARATUS AND PROCEDURE: Water flows in the thin (0.020 in.) gap between parallel transparent plates (6 in. x 20 in.). The camera looks normal to the plates; the side gaskets which establish the plate separation are visible along the top and bottom of the field of view. The water can be made to enter or leave the flow gap uniformly from the left or right ends, as well as from any of five small holes (sources and sinks) which pierce one plate and are located on a line midway between the side gaskets. The left-most hole is surrounded by eight dye-injection holes. Dye holes (out of view) mark the uniform flow when it enters from the left.

DISCUSSION: Because of very small speed and gap width, the Reynolds Number is low enough to make viscous forces dominant and inertial forces negligible. Hence the velocity averaged across the gap is proportional to the gradient of the pressure, and the pressure obeys Laplace's equation. (See references.) Thus the Hele-Shaw flows are formally analogous to two-dimensional fields in (i) irrotational flow of an incompressible liquid, (ii) electrostatics, (iii) magnetostatics, and (iv) steady heat flow in a medium of uniform conductivity; and the pressure in the Hele-Shaw experiment is the potential, corresponding respectively to (i) the velocity potential, (ii) the electrostatic potential, (iii) the vector magnetic potential, and (iv) the temperature.

SEQUENCES: (1) Only the left hole is open, and both ends are open equally to the drain. A source flow marked by dye is suddenly started. Initially the dye front is circular, but as it enlarges, the influence of the horizontal walls distorts it. (2) Dye is emitted from the holes around the source, thus showing the streaklines (and streamlines) of a source flow centered between two walls. (3) Initially, there is a steady source flow from the left hole, marked by dye. The right hole is opened to drain, thereby adding a sink which is weaker than the source. (4) There is steady flow from the source on the left to the sink on the right; the ends are closed to flow. Dye is then emitted from the holes around the source and traces out the streaklines of the flow. Near the source-sink, the flow approximates a source-sink pair in an infinite medium. Far from the source-sink, the walls at top and bottom of the picture strongly affect the flow. (5) A flow similar to (4), but here the spacing between source and sink is much smaller than the channel width, and the effect of the walls is reduced. (6) Initially a source flow with streamlines marked by dye. The channel ends pass equal flows. A second source, equal in strength to the first, is started. The dividing streamline is a nearly straight line, perpendicular to a line joining the sources, and midway between them. The source and "image" source pair thus represents a source near a wall. A solid wall replaces the dividing streamline of sequence (6).

REFERENCES: (1) Hele-Shaw, H. S., Trans. Inst. Nav. Arch. Vol. XL. 1898. (2) Schlichting, H., Boundary Layer Theory, McGraw-Hill, 1960.

CREDITS: Experiments and script designed by C. Conn; Experimental technician, G. Fardy; Director, A. Pesetsky; Cinematography by A. Morochnick; Editor, W. Hansard. Produced by the National Committee for Fluid Mechanics Films and Educational Services Incorporated, with the support of the National Science Foundation.

FM-82
Water Jet Instability in Electric Field

PURPOSE: To show wave amplification with distance on a sinusoidally-excited water jet in the presence of an electric field.

APPARATUS: A horizontal jet of ordinary tap water issues from a nozzle. At the nozzle exit, a small-amplitude sinusoidal force produces horizontal disturbances of the jet. The jet is grounded electrically and passes between a pair of vertical plane-parallel electrodes that are at a common d-c potential. The jet and plates are viewed from above. Both continuous and stroboscopic light are used to view the jet.

SEQUENCES: (1) *Overall View*. Relative sizes of nozzle, jet and plates are shown. (2) *Schematic of Experiment Without Electric Field*. A sketch shows the nozzle, the sinusoidal excitation, and the jet without electric field. (3) *Ordinary Waves, Stroboscopic Light*. There is no electric field. The stroboscopic light flashes at the exciter freqency. Surface tension causes the jet to behave like a moving stretched string, and waves propagate upstream and downstream, relative to the jet. However, since the jet velocity far exceeds the velocity of these waves, both waves propagate downstream with respect to the viewer. Because they have differing wavelengths, they beat with each other and form a standing wave pattern, a portion of which is seen in this sequence. (4) *Ordinary Waves, Continuous Light*. Repeat of (3) with ordinary lighting. (5) *Experiment with Electric Field*. The sketch of the apparatus is shown with the plates and source of potential added. (6) *Jet Instability, Stroboscopic Light*. The jet is shown under stroboscopic light as the voltage is slowly increased from 0 to 15 kilovolts. With no voltage, the waves simply interfere as before to form a standing wave pattern. With the electric field present, the waves grow with distance along the jet, i.e. they amplify. To understand why, we recognize first that the water is sufficiently conductive for the jet to remain at a uniform potential. When the jet is centered between the plates, the electric forces of attraction toward each of the plates cancel. However, when it is deflected slightly toward one plate, there is an increase in the amount of charge on that side and hence a force tending to deflect the jet further toward that plate. The wave amplitude grows with distance along the jet instead of with time at a fixed point in the jet, because the longitudinal velocity of the jet exceeds the velocity of waves relative to the jet. (7) *Jet Instability, Continuous Light*. Repeat of (6) with ordinary lighting, and then with stroboscopic lighting added.

REFERENCE: Ch. 10, "Electromechanical Dynamics", H. H. Woodson and J. R. Melcher, Wiley 1968.

CREDITS: Scenes from the film "Complex Waves II: Instability, Convection, and Amplifying Waves" (a production of the National Committee for Electrical Engineering Films and E.D.C.) by J. R. Melcher, M.I.T.; Consultants, A. H. Shapiro, M.I.T., and J. R. Melcher, M.I.T.; Producer, V. Komow, E.D.C.; Editor, E. Carini, E.D.C.; Art Director, R. P. Larkin, E.D.C. Produced by the National Committee for Fluid Mechanics Films and Education Development Center, Inc. (formerly Educational Services Incorporated), with the support of the National Science Foundation.

FM-83
Induced J X B Forces in Solids and Liquids

PURPOSE: To demonstrate the electromagnetic forces that act on an electrical conductor moving through a magnetic field.

SEQUENCES: (1, 2) A metallic ring is suspended on a pendulum. The pivot axis of the pendulum is parallel to the axis of the ring. At the bottom of the ring's path is a permanent magnet whose field **B** lies along the axis of the ring. When the ring is released from an elevated position it quickly comes to rest as it passes into the magnetic field. A circumferential e.m.f. is generated as the ring begins to link magnetic flux. Since the ring is conducting, the e.m.f. produces a circumferential current, **J**. In the region where the axial field **B** is present, the circumferential current **J** produces a Lorentz force **J** X **B** per unit volume acting perpendicular to both **J** and **B**, i.e. along the line of motion of the ring. The force thus generated by induced currents always opposes the motion of the conductor (Lenz' law), and the motion of the ring is quickly damped. The kinetic energy of the ring is dissipated in Joulean losses. (3) A large flat plate of aluminum, suspended on strings, is allowed to swing into the gap of the magnet. Its motion is rapidly damped as it passes into the field region. This shows that the impeding **J** X **B** forces do not depend on a wire-shaped geometry, but arise in any electrical conductor whose geometry allows closed current paths under the action of the induced e.m.f. (4) A low-friction rotary bearing carries four pairs of permanent magnets that produce fields parallel to the axis of rotation. These fields are of opposite sense in each quadrant. A flat sheet of copper is passed several times through the field region. This sets the device into rotation, even though there is no physical contact between the copper sheet and the apparatus. The **J** X **B** forces acting on the copper sheet as a result of the induced currents act equally on the permanent magnets, but in opposite sense (Newton's law of action and reaction). (5) The same rotary device is used, but a pipe carrying liquid metal passes through the field region. When the liquid metal begins to flow through the pipe the induced forces "push" on the magnets, and the rotor turns. Such a device could in principle be used as an MHD turbine or an MHD flowmeter. Its principle of operation is similar to that for the preceding experiment. (6, 7, 8) A long, horizontal trough of rectangular cross-section, containing mercury, lies between the poles of a permanent magnet. The magnet has a vertical field and may be moved parallel to the trough. The mercury is initially stationary. When the magnet moves, thus inducing electric currents in the mercury, the mercury is set into motion. Unlike the solid bodies, the liquid medium deforms greatly under the action of **J** X **B** forces. Swirls and vortices are developed in the mercury. Since the **J** X **B** force is, in general, a rotational force (i.e., curl **J** X **B** ≠ 0), induced **J** X **B** forces usually alter the vorticity of a fluid motion.

CREDITS: Scenes from the NCFMF film "Magnetohydrodynamics" by J.A. Shercliff, Warwick University; Consultants, J.A. Shercliff and A.H. Shapiro, M.I.T.; Producer/Director, J. Friedman, E.D.C.; Editor, W. Tannebring, E.D.C.; Art Director, R.P. Larkin, E.D.C. Produced by the National Committee for Fluid Mechanics Films and Education Development Center, Inc., with the support of the National Science Foundation.

FM-84
An MHD Pump

PURPOSE: To demonstrate the operation of a simple magnetohydrodynamic pump constructed by applying crossed **J** and **B** fields to an electrically-conducting liquid.

APPARATUS: A mercury flow system comprises an open, vertical reservoir that leads into a horizontal pumping duct of rectangular cross-section, followed by a vertical tube which then bends into a horizontal return leg whose highest level is above that of the mercury in the reservoir. At the exit of the horizontal return leg is a nozzle, open to the atmosphere, and directed downwards toward the reservoir. Two open, vertical manometers are spaced longitudinally along the rectangular pumping duct. The system is made of non-conducting plastic. Between the manometers, in the top and bottom walls of the pumping duct, are plate electrodes connected to an external battery which passes a current **J** vertically through the mercury in the duct. A permanent magnet is arranged so that a horizontal **B**-field can be applied to the mercury in the region of current flow. The electrodes and magnet, together with the current source, comprise the MHD pump.

SEQUENCES: (1) A general view of the apparatus. (2) A schematic sketch of the apparatus. (3) The current is on, but the **B**-field is at first not present. There is no motion of the mercury, and the manometers are at the same level. When the magnet is moved to the current region, the mercury flows around the circuit. The net head rise produced by the MHD pump is the kinetic-energy head at the nozzle exit plus the gravitational head between the reservoir level and the nozzle exit. (4) A close-up of the manometers shows that the pressure *rises* in the direction of flow. At steady state, there are no fluid accelerations in the pumping duct. Moreover, viscous forces are very small. The magnitude of the pressure gradient is, therefore, that which will balance the longitudinal **J** X **B** force arising from the vertical current in the horizontal, transverse magnetic field. The **J** X **B** force is from right to left, requiring that the balancing pressure gradient be one that rises in the flow direction. The pressure head thus developed by the **J** X **B** force accounts for the net head rise referred to above, plus whatever frictional head losses are present. (5) A close-up of the manometers and the pumping duct as the magnet moves into place, showing the development of the pressure gradient. (6, 7) Several repeats of the entire experiment, with the magnet alternately in and out of place.

CREDITS: Scenes from the NCFMF film "Magnetohydrodynamics" by J.A. Shercliff, Warwick University; Consultants, J.A. Shercliff and A.H. Shapiro, M.I.T.; Producer/Director, J. Friedman, E.D.C.; Editor, W. Tannebring, E.D.C.; Art Director, R.P. Larkin, E.D.C. Produced by the National Committee for Fluid Mechanics Films and Education Development Center, Inc., with the support of the National Science Foundation.

FM-86
Suppression of Vorticity by MHD Forces

PURPOSE: To show how the electric currents induced by fluid motion in a magnetic field can produce electromagnetic forces that damp out vorticity.

A PRELIMINARY MECHANICAL EXPERIMENT: A metallic ring is suspended on bearings so that it can spin freely about a diameter. The ring is set spinning while between the pole faces of a permanent magnet. When the axis of spin is aligned with the direction of the **B**-field, the presence of the **B**-field does not damp out the motion. However, when the device is turned so that the axis of spin is normal to the **B**-field, the rotation of the ring is rapidly damped out. In this case, the magnetic flux linked by the ring changes as the ring turns. The changing of the linked flux generates an electromotive force around the ring, and this in turn drives a current **J** around the ring. The electromagnetic force (**J** X **B**) of the induced current in the **B**-field of the magnet produces in the material of the ring a torque which reduces the angular momentum of the ring. This suggests that if a magnetic field has a component perpendicular to the angular momentum of a conducting medium, the induced currents may lead to forces that reduce the angular momentum.

A FLUID-MECHANICAL EXPERIMENT: A long horizontal trough of rectangular cross-section contains mercury. The trough lies between stationary pole pieces that produce a horizontal magnetic field in the mercury over a certain horizontal length. A vertical flat plate, held broadside, is moved horizontally through the trough. An overhead camera views the surface of the mercury. Reflection of a grid pattern in the surface makes visible distortions in that surface due to the dimples of vertical vortices that are shed from the vertical edges of the plate.

SEQUENCES: (1) *No **B**-Field*. The magnet is not on. A Karman vortex street is left in the wake of the plate. Each vertical vortex decays slowly as it is left behind. (2) *Comparison of Zero **B**-Field with Low **B**-Field*. In a split-screen view, the vortices are seen to decay much more rapidly when a low **B**-field is applied than when there is no **B**-field at all. Note that the **B**-field is perpendicular to the axis of vorticity. (3) *High **B**-Field*. No vortices are clearly visible. Either the shed vortices are damped out extremely quickly, or the boundary layer is stabilized against the shedding of vortices. (4) The camera now moves with the plate. The plate starts outside the **B**-field region, and at first strong vortices are shed. Then the plate enters the **B**-field region and the vortices are suppressed. When the plate leaves the **B**-field region vortices reappear in the wake.

CREDITS: Scenes from the NCFMF film "Magnetohydrodynamics" by J.A. Shercliff, Warwick University; Consultants, J.A. Shercliff and A.H. Shapiro, M.I.T.; Producer/Director, J. Friedman, E.D.C.; Editor, W. Tannebring, E.D.C.; Art Director, R.P. Larkin, E.D.C. Produced by the National Committee for Fluid Mechanics Films and Education Development Center, Inc., with the support of the National Science Foundation.

FM-87
The Hartmann Layer

PURPOSE: To show how a magnetic field normal to a wall produces a Hartmann-type boundary layer in a conducting fluid flowing along the wall.

APPARATUS: Mercury is contained in the annular space between a pair of concentric, vertical cylinders. The cylinders and the bottom of the annulus are electrical insulators, while the top is open. The inner cylinder rotates clockwise; the outer cylinder is stationary. An electromagnet applies a radial **B**-field between the inner and outer cylinders. A non-conducting vorticity float is free to rotate at the end of an arm that is itself pivoted about the vertical axis of the cylinders. It shows the fluid velocity and the vertical vorticity in the center of the annular gap.

SEQUENCES: (1) A schematic of the apparatus. (2) Installation of the vorticity float, showing its construction. (3) With no **B**-field present, the float shows that the fluid velocity is clockwise and the vorticity anti-clockwise. The latter is associated with the velocity distribution produced by viscosity and the no-slip conditions at the inner and outer cylinder walls. (4) When the radial **B**-field is applied, the vorticity in the center of the annular gap quickly becomes clockwise. The rotation of the float shows that the mercury there is in solid-body rotation. (5) A mechanical analog experiment. A conducting loop of metal is suspended in the empty annulus so that it can be spun about a vertical axis. It simulates a conducting loop of mercury. The loop is first given an anti-clockwise spin while the tethering arm is rotated clockwise. When the **B**-field comes on, the loop is locked into a clockwise solid-body rotation. Currents are induced in the spinning loop as it cuts the radial field lines. The resulting **J** X **B** forces slow it down until it no longer links a changing flux (see FM-86). (6, 7) A repeat of (4), with the circumferential velocity distribution superimposed, showing that the solid-body velocity distribution cannot fit the no-slip boundary conditions at the walls. By inference, there are thin layers (Hartmann layers) near the walls in which the necessary velocity adjustment occurs.

DISCUSSION: The strength of the radial field **B** is inversely proportional to radius. Except near the walls of the inner and outer cylinders, the fluid velocity **V** is also proportional to radius when the **B**-field is on. Thus the vertical induced electric field **V** X **B** is constant with radius, and there is no tendency to drive vertical currents. Near the walls **V** X **B** is not constant and there are vertical currents that are closed by horizontal current flows at top and bottom of the annulus. The thickness of the Hartmann layer is governed by a balance between the viscous forces and the **J** X **B** forces developed by the induced currents.

CREDITS: Scenes from the NCFMF film "Magnetohydrodynamics" by J.A. Shercliff, Warwick University; Consultants, J.A. Shercliff and A.H. Shapiro, M.I.T.; Producer/Director, J. Friedman, E.D.C.; Editor, W. Tannebring, E.D.C.; Art Director, R.P. Larkin, E.D.C. Produced by the National Committee for Fluid Mechanics Films and Education Development Center, Inc., with the support of the National Science Foundation.

FM-88
Laminar Boundary Layers

PURPOSE: To illustrate some characteristics of laminar boundary layers.

APPARATUS: Water flows in a channel of rectangular cross-section. The flow is made visible by hydrogen bubbles electrolyzed at a wire normal to the flow. By pulsing the current, fluid lines (time lines) normal to the flow are released periodically from the wire.

SEQUENCES: (1) Time lines are released upstream of a flat plate installed in the channel. Each time line remains virtually straight until it reaches the plate. The fluid in contact with the plate is at rest, and a region of decelerated flow diffuses outward from the plate. (2) Time lines are released from a wire passing through the plate at a location downstream of the leading edge. The fluid near the plate moves more slowly than the undisturbed fluid far from the plate. The boundary layer is that region adjacent to the plate in which the velocity gradient is sufficiently large for viscous forces to be significant. (3) The flat-plate experiment is performed with zero longitudinal pressure gradient. By following a single time line, it can be seen that the boundary layer thickness grows with distance from the leading edge. This is also observed by examining time lines simultaneously released at different distances from the leading edge. The increasing thickness may be ascribed to the diffusion of vorticity through the action of viscosity. (4) A close-up of a time line released downstream of the leading edge shows that there is no fluid velocity at the plate, i.e. no slip between the fluid and the wall. (5) The channel is now convergent, with one straight wall. The pressure falls along this wall in the flow direction. A streakline near the straight wall serves as a reference for relative transverse distances. Examination of time lines shows that the boundary layer thickness decreases in the zone of decreasing pressure. Although viscous diffusion increases the amount of flow in the boundary layer, the increase of flow speed and the fuller form of the velocity profile make it possible for the increased flow to be carried in a layer of smaller thickness. (6) The boundary layer is observed on the wall of a diffuser, in which the pressure increases with longitudinal distance. The boundary layer thickness now increases rapidly, owing to the additive action of three effects: (i) viscous entrainment of additional mass flow, (ii) reduction in velocity level, and (iii) a less full form of the velocity profile. A close-up also shows back flow in the boundary layer very close to the wall.

CREDITS: Scenes from the NCFMF film "Fundamentals of Boundary Layers" by F. H. Abernathy, Harvard University; Consultants, F. H. Abernathy and A. H. Shapiro, M.I.T.; Producer/Director, J. Friedman, E.D.C.; Editor, M. Chalufour, E.D.C.; Art Director, R. P. Larkin, E.D.C. Produced by the National Committee for Fluid Mechanics Films and Education Development Center, Inc., with the support of the National Science Foundation.

FM-89
Turbulent Boundary Layers

PURPOSE: To illustrate some characteristics of turbulent boundary layers, especially by comparison with laminar boundary layers.

APPARATUS: Water flows in a channel of rectangular cross-section. The flow is made visible by hydrogen bubbles electrolyzed at a wire normal to the flow. By pulsing the current, fluid lines (time lines) normal to the flow are released periodically from the wire.

SEQUENCES: (1) A flat plate is installed in the channel. On the lower side the flow is laminar, while roughness near the leading edge on the upper side induces the boundary layer flow to become turbulent. The camera views a region well downstream of the leading edge. Time lines normal to the plate are released simultaneously on the upper and lower sides. (2) A succession of still photos shows the positions of the time lines at a fixed interval of time after they are released. Each time line approximately represents an average velocity profile during that time interval. On the lower (laminar) side the flow is essentially steady. On the upper (turbulent) side the differing appearance of successive time lines shows that the flow is inherently unsteady. (3) The additive optical super-position of the time lines of the previous sequence suggests a procedure by which average velocities may be defined in an unsteady turbulent flow. (4) The laminar velocity profile is compared with the time-mean turbulent velocity profile. The turbulent layer is thicker, indicating that the diffusive transport of momentum is more rapid. Also, the velocity gradient at the wall is greater for the turbulent case, indicating a greater wall skin-friction stress. (5) The laminar boundary layer is observed on the divergent walls of a diffuser, where the pressure gradient is adverse, i.e. rising in the flow direction. The flow separates along both walls. (6) In the same diffuser the flow on the bottom wall is made turbulent by trip wires. The turbulent layer does not separate, indicating that turbulent layers can withstand higher adverse pressure gradients than laminar layers before separation occurs. This results from the greater ability of the turbulent boundary layer, with its transverse mass interchange, to transfer momentum from the free stream into the slow-moving layers near the wall. (7) In the same experiment, it is seen that the laminar layer on the top wall continues to separate while the turbulent layer on the bottom wall remains attached.

CREDITS: Scenes from the NCFMF film "Fundamentals of Boundary Layers" by F. H. Abernathy, Harvard University; Consultants, F. H. Abernathy and A. H. Shapiro, M.I.T.; Producer/Director, J. Friedman, E.D.C.; Editor, M. Chalufour, E.D.C.; Art Director, R. P. Larkin, E.D.C. Produced by the National Committee for Fluid Mechanics Films and Education Development Center, Inc., with the support of the National Science Foundation.

FM-90
Supersonic Flow Past Diamond Airfoil

PURPOSE: To show the wave pattern for supersonic flow past a diamond-shaped (symmetrical double-wedge) airfoil at zero incidence; also to show the establishment of the steady supersonic flow over the airfoil in a wind tunnel.

APPARATUS: The two-dimensional airfoil is in a two-dimensional, supersonic wind tunnel with a test-section Mach number of 1.6. High-speed schlieren photography is used to observe the flow. With the schlieren knife edge parallel to the flow (horizontal), density gradients transverse to the flow direction are shown. With the knife edge transverse to the flow (vertical), density gradients in the flow direction are shown.

SEQUENCES: (1) Starting of the supersonic flow. (2) Steady supersonic flow at $M \infty = 1.6$; seen first with knife edge horizontal, then vertical. Oblique shocks spring from the leading edge and trailing edge, and centered rarefaction waves (Prandtl-Meyer expansion fans) from the sharp shoulders. Between the shock and rarefaction waves, the flows are locally uniform. The differences in light intensity between the leading and trailing edge waves on the one hand and the shoulder waves on the other hand show that they are respectively of opposite sense (compression and rarefaction, respectively). Although the pictures with the two different orientations of the knife edge look different, they yield consistent interpretations. With the knife edge horizontal, the wake (with strong vertical density gradients) is clearly seen; with the knife edge vertical, it is barely visible. The geometry of the three wave systems (leading-edge shock, rarefaction wave, trailing-edge shock) springing from either surface shows that the shocks will interact with the rarefaction fan far from the airfoil, tending toward mutual cancellation. The shocks are actually very thin, but appear to broaden because they interact with the viscous boundary layers on the glass side walls of the tunnel. (3) Establishment of the supersonic flow over the airfoil, as in sequence (1), but in extra-slow motion. Observed first with knife edge horizontal, then vertical. As the tunnel flow begins, the flow is at first subsonic; no waves are present, and the maximum density gradients are localized at the leading edge (which is a stagnation point) and at the shoulders (where the velocity tends to be very large). At higher subsonic speeds in the test section, the flow at the sharp shoulders becomes sonic and is followed by a region of supersonic flow terminated by a shock wave. When the nearly-normal "starting shock" of the tunnel moves downstream to the airfoil nose, it curves, then momentarily becomes an oblique shock of "strong" type, then becomes an oblique shock of "weak" type. As the now-distorted starting shock wave passes over the model, complex and interesting interactions with the waves already existing on the model are produced. As soon as the starting shock leaves the trailing edge, the steady supersonic wave pattern near the airfoil is fully established.

CREDITS: Experiments and film by the National Physical Laboratory, England; Consultant, A. H. Shapiro, M.I.T.; Producer, R. Bergman, E.S.I.; Editor, W. Tannebring, E.S.I.; Art Director, R. P. Larkin, E.S.I. Produced by the National Committee for Fluid Mechanics Films and Educational Services Incorporated, with the support of the National Science Foundation.

FM-91
Modes of Sloshing in Tanks

PURPOSE: To show how horizontal oscillation of an open tank of water produces different modes of sloshing at different frequencies.

APPARATUS: An open cylindrical tank of dyed water is mounted on a frame which is translated to-and-fro by a drive mechanism in a horizontal, simple harmonic motion. The amplitude of horizontal displacement is maintained constant, but the frequency is varied from one experiment to the next.

SEQUENCES: (1) View of the tank and its supporting frame. (2) View of the drive mechanism. (3) Chart with ordinate "wave height" and abscissa "exciting frequency", showing that there is a particular resonant frequency for planar waves. In this mode of oscillation or sloshing the free surface of the water remains essentially planar and oscillates about that horizontal diameter in the free surface which is perpendicular to the direction of excitation displacement. (4) An experiment with low exciting frequency, well below resonance. The free surface oscillates in a planar fashion, the fluid motion at the free surface being essentially vertical. (5) Excitation above the resonant frequency. Again the planar mode of wave motion is observed. (6) Excitation above the resonant frequency, at the same frequency as in (5), but the water is first stirred to create an initial rotary motion. The resulting steady wave motion is a rotary sloshing mode. In this rotary mode the free surface remains planar, but the horizontal axis about which the surface tilts is itself rotating. The rotary mode may be regarded as the superposition of two planar modes of equal amplitude but 90° apart. It is common experience that the rotary mode may be produced easily in a hand-held glass of liquid which is translated back-and-forth in a straight line. (7) There is a range of subresonant frequencies, from about ⅔ of resonant frequency to the resonant frequency itself, for which the simple planar mode does not persist even when started. The experiment shows that there is a complicated back-and forth transfer of energy between the planar mode, the rotary mode, and a more complicated assymmetrical mode in which there are two antinodes located about midway between the center and the vessel wall. The antinodes change in amplitude and the line connecting them rotates.

CREDITS: Edited from a film by R. E. Hutton, TRW Systems, Redondo Beach, California, which was supported by the National Aeronautics and Space Administration. Consultant, A. H. Shapiro, M.I.T.; Producer, R. Bergman, E.D.C.; Editor, W. Tannebring, E.D.C. Produced by the National Committee for Fluid Mechanics Films and Education Development Center, Inc., with the support of the National Science Foundation.

FM-92
Stages of Boundary-Layer
Instability and Transition

PURPOSE: To show the successive steps of instability and transition by which a laminar boundary layer becomes turbulent.

APPARATUS: A long circular cylinder with a pointed ogive nose is aligned with the flow in a low-turbulence wind tunnel. The boundary layer on the cylinder is thin compared with the diameter; hence the boundary layer is very much like that of a planar two-dimensional flow. Events in the boundary layer are made visible by a streamtube of smoke which impinges on the pointed nose and then flows along the cylinder as a sheath thinner than the boundary layer. Except for the final experiment, which is seen in silhouette by means of back lighting, the test body is seen by means of front lighting. High-speed photography is used.

SEQUENCES: (1) Sketch of apparatus. (2) A preliminary view of the flow. (3) An experiment with nearly zero longitudinal pressure gradient. Various stages of instability and transition are seen, which shift forward and backward in an unsteady manner. This is characteristic of instability phenomena which depend for their initiation upon accidental perturbations. (4) A much-simplified sketch identifying four stages of the process by which a laminar boundary layer becomes turbulent. Close examination of the experimental scenes reveals these four stages, although as explained the positions of the four steps varies with time. The stages are (i) growth of two-dimensional Tollmien-Schlichting waves, corresponding to a small-perturbation instability in the laminar boundary layer; (ii) growth of a secondary instability upon the combination of the original undisturbed flow with the Tollmien-Schlichting motion, and involving the three-dimensional distortion of the original transverse vortex lines such that streamwise vorticity develops; (iii) growth of turbulent spots in the regions containing streamwise vorticity; (iv) development and merging of the spots into fully developed turbulence. (5) Repeat of the experiment with nearly zero pressure gradient. (6) An experiment with a positive (adverse) pressure gradient. The events are similar to those with zero pressure gradient, except that the four stages do not shift about as much and the Tollmien-Schlichting instability is more pronounced. (7) A repeat of the sketch identifying the four stages. (8) A repeat of (6). (9) The experiment seen with silhouette lighting. The Tollmien-Schlichting instability occurs occasionally and is seen as exponentially-growing sine waves. Each such growth, after reaching a certain amplitude, develops strong motions normal to the surface followed rapidly by a "bursting" of the flow that corresponds to stages (iii) and (iv) as described previously.

CREDITS: From a film by F. N. M. Brown, University of Notre Dame; Consultant, A. H. Shapiro, M.I.T.; Producer, R. Bergman, E.D.C.; Editor, W. Tannebring, E.D.C.; Art Director, R. P. Larkin, E.D.C. Produced by the National Committee for Fluid Mechanics Films and Education Development Center, Inc. (formerly Educational Services Inc.), with the support of the National Science Foundation.

FM-93
Supersonic Spike Inlet
With Variable Geometry

PURPOSE: To show the various flow patterns near the inlet as the central spike is moved inwards and outwards.

APPARATUS: A supersonic wind tunnel produces a flow with Mach number 3.0 in the test section. The front end of the model containing the axisymmetric air inlet is seen through a circular window. The inlet is annular, with a central spike that may be moved axially. This spike is conical at its forward end. At the design location of the spike for M = 3, the conical shock attached to the spike is supposed to intersect the sharp lip of the annular inlet. The flow is observed by schlieren photography, with the knife edge perpendicular to the flow (i.e. vertical).

SEQUENCES: (1), (2) Sketches of the apparatus. (3) Preliminary experiment, with the spike forward of the design location. The conical shock, which is of the "weak" family, passes outside the lip of the annulus. (4) The spike is gradually retracted until it reaches the design location. This position is marked for reference by an optical superimposure which is seen again in later experiments. (5) The spike is retracted through and beyond the design location. At a certain position, the stagnation pressure losses in the inlet are so great that the throat of the inlet can no longer pass all the flow in the "capture cross-section" of the lip. The shock system suddenly assumes a different configuration: the attached conical shock is of the "strong" family and passes outside the annulus lip; further, there is a normal shock near the lip. This configuration allows some of the flow in the capture cross-section to spill around the lip. The flow pattern is also unstable, with the result that a high frequency oscillatory transition between different stages occurs, called "buzz." (6) Starting with the "buzz" condition, the spike now reverses and moves forward. The "buzz" instability soon ceases but the above-described shock system allowing spillage persists even when the spike passes again through the design location. After the spike moves well forward of the latter the shock pattern changes to a simple weak conical shock from the nose which, however, passes outside the lip. (7) The foregoing sequences of spike retraction followed by spike advance are repeated, with the superimposure of the design location as a guide. The hysteresis in the flow patterns is shown by the fact that, for the same spike location, the wave patterns are different for the retraction phase and the advancing phase. This comes about because the air-swallowing capacity of the annulus is less when the shock system produces large losses in stagnation pressure than when it produces small losses.

CREDITS: From a film by Office National d'Études et de Recherches Aérospatiales (O.N.E.R.A.), France; Consultant, A. H. Shapiro, M.I.T.; Producer, R. Bergman, E.D.C.; Editor, W. Tannebring, E.D.C.; Art Director, R. P. Larkin, E.D.C. Produced by the National Committee for Fluid Mechanics Films and Education Development Center, Inc. (formerly Educational Services Inc.), with the support of the National Science Foundation.

FM-97
Flow Through Fans and Propellers

PURPOSE: To illustrate the flow patterns through a propeller or through an unshrouded fan.

APPARATUS: A rotating model propeller is supported in a large water tunnel. Water flowing through the tunnel past the fixed propeller produces a flow relative to the propeller that is essentially the same as though the propeller were itself advancing through stationary fluid. The flow is made visible by light reflected from numerous small air bubbles entrained in the water.

SEQUENCES: (1) An establishing scene showing the general character of the flow. (2) An experiment in which there is no net water flow through the tunnel. This shows the flow in the neighborhood of a propeller on an airplane when the latter is stationary on the ground. It also simulates an unshrouded ventilating fan in a large room. The flow upstream of the fan enters from all directions as though the disk swept by the fan were a sink of fluid. Through the fan disk itself the principal motion is a jet-like flow which propels the fluid. This jet remains fairly well-defined downstream of the fan, but becomes narrower, indicating that the jet is accelerating. The sink-like flow upstream accelerates as it approaches the fan; thus the pressure is below ambient just ahead of the fan disk. The jet leaves the fan disk at a pressure above ambient; as it readjusts to ambient pressure far downstream, it accelerates, thus explaining the contraction of the jet wake. The pressure rise across the fan disk results in a forward thrust on the fan. (3) Repeat of foregoing scene. (4) An experiment in which there is net water flow through the tunnel, simulating a propeller advancing through stationary fluid. The stream tube passing through the propeller disk contracts as it approaches the disk and further contracts as it leaves the disk. Corresponding to these area changes are velocity changes which make the pressure on the forward side of the disk less than ambient, and that on the aft side greater than ambient. Again, the pressure rise across the disk yields a forward thrust. (5) Similar to scene (4), but dye is emitted through the tips of the propeller blades. The dye moves downstream on helical paths. This shows that swirl is put into the fluid. The corresponding angular momentum is related to the driving torque that must be applied to the propeller.

CREDITS: Experiments and film by Office National d'Études et de Recherches Aérospatiales (O.N.E.R.A.), France; Consultant, A. H. Shapiro, M.I.T.; Producer, R. Bergman, E.S.I.; Editor, W. Tannebring, E.S.I.; Art Director, R. P. Larkin, E.S.I. Produced by the National Committee for Fluid Mechanics Films and Educational Services Incorporated, with the support of the National Science Foundation.

FM-98
Aerodynamic Heating and Ablation of Missile Shapes

PURPOSE: To show the patterns of aerodynamic heating on missiles, and how these patterns are affected by ablation and by missile shape.

APPARATUS: The high stagnation temperature relative to a vehicle in hypersonic flight is simulated by placing models in a free supersonic jet discharged from the nozzle of a rocket engine. Three axisymmetric models are tested, all made of stainless steel. Each run starts with the model cold, and is seen twice. The distributions of temperature and of rate of heat transfer may be observed qualitatively from the brightness of the model as it heats up to the point where it radiates visible light.

SEQUENCES: (1) Long view of rocket engine and model. (2) Closeup of same. (3), (4), (5) *Cylindrical Body with Relatively Sharp Ogive Nose.* After a sketch of the shape, the experiment is seen twice. The tip of the nose heats up first. After some time it begins to melt, and then to evaporate. The evaporated gas emitted from the nose region produces a different equivalent shape around which the supersonic jet must be diverted. This causes a drastic change in flow pattern, with a detached bow wave (seen by luminosity of the gas) such as would occur with a blunt nose. The ablating and evaporating material flowing along the cylinder creates an insulating buffer layer which reduces the rate of heat transfer. Thus, except for the nose region, the body does not become incandescent. (6), (7), (8) *Cylindrical Body with Relatively Blunt Ogive Nose.* The tip of the nose becomes incandescent very rapidly and the nose melts back, thus becoming more blunt. Emitted vapor from the nose region interacts with the oncoming flow to produce a detached bow shock, but this is not as strong as with the first model. The shielding effect of the vapor protects the model for some distance back of the nose, beyond which the cylindrical section heats up and becomes incandescent, although not so rapidly as does the nose. (9), (10), (11) *Sharp Ogive Nose Followed by Downstream-Facing Step.* Compared with the first model, the flow in the step region, coupled with its influence on the upstream boundary layer, changes the whole pattern of heating and ablation. Except for the very tip of the nose, which does not melt appreciably, the model does not become incandescent in the region forward of the step. Downstream of the step, incandescence occurs only after about 2 or 3 diameters beyond the step.

CREDITS: From a film by Office National d'Études et de Recherches Aérospatiales (O.N.E.R.A.), France; Consultant, A. H. Shapiro, M.I.T.; Producer, R. Bergman, E.D.C.; Editor, W. Tannebring, E.D.C.; Art Director, R. P. Larkin, E.D.C. Produced by the National Committee for Fluid Mechanics Films and Education Development Center, Inc. (formerly Educational Services Inc.), with the support of the National Science Foundation.

FM-99
Passage of Shock Waves Over Bodies

PURPOSE: To show the wave patterns and flows when a normal shock wave passes over a wedge, a circular cylinder, and a sphere.

APPARATUS: A pressure-driven shock tube is used. Initially, the high-pressure driver gas is confined at the left behind a diaphragm. When the diaphragm is burst, a shock wave forms and moves into the test gas with a speed about 1.5 times the sound speed in the original test gas. The flow behind the shock is subsonic relative to the models. A shadowgraph optical system makes the flow visible. Many spark pictures from successive identical runs were pieced together to make a motion picture with an effective framing speed of several hundred thousand frames per second. Some jumpiness is evident due to the techniques used. Each experiment is shown twice.

SEQUENCES: (1) Schematic of apparatus. (2) *Wedge.* The flow and wave patterns are self-similar until the main shock reaches the base of the wedge. A cylindrical shock reflects from the nose. The wedge angle is too small to permit regular reflection (two shock waves intersecting at the surface). Thus the cylindrical shock, the original shock, and a short length of shock normal to the wedge surface join at a triple-point ("Mach reflection"), leaving behind an entropy sheet (or contact discontinuity). When the normal shock reaches the sharp shoulder at the base of the wedge, the subsonic flow behind the shock separates and produces a vortex sheet that rolls up into a vortex. The waves reflect from the side walls of the tunnel and from the entropy sheets. An increasingly complex pattern develops: some of the lines represent pressure waves traveling at or above the sound speed; some represent entropy sheets moving with the fluid. (3) A close-up of the shock passing over the wedge. (4) *Circular Cylinder.* The reflection is not self-similar, because the angle of the main shock to the surface of the cylinder changes. It starts out as a regular reflection, but then becomes a Mach reflection. (5) *Sphere,* suspended on vertical wires. Compared to the circular cylinder, the sphere produces a much smaller disturbance, the reason being that it constitutes a much smaller blocking target. In particular, the main shock passes the sphere with relatively little change in strength and shape.

DISCUSSION: The experiments exhibit a wide variety of gas, dynamical phenomena which may be seen on careful repeated study. These are: (a) Regular reflection. (b) Mach reflection with production of entropy sheets. (c) Reflection of shock from shock, both of regular type and Mach type. (d) Reflection and transmission of shock at entropy sheet. (e) Shocks overtaking all waves ahead. (f) Shocks being overtaken by all waves behind.

CREDITS: Pictures taken by Prof. Dr. -Ing. H. Schardin, German-French Research Institute, Saint-Louis, France; Consultant, A. H. Shapiro, M.I.T.; Producer, R. Bergman, E.D.C.; Editor, E. Carini, E.D.C.; Art Director, R. P. Larkin, E.D.C. Produced by the National Committee for Fluid Mechanics Films and Education Development Center, Inc., with the support of the National Science Foundation.

FM-100
Passage of Shock Waves Through Constrictions

PURPOSE: To show the wave patterns and flows when a normal shock wave passes through various constrictions in a duct.

APPARATUS: A pressure-driven shock tube is used. Initially, the high-pressure driver gas is confined at the left behind a diaphragm. When the diaphragm is burst, a shock wave forms and moves into the test gas with a speed about 1.5 times the sound speed in the original test gas. The flow behind the shock is subsonic relative to the constrictions. A shadowgraph optical system makes the flow visible. Many spark pictures from successive identical runs were pieced together to make a motion picture with an effective framing speed of several hundred thousand frames per second. Some jumpiness is evident due to the technique used.

SEQUENCES: (1) A schematic drawing of the apparatus. (2) The shock is incident on a sharp-edged circular orifice (the orifice itself is just out of view at the left of the frame). (3) The shock is incident on a sharp-edged two-dimensional slit. (4) The shock is incident on a two-dimensional Laval nozzle. (5) The shock passes over a two-dimensional ramp-shaped obstacle in the duct. (6) Same as (5), but viewed at positions successively further downstream. The overtaking of shocks by shocks causes the main propagating shock, which was greatly altered in form as it passed over the ramp, to tend to become planar again. (7) Same as (5), but with the viewing position attached to the upper part of the incident shock.

DISCUSSION: The patterns produced become increasingly complicated through a succession of interactions and the creation of new reflected waves and entropy surfaces. Careful study reveals that the complicated patterns are the result of a few basic gas dynamical phenomena: (a) Regular shock reflection. (b) Mach-type shock reflection, with production of entropy surfaces. (c) Regular and Mach-type impingement of a shock on another shock. (d) Reflection and transmission of a shock at an entropy sheet. (e) Shocks overtake all waves ahead. (f) Shocks are overtaken by all waves behind.

CREDITS: Pictures taken by Prof. Dr. -Ing. H. Schardin, German-French Research Institute, Saint-Louis, France; Consultant, A. H. Shapiro, M.I.T.; Producer, R. Bergman, E.D.C.; Editor, E. Carini, E.D.C.; Art Director, R. P. Larkin, E.D.C. Produced by the National Committee for Fluid Mechanics Films and Education Development Center, Inc., with the support of the National Science Foundation.

FM-101
Reflections of Shock Waves

PURPOSE: To show reflection and diffraction patterns when a normal shock wave impinges upon inclined walls, a knife edge, and a 90° corner.

APPARATUS: A pressure-driven shock tube is used to drive a normal shock wave from left to right. A shadowgraph optical system makes the flow visible. Many spark pictures from successive identical runs were pieced together to make a motion picture with an effective framing speed of several hundred thousand frames per second. Some jumpiness is evident due to the technique used.

SEQUENCES: (1) A schematic drawing of the apparatus. (2) The normal shock wave impinges on an end wall that is parallel to the plane of the shock. The incident shock reflects as a normal shock that brings the flow behind it to rest. Weak reflected shocks are generated at the corners of the duct. These arise from the boundary layers formed behind the incident shock and from the lack of perfect sharpness of the corners. The entire region of stagnant flow behind the reflected shock appears grainy; this is the result of the turbulence in the side-wall boundary layers through which the flow is viewed. (2) The normal shock impinges on an oblique end wall, with the angle between shock and wall small enough so that there is a regular shock reflection at the impingement point. (3) When the oblique end wall makes too large an angle with the original shock, regular reflection is impossible. A Mach triple-shock reflection occurs, with one leg a shock that is normal to the wall. An entropy sheet (contact discontinuity) is formed at the triple-junction point. (4) The normal shock wave is incident on a knife edge parallel to the shock. The reflection and diffraction pattern at the edge is initially circular and self-similar as the waves move away from the edge. (5) The normal shock wave diffracts around a 90°- corner. The wave pattern is initially cylindrical and self-similar. In this experiment the flow behind the shock is subsonic, and separates in typical subsonic fashion at the corner, producing a vortex sheet that rolls up into a vortex. (6) Similar to (5), but the normal shock is now stronger, and the flow behind it is supersonic. The supersonic flow behind the shock expands around the corner in a Prandtl-Meyer centered rarefaction fan. When the pressure behind the rarefaction fan is suddenly increased by a reflected shock wave that penetrates this region, the Mach angles of the rarefaction fan are abruptly changed.

CREDITS: Pictures taken by Prof. Dr. -Ing. H. Schardin, German-French Research Institute, Saint-Louis, France; Consultant, A. H. Shapiro, M.I.T.; Producer, R. Bergman, E.D.C.; Editor, E. Carini, E.D.C.; Art Director, R. P. Larkin, E.D.C. Produced by the National Committee for Fluid Mechanics Films and Education Development Center, Inc., with the support of the National Science Foundation.

FM-102
Passage of a Shock Wave Through a Circular Orifice

PURPOSE: To show the wave patterns and flows as a normal shock wave passes through a circular orifice plate mounted in a shock tube.

APPARATUS: A pressure-driven shock tube is used. Initially, the high-pressure driver gas is confined at the left behind a diaphragm. When the diaphragm is burst, a shock wave forms and moves to the right into the test gas with a speed 1.52 times the sound speed in the gas. The shock wave impinges on a thin flat plate (normal to the tube axis) in which is a sharp-edged circular hole. Part of the shock wave passes through the orifice into the part of the shock tube beyond the plate. The camera views the region just downstream of the plate; the plate lies along the left edge of the frame. That portion of the shock reflecting from the solid part of the plate leaves between it and the plate a growing region of high-pressure gas that acts as a reservoir for continued flow through the orifice. The flow is made visible by a shadowgraph optical system, which is sensitive to the second spatial derivative of the gas density. Many spark pictures from successive experiments were pieced together to make a motion picture with an effective framing speed of several hundred thousand frames per second. Some jumpiness is evident due to the technique used.

SEQUENCES: The same scene is shown many times over. Different aspects of the flow may be concentrated upon one at a time. Important features: (1) The plane shock incident on the orifice remains plane just as it emerges. Transverse wave propagation near the edges of the orifice causes the shock rapidly to become approximately spherical, with its center at the middle of the orifice. As it spreads radially, reflections from the walls of the shock tube, and the interactions of these reflections and the associated entropy sheets with the main transmitted shock cause the latter to become more plane. (2) Small-amplitude waves behind the main shock overtake the latter, in accord with the general rule that a shock wave is overtaken by any wave behind it, whether it be small or large in amplitude, whether it be a rarefaction or a compression. (3) The lines seen by shadowgraph are either waves or entropy sheets (contact surfaces). The latter separate parcels of gas having the same pressure and velocity but different temperatures. The waves move with speeds of the order of the sound speed, while the entropy surfaces move with exactly the gas speed, which is much lower. (4) The gas passing through the orifice just behind the emerging main shock form a vortex ring because of viscous separation at the sharp edge. This vortex ring moves downstream ahead of the gas jet. (5) The gas jet, emerging behind the main shock and vortex ring, rapidly assumes the pattern of steady supersonic jet flow. It has the characteristic diamond-shaped wave pattern of underexpanded supersonic jets. (6) Turbulence at the edge of the jet generates waves which radiate sound energy in the form of jet noise.

CREDITS: Pictures taken by Professor Dr. Ing. H. Schardin, German-French Research Institute, Saint-Louis, France; Consultant, A. H. Shapiro, M.I.T.; Producer, R. Bergman, E.S.I.; Editor, J. Hirschfeld, E.S.I.; Art Director, R. P. Larkin, E.S.I. Produced by the National Committee for Fluid Mechanics Films and Educational Services Incorporated, with the support of the National Science Foundation.

FM-103
Effect of Knudsen Number on Flow Past a Blunt Body

PURPOSE: To show how the flow field around a body changes in character as the ratio of molecular mean free path to body size decreases from a very large value (free molecule flow) to a very small value (continuum flow).

APPARATUS: Argon gas, seeded with sodium vapor, escapes from a reservoir through a small circular orifice into a chamber where the pressure is kept as low as possible. The mean free path in the argon/sodium jet is controlled by varying the upstream reservoir pressure. At the lowest reservoir pressures, the escaping flow is of free-molecule type: the molecules move in straight rays, at molecular speeds, and with a cosine-law distribution of flux with angle. A washer-like disk, with a hole through its center, is positioned co-axially in the jet. The flow is made visible by a narrow sheet of light from a sodium vapor lamp which shines vertically downward through the symmetry axis of the flow. The light is absorbed by the sodium molecules in the jet, and then re-emitted with an intensity proportional to the number density of sodium molecules. Still photographs were taken at various flow conditions and later composed into a motion picture.

SEQUENCES: (1) A view of the orifice and the disk. (2) A single still picture, showing the optical shadow cast vertically downward by the disk. (3) Another still picture showing how the pressure level is characterized by a horizontal white line, optically superposed, that shows in correct scale the mean free path in the jet. Also shown is the value of the Knudsen number, λ/d, where λ is the mean free path in the jet and d is the disk diameter. (4) A still picture for a large value of Knudsen number, $\lambda/d = 68$. Since the mean free path is extremely large compared with the disk dimensions, the flow relative to the disk is free-molecular. Collisions between molecules are negligible. Thus the straight-line molecular trajectories from the orifice are interrupted only by the disk, which casts a sharp molecular shadow. Molecules reaching the hole in the disk pass through in a collimated beam, with little attenuation. (5) A succesion of three still pictures, all at very low Knudsen number, $\lambda/d \cong 0.026$. The molecules which are incident on the disk are re-emitted and then collide with incoming molecules to form a relatively dense layer in front of the disk. The important features of the flow relative to the disk are of continuum type. Since the molecular speed exceeds the sound speed, the flow is supersonic, and a detached shock wave is formed ahead of the blunt disk. (6) By increasing the pressure in the reservoir, the Knudsen number is reduced in many small steps from a value of 68 to 0.024. The succession of pictures shows the intermediate steps in the transition from free-molecule to continuum flow. The sharp molecular shadows become fuzzy at $\lambda/d \cong 2$, while the first indication of a dense collision region ahead of the body occurs at $\lambda/d \cong 0.3$.

CREDITS: Scenes from the film "Rarefied Gas Dynamics" by F.C. Hurlbut and F.S. Sherman, University of California at Berkeley; Consultants, F.C. Hurlbut, F.S. Sherman, and A.H. Shapiro, M.I.T.; Producer/Director, R. Bergman. E.D.C.; Editor, W. Tannebring, E.D.C. Produced by the National Committee for Fluid Mechanics Films and Education Development Center, Inc., with the support of the National Science Foundation.

FM-104
Effect of Knudsen Number on a Jet

PURPOSE: To show the change in character of a jet issuing from an orifice when the ratio of molecular mean free path to jet size increases from a very low value (continuum flow) to a high value (free-molecule flow).

APPARATUS: Argon gas, seeded with sodium vapor, escapes through a circular orifice from a region of high pressure to a region of low pressure. Both pressure levels are adjustable, but the pressure *ratio* is maintained constant. The flow is made visible by a narrow sheet of light from a sodium vapor lamp that shines vertically downward through the symmetry axis of the flow. The light is absorbed by the sodium molecules of the flow, then re-emitted with an intensity proportional to the number density of sodium molecules. Still pictures were taken at many pressure levels, and were later composed into a motion picture.

SEQUENCES: (1) A still photo to show the light re-emitted by the jet. (2) A still photo shows how the pressure level is characterized by a horizontal white line, optically superposed, that exhibits in correct scale the mean free path λ in the jet. Also shown is the value of the Knudsen number, λ/d, where d is a characteristic dimension of the jet. Here d is taken to be the distance from the orifice to the location of the Mach disk as observed under continuum-flow conditions. (3) A still photo showing the continuum, supersonic type of jet that exists when $\lambda \ll d$. Clearly evident are the barrel shock and Mach disk of a greatly underexpanded supersonic jet. (4) A still photo showing the free-molecule type of jet that exists when $\lambda \gg d$. The molecules travel along straight-line trajectories. The approximately-circular illumination pattern illustrates the cosine law for distribution of number flux of molecules with angle. In interpreting the illumination pattern, note that the light beam is of constant width normal to the plane of the picture, while the jet spreads in axi-symmetrical fashion. (5) A succession of many stills showing the stages of transition from continuum flow ($\lambda/d = 0.006$) to free-molecule flow ($\lambda/d = 2.4$). Clarity in the shock pattern is lost at $\lambda/d \cong 0.01$, while the characteristic free-molecule pattern begins to appear at $\lambda/d \cong 0.1$.

CREDITS: Scenes from the film "Rarefied Gas Dynamics" by F.C. Hurlbut and F.S. Sherman, University of California at Berkeley; Consultants, F.C. Hurlbut, F.S. Sherman, and A.H. Shapiro, M.I.T.; Producer/Director, R. Bergman. E.D.C.; Editor, W. Tannebring, E.D.C. Produced by the National Committee for Fluid Mechanics Films and Education Development Center, Inc., with the support of the National Science Foundation.

FM-105
Ripple-Tank Radiation Patterns of Source, Dipole, and Quadrupole

PURPOSE: To illustrate radiation fields relevant to sound production by source, dipole, and quadrupole emitters.

APPARATUS: A glass-bottomed tank contains water about ⅜-inch deep. Long gravity waves (ripples) are produced by various oscillating exciters which disturb the water surface. These waves move outward at about 1 ft./sec.; they are analogous to small-amplitude pressure waves in a two-dimensional gas flow. Nearly-parallel light coming through the bottom is focused on a screen by the varying slopes of the waves.

SEQUENCES: (1) *Single Source*. A cone-shaped plunger oscillates vertically. (2) On the viewing screen, the circular waves move with fixed speed and have uniform wavelengths. (3) *Multiple Sources*. Three simple sources located near each other are oscillated out of phase, such that there is a net algebraic source strength. (4) On the viewing screen, the wave field is irregular in the near field (i.e., at distances not large compared with the size of the multisource region). In the far field the irregularities have smoothed out; the field is virtually that of a single source having the same net source strength as the multi-source system. (5, 6) *Dipole*. Two nearby sources of equal strength, situated on a north-south axis, are oscillated 180 degrees out of phase. Although the net source strength is zero at every instant, the far field is not zero, because the positive and negative sources are at slightly different distances from each field point. The pattern is that of a dipole. The intensity is a maximum on the north-south axis, and is zero on the east-west axis where the wave cancellation is perfect. (7) A vertical circular rod oscillates horizontally in the north-south direction. It has a dipole field like that of the oscillating source-sink pair of the previous experiment. (8, 9) *Quadrupole*. Two nearby circular rods, situated on an east-west axis, are oscillated horizontally in the north-south direction. They have equal amplitude, but they are 180 degrees out of phase. Each produces a dipole field. In the far field the two dipole emitters nearly cancel, the residue constituting a quadrupole field. The wave intensity is strongest in the directions NE, NW, SE, SW, and is exactly zero along N, E, S, W, where the cancellation is perfect. (10) *Scattering from a Fixed Obstacle*. A small obstacle is held stationary far enough from a simple source so that it senses nearly-plane radiation. The waves scatter from the fixed obstacle in a dipole pattern. (11) In a similar experiment, but with a very light obstacle which is free to move, there is no scattering.

CREDITS: Scenes from the NCFMF film "Aerodynamic Generation of Sound" by M. James Lighthill and John E. Ffowcs-Williams, Imperial College of Science & Technology, University of London; Consultants, A.H. Shapiro, M.I.T. and E. Mollo-Christensen, M.I.T.; Producer/Director, J. Friedman, E.D.C.; Editor, W. Tannebring, E.D.C. Produced by the National Committee for Fluid Mechanics Films and Education Development Center, with the support of the National Science Foundation.

FM-107
Deformation in Fluids Illustrated by a Rectilinear Shear Flow

PURPOSE: To illustrate experimentally the mathematical descriptions of deformation of a continuous medium as it is classically analyzed, and to show some properties of the deformation tensor.

APPARATUS AND PROCEDURE: A rectilinear shear flow is produced in a tank of glycerine by two parallel belts moving at equal speeds in opposite directions. Patterns are marked in powder on the surface of the glycerine, and their evolution followed by stationary and moving cameras.

SEQUENCES: (1) Schematic drawing of the apparatus. (2) The linear velocity profile is demonstrated by a straight line of crosses marked on the fluid. (3) Straight lines are marked in the fluid at several successive angles. They remain straight during deformation, but rotate at different rates depending on initial orientation. (4) A circle is marked. It develops into a strain ellipse. The axes of symmetry of the strain ellipse are the principal axes of strain. (5) All sequences from here on are photographed with a camera rotating at a speed that removes the average vorticity, thus leaving visible only the "pure deformation." The first illustration is of the circle deforming into the strain ellipse. The principal axes do not rotate in this frame of reference. (6) Similar to (5), and showing that during a time in which the deformation is small the perpendicular fluid lines which initially mark the principal axes *remain* perpendicular. (7) Again similar to (5), and showing that the principal axes mark the directions of maximum and minimum strain rate. (8) Repeat of (7); the deformation of the strain ellipse shows that the strain is purely lineal in principal-axis coordinates.

DISCUSSION: Deformation can be analyzed into rotation plus strain, and deformation rate into rotation rate plus strain rate. With the rotation removed, strain or strain rate produces an elliptical deformation of an initial circle. The axes of symmetry of the ellipse are the directions of extreme strain (or strain rate), and they rotate at the average rate of lines in all directions. The principal axes of strain are perpendicular before and after deformation; those of strain rate remain orthogonal for small deformations. In principal-axis coordinates the strain or strain rate in one axis direction is not a function of position on the other; this is a way of saying that the strain (or strain rate) tensor is diagonal.

CREDITS: Scenes from the film "Deformation of Continuous Media" by John Lumley, Pennsylvania State University; Consultants, J. Lumley and A. H. Shapiro, M.I.T.; Producer, R. Bergman, E.S.I.; Editor, W. Hansard, E.S.I.; Art Director, R. P. Larkin, E.S.I. Produced by the National Committee for Fluid Mechanics Films and Educational Services Incorporated, with the support of the National Science Foundation.

FM-108
Small-Amplitude Waves

PURPOSE: To show several aspects of the generation, propagation, and reflection of small-amplitude waves in a fluid.

APPARATUS: A free-surface water table is used. A layer of water about 1/4″ thick rests on a horizontal glass floor. Gravity waves of small amplitude are produced by various means. The wave length is large compared with the depth, and thus the wave propagation is analogous to two-dimensional wave propagation in a gas. The wave propagation speed, however, is only about 1 ft./sec. Light from below passes through the glass floor and the water layer; it is refracted at the disturbed water surface, and patterns of light and dark are produced on a screen held horizontally above the surface, thus making the waves visible. The screen is photographed from above.

SCENES: (1) *A Point Source*. A drop of water falls onto the surface. The disturbance propagates outward with circular symmetry and at constant speed. (2) *A Plane Source*. A long bar in the water is moved perpendicular to its axis. The disturbance propagates as a plane wave at constant speed. (3) *Reflection of a Plane Wave*. The plane wave is incident on a straight wall held first parallel to the wave and then at an angle. In each case there is a reflected wave of like sense, with the angle of incidence equal to the angle of reflection, as in the mirror-like reflection of a light ray. (4) *Reflection of Circular Wave from Straight Wall*. The reflection is also a circular wave. An animation scene shows that the reflected wave can be represented as arising from an image source (in phase with the real source) symmetrically placed on the other side of the wall. (5) *Circular Wave Reflected from Concentric Circular Wall*. A circular wave is generated at a point source at the center of a circular boundary. The reflected wave is also circular and comes to a focus at the center, where it reflects outward again. (6) *Waves from a Point Source Within an Elliptical Boundary*. The point source, which is placed at various positions, emits circular waves which reflect from the walls. When the point source is at one focus of the ellipse, the reflected wave concentrates at the other focus. (7) *Waves from a Point Source Within a Parabolic Boundary*. The point source, which is placed at various points, emits circular waves which reflect from the parabolic wall. When the source is at the focus of the parabola, the reflected wave leaving the parabola is plane. (8) *Plane Waves Incident on a Parabolic Reflector*. This is the inverse experiment of the foregoing. Incoming plane waves are concentrated at the focus of the parabola.

CREDITS: Scenes from the film "Reflection and Refraction" (produced by the College Physics Film Program of E.S.I.) by James Strickland, E.S.I.; Consultant, A. H. Shapiro, M.I.T.; Producer, R. Bergman, E.S.I.; Editor, E. Carini, E.S.I.; Art Director, R. P. Larkin, E.S.I. Produced by the National Committee for Fluid Mechanics Films and Educational Services Incorporated, with the support of the National Science Foundation.

FM-109
Source Moving at Speeds Below and Above Wave Speed

PURPOSE: To show the types of disturbance fields produced by a disturbance source that moves with respect to the fluid.

APPARATUS: A free-surface water table is used, with a constant depth of about 1/4-inch, for which the speed of long gravity waves is about 1 ft./sec. Such waves behave much like two-dimensional compressibility waves in a gas. A vertically-oscillating sphere which penetrates the water surface slightly produces gravity waves that radiate cylindrically from the source. The periodic succession of waves gives a picture of the disturbance field. The sphere also translates horizontally. Diffuse light from below is transmitted upwards through the water layer, and is refracted by the wave disturbances at the water surface, producing patterns of light and dark on a horizontal screen above the surface. The screen is photographed. The wave speed V_w is held constant, while the source speed V_S is varied.

SEQUENCES: (A) *Subsonic*. (1) Establishing scene. (2) $V_s/V_w = 0$. Disturbances propagate symmetrically in all directions. (3) $V_s/V_w = 0.4$. The waves propagate in all directions, but not symmetrically upstream and downstream. (4) $V_s/V_w = 0.7$. Propagation in all directions, but with even more asymmetry. (B) *Transonic*. (5) $V_s/V_w = 0.95$. The source and wave speeds are nearly the same. Hardly any disturbances proceed ahead of the source. (6) $V_s/V_w = 1.05$. Similar, with an even greater localization of the disturbance field behind the source. (C) *Supersonic*. (7) $V_s/V_w = 1.6$. All the disturbances are confined within a Mach wedge behind the source. The disturbances are concentrated on the Mach lines. (8) Animation shows that when $V_s > V_w$ the waves spreading from a moving source never extend beyond the downstream-running Mach wedge, and generate the Mach wedge. (9) Repeat of (7). (10) $V_s/V_w = 1.8$. The Mach wedge is narrower.

DISCUSSION: The compressible flow field past a thin body may be represented as the sum of the disturbances initiated at all previous times from all the elements of the body surface. Each such element, as it pushes fluid away or sucks it in, generates a pressure wave that propagates outward through the fluid from the point of origin. Thus a single disturbance source gives insight into more complicated flows. When the source moves with a speed less than the wave speed, disturbances can propagate in all directions. Such subsonic flows are influenced by changes in the boundary conditions anywhere; their governing equations are elliptic. When the source moves with a speed greater than the wave speed, disturbances cannot propagate ahead of the downstream Mach cone or wedge. Within the Mach cone or wedge is the "zone of action", outside it is the "zone of silence", and the Mach surface is the "zone of concentrated action". Such supersonic flows are not affected by the boundary conditions everywhere; their governing equations are hyperbolic.

CREDITS: Scenes from the film "Doppler Effect" (produced by the College Physics Film Program of E.S.I.) by James Strickland, E.S.I.; Consultant, A. H. Shapiro, M.I.T.; Producer, R. Bergman, E.S.I.; Editor, E. Carini, E.S.I.; Art Director, R. P. Larkin, E.S.I. Produced by the National Committee for Fluid Mechanics Films and Educational Services Incorporated, with the support of the National Science Foundation.

FM-112
Examples of Low-Reynolds-Number Flows

PURPOSE: To illustrate a wide variety of flows in which the Reynolds number is very small.

INTRODUCTION: In flows at very low Reynolds number the inertial forces associated with the motion are virtually negligible, and a near balance prevails between the pressure gradient and the viscous forces, A very low Reynolds number may be achieved by very high viscosity, very low density, very small scale of size, very low scale of speed, or by some combination of these.

SEQUENCES: (1) An annular tank contains viscous liquid, on the surface of which the statement "R≪1" is written with dye. The inner cylinder is slowly turned clockwise until the dye seems all smeared out. Then the inner cylinder is slowly turned anti-clockwise exactly the same number of turns, after which the statement "R≪1" is fully reconstructed. This illustrates the kinematic reversibility of inertia-free flows. (2) A knife is withdrawn vertically from a pot of honey. A column of honey is formed between the knife and the pot. The high viscous stresses in the honey arising from tangential sheer at the knife surface allow the development of tensile stresses over the cross-section of the stretching column. (3) A jet of dyed corn syrup is discharged into clear corn syrup. The jet's directed momentum is lost almost immediately, and the injected fluid advanced spherically into the surroundings. (4) A steel sphere whose surface has been coated with dyed glycerine falls slowly, at terminal speed, through clear glycerine. All the dye swept off the surface collects into a thread-like wake at the rear. At very low Reynolds number, the flow does not separate from the rear of a blunt body. (5) A Hele-Shaw cell. Water flows in the very thin gap between parallel plates. This inertia-free flow is a potential flow with the pressure acting as the potential for the mean velocity; the flow pattern simulates flow of an ideal liquid. The combination of a parallel flow with a source is shown. This is followed by the addition of a sink to make a closed Rankine oval. (6) Bull spermatozoa swim by sending transverse waves of bending down their tails. The Reynolds number, however, is so low that the forward thrust comes not from imparting momentum to the liquid, but rather through the generation of viscous forces that have a component in the thrust direction. (7) A glacier flows like a very viscous liquid. It flows· because the stresses resulting from the weight of the ice very much exceed the yield stress of ice. Although the scale of size is large, the viscosity is so great and the speed so low that inertial forces are inconsequential.

CREDITS: Scenes from the NCFMF film "Low Reynolds Number Flows" by G.I. Taylor, Cambridge, England. Consultants, A.H. Shapiro, M.I.T., and G.I. Taylor; Producer/Director, J.R. Friedman, E.D.C.; Editor, W. Hansard, E.D.C.; Art Director, R.P. Larkin, E.D.C. Produced by the National Committee for Fluid Mechanics Films and Education Development Center, Inc., with the support of the National Science Foundation.

FM-113
Hydrodynamic Lubrication

PURPOSE: To illustrate the principles by which viscous stresses may be used to generate high pressures that can support large loads in bearings.

SEQUENCES: (1) A sheet of light cardboard is slid along the tabletop. It moves with very little apparent friction. In falling toward the table, the board must squeeze out the air between itself and the table. The pressure gradient that is established to balance the viscous forces of the escaping air makes the pressure fall from an above-atmospheric value under the center of the board to atmospheric at its edges. When the air gap is very small, the viscous stresses act over the area of the board, which is much larger than the gap area on which the pressure difference acts. Thus small viscous stresses can generate large excess pressures. The above-atmospheric pressures induced under the board support the latter on a film of air and prevent solid-solid friction. When a board with punched holes is tried, the air quickly escapes and solid-solid friction rapidly decelerates the sliding board to rest. (2) A "teetotum" is a spinning toy with flat-plate blades that are slightly inclined to the tabletop. When spun so that the high side of the blade is the leading edge, the teetotum spins with almost imperceptible friction. A schematic sketch shows the flow and the induced pressure distribution when a flat surface moves past an inclined plate. Fluid is swept into the wedge, and an excess pressure is developed under the plate. Here again, when the gap is small compared with the length of the plate, small viscous stresses can generate large load-sustaining pressures. When the teetotum is spun in the wrong direction, negative pressures are induced in the air gap; the teetotum touches the tabletop and quickly stops. (3) A model of an air-lubricated thrust bearing, based on the tapered wedge principle, is shown. (4) A model illustrating the principle of a journal bearing is shown. The shaft is located in a constrained position eccentric to the bearing, with the minimum gap of the annulus at the 12 o'clock position. Manometers show the pressures at 11, 12, and 1 o'clock. When the shaft rotates clockwise, the liquid is first driven into a narrowing gap, then into a widening gap. The pressure, however, rises all the way from 11 to 1 o'clock.

CREDITS: Scenes from the NCFMF film "Low Reynolds Number Flows" by G.I. Taylor, Cambridge, England. Consultants, A.H. Shapiro, M.I.T., and G.I. Taylor; Producer/Director, J.R. Friedman, E.D.C.; Editor, W. Hansard. E.D.C.; Art Director, R.P. Larkin, E.D.C. Produced by the National Committee for Fluid Mechanics Films and Education Development Center, Inc., with the support of the National Science Foundation.

FM-114
Sedimentation at Low Reynolds Number

PURPOSE: To illustrate some features of sedimentation flows in which particles settle slowly, at very low Reynolds number, through a liquid.

SEQUENCES: (1) Two steel spheres, one having twice the diameter of the other, fall slowly at terminal speed through corn syrup. At low Reynolds number, the gravity force (proportional to D^3) balances the drag (proportional to $U \cdot D$). Hence U/D^2 is constant, and the terminal speed U is proportional to the square of the diameter D. This is confirmed by the experiment. (2) A sphere settles in a viscous liquid in which many smaller, colored spheres are dispersed. The disturbance field is seen to extend far from the falling sphere. Behind the sphere is a broad viscous wake of fluid that has acquired a velocity in the same direction as the sphere. (3) A cloud of solid particles, having a range of sizes, is dispersed in liquid in a transparent box. After settling begins, a rather sharp edge (a "kinematic shock wave") is formed between the lower region containing particles and the upper clear region of which particles have settled. Note that while a cluster of dispersed particles in an otherwise clear fluid falls faster than a single particle, a cloud of particles dispersed completely throughout a box of liquid appears to fall more slowly than a single particle. Actually they fall more rapidly with respect to the liquid in which they are moving, but the liquid itself has a speed upward relative to the box as it is displaced by the falling cloud. The edge region between the clear liquid and the cloudy liquid is quite sharp, because smaller particles (which, alone, would fall more slowly) get caught in the wake of the larger particles; and, moreover, they fall through liquid at rest with respect to the box, not liquid moving upwards. (4) The top edge of the cloud has a characteristic thickness that depends on the number concentration of particles. When the box contains a smaller number of particles, the top edge is less crisp and the cloud falls more rapidly.

CREDITS: Scenes from the NCFMF film "Low Reynolds Number Flows" by G.I. Taylor, Cambridge, England. Consultants, A.H. Shapiro, M.I.T., and G.I. Taylor; Producer/ Director, J.R. Friedman, E.D.C.; Editor, W. Hansard, E.D.C.; Art Director, R.P. Larkin, E.D.C. Produced by the National Committee for Fluid Mechanics Films and Education Development Center, Inc., with the support of the National Science Foundation.

FM-115
Kinematic Reversibility of Low-Reynolds-Number Flows

PURPOSE: To illustrate the property of kinematic reversibility that is characteristic of flows at very low Reynolds number.

INTRODUCTION: When the Reynolds number is small enough for inertial forces to be negligible, the equations of quasi-steady motion may be expressed as $\nabla^2 \omega = 0$, where ω is the vorticity. This is linear and does not contain time explicitly. Hence, given a certain motion up to a certain point, a reversal of all the boundary conditions through the original sequence causes the motions to be retraced in reversed sequence. The original displacement field of the fluid can thus be completely restored: the motion is *kinematically* reversible.

SEQUENCES: (1) The annular space between two concentric circular cylinders contains corn syrup. A small square is drawn on the liquid surface with dyed corn syrup. The inner cylinder is then rotated *clockwise* through 1.5 turns, and the square is sheared into a highly elongated parallelogram. When the inner cylinder is next rotated *anti-clockwise* through 1.5 turns, the dyed figure returns to its original location and shape. A similar result would be observed for any fluid curve of fixed identity. (2) In another concentric cylinder apparatus, a dyed blob of liquid is placed in the annulus, and is viewed from the side. The inner cylinder is rotated *clockwise* through 3.5 turns, at which time the blob is so sheared out as to be barely visible. Then the inner cylinder is rotated *anti-clockwise* through 3.5 turns. The blob returns to its original location and shape, except for a slight fuzziness due to molecular diffusion. (3) A circular rigid ring of plastic, with an open cut at 12 o'clock, is placed at the 12 o'clock position in the annulus of sequence (1). Turning of the inner cylinder causes the plastic ring to rotate and also to translate in a circular path. Turning the inner cylinder back to its original position causes the plastic ring to return to its original position and orientation. (4) When the same experiment is repeated with a piece of yarn, kinematic reversibility is *not* achieved, mainly because the yarn deforms differently in tension and in compression. With a *deformable* body in the fluid, kinematic reversibility occurs only if the sequence of boundary *speeds* is reversed and if the body has the same deformation under reversed forces. (5) A "mechanical fish" has a rubber-band motor that oscillates a rudder-like tail at the rear. In water (i.e. at high Reynolds number) it swims by imparting net rearward momentum to the water. In corn syrup (i.e. at very low Reynolds number) it does not move forward at all. Because of kinematic reversibility the forces due to the oscillating tail are exactly reversed for the to-and-fro strokes, and the time-average thrust is zero.

CREDITS: Scenes from the NCFMF film "Low Reynolds Number Flows" by G.I. Taylor, Cambridge, England. Consultants, A.H. Shapiro, M.I.T., and G.I. Taylor; Producer/ Director, J.R. Friedman, E.D.C.; Editor, W. Hansard, E.D.C.; Art Director, R.P. Larkin, E.D.C. Produced by the National Committee for Fluid Mechanics Films and Education Development Center, Inc., with the support of the National Science Foundation.

FM-116
Swimming Propulsion
at Low Reynolds Number

PURPOSE: To illustrate and explain one form of propulsion by which microscopic organisms swim at very low Reynolds number, under circumstances where viscosity rather than momentum must be the source of thrust.

SEQUENCES: (1) Bull spermatozoa are seen under the microscope to swim by sending progressive waves of bending down their tail. This action resembles that of fish, who swim by imparting rearward momentum to the water. However, this inertia mechanism is not available to spermatozoa because the Reynolds number is too small for inertial forces to be significant. Microscopic organisms must use viscous forces for generating thrust, as explained below. (2) An experiment with two identical rods (one horizontal, the other vertical) falling through corn syrup demonstrates the general proposition that, for inertia-free flows, the drag per unit speed of any long, slender body is twice as great for flow perpendicular to the body as for flow along the body. (3) A rod oriented at about 45° to the horizontal is observed to fall in a direction about 15° or 20° to the vertical. From the experimental result of sequence (2), and noting that forces and velocities may each be resolved along and perpendicular to the axis of a long rod, it may be proved that a long obliquely-oriented rod does not fall vertically under gravity, but slips sideways. *Exercise:* show that $\beta = [\text{arc tan } (2 \tan \theta)] - \theta$, where θ is the angle of the rod to the horizontal, and β is the angle between the vertical and the direction of fall; and, furthermore, that β is a maximum when $\theta \cong 35°$. (4) An animated sequence shows that if a vertical force on a long oblique rod causes a horizontal component of motion, a vertical motion of an oblique rod will induce a horizontal component of drag force. The animation proceeds to show that each element of a rotating spiral of wire acts in the way described: it moves normal to the axis of the spiral and creates a viscous drag force along the axis. This produces a swimming thrust. The sketch shows a mechanical swimming model: left and right-handed spirals are turned in opposite directions by a wound rubber-band untwisting inside the central body. (5) The mechanical model swims through corn syrup. (6) A spermatazoon swims in a similar way by sending progressive, transverse waves down its tail.

REFERENCES: (1) Hancock, G.J., 1953: The Self-Propulsion of Microscopic Organisms Through Liquids., Proc. Roy. Soc. A, 217. (2) Taylor, G.I., 1952: Analysis of the Swimming of Long and Narrow Animals, Proc. Roy. Soc. A, 214.

CREDITS: Scenes from the NCFMF film "Low Reynolds Number Flows" by G.I. Taylor, Cambridge, England. Consultants, A.H. Shapiro, M.I.T., and G.I. Taylor; Producer/Director, J.R. Friedman, E.D.C.; Editor, W. Hansard, E.D.C.; Art Director, R.P. Larkin, E.D.C. Produced by the National Committee for Fluid Mechanics Films and Education Development Center, Inc., with the support of the National Science Foundation.

FM-117
Subsonic Flow Patterns and Pressure
Distributions for an Airfoil

PURPOSE: To show, for a symmetrical subsonic airfoil, the flow fields and surface pressure distributions over a wide range of incidence.

APPARATUS: A two-dimensional low-speed wind tunnel is used. Smoke streamers introduced upstream show the streak lines of the flow, and, when the flow is steady, the streamlines as well. A symmetrical airfoil of conventional subsonic type has pressure taps on the upper and lower surface which are connected to alternate tubes on an external manometer board. The spacing of the pressure taps corresponds to that of the tubes on the manometer board; thus the board exhibits the curves of pressure vs. chordwise distance. The manometer tubes are connected so that the mercury level rises as the pressure falls: a "positive" pressure is shown by a low level, and a "negative" pressure by a high level.

SEQUENCES: (1) A preliminary experiment with smoke lines, in which the incidence increases from zero to about 30°. (2) The same experiment, showing the manometer board. (3) Repeat of (2). (4) As the airfoil is adjusted to zero incidence, the manometer board shows the same pressure distribution on the upper and lower surfaces, as required by symmetry. (5) The smoke lines, for zero incidence, are symmetrical above and below the airfoil. (6) As the incidence increases from zero, the smoke lines show the whole flow field to change (top and bottom; upstream and downstream). The streamlines above the airfoil become more closely crowded than those below, thus indicating a higher speed, a lower pressure, and a net upward lift. (7) As the incidence increases from zero, the manometer board shows the asymmetry between the pressure distributions on the upper ("negative" pressure) and lower ("positive" pressure) surfaces. The area contained between the two pressure-distribution curves is proportional to the lift. On the upper surface the pressure starts at the stagnation value, decreases to a minimum, and rises again toward the trailing edge. As the incidence increases the point of minimum pressure moves forward, the minimum pressure decreases, and the adverse pressure gradient aft of the point of minimum pressure becomes stronger. (8) Repeat of (7), with a closer view of the smoke lines, showing that as the incidence increases, the point of transition from a laminar to a turbulent boundary layer on the upper surface moves forward. (9) As the incidence is increased beyond a certain value, the smoke-flow pictures show that the boundary layer separates. This is due to the very large adverse pressure gradient. The entire external flow is thereby disrupted. It no longer follows the upper surface of the airfoil, and there is a low-speed separated zone on the upper surface. The airfoil is "stalled", and produces less lift than when unstalled.

CREDITS: Scenes from the film "Boundary Layer Control" by D. C. Hazen, Princeton University; Consultants, A. H. Shapiro, M.I.T., and D. C. Hazen, Princeton University; Producer, R. Bergman, E.D.C.; Editor, E. Carini, E.D.C. Produced by the National Committee for Fluid Mechanics Films and Education Development Center, Inc., with the support of the National Science Foundation.

FM-118
Laminar-Flow Versus Conventional Airfoils

PURPOSE: To show the difference in pressure distributions on a "conventional" airfoil and a "laminar-flow" airfoil. The latter is designed to maintain a long run of laminar flow on the suction surface. Also, to show the difference in wake-momentum-defect between the two airfoils.

APPARATUS: The two-dimensional airfoils, both of which are symmetrical, are tested in a low-speed wind tunnel. Pressure taps on the upper and lower airfoil surfaces are connected to alternate tubes of a manometer board. The manometer board exhibits the curves of pressure vs. chordwise distance. The manometer tubes are connected so that the liquid level rises as the pressure falls. An array of total-head tubes (a "rake") downstream of the airfoil shows the momentum defect in the wake by means of the rise in levels of the manometer tubes to which they are connected.

SEQUENCES: (1) A preliminary experiment. The conventional airfoil is initially at zero incidence, and the pressure distributions are the same on the upper and lower surfaces. As the incidence is increased, the pressure distributions become different, with substantial "negative" pressures appearing on the upper (suction) surfaces. (2) Repeat. (3) A drawing compares the two airfoils, which have the same thickness ratio. The laminar-flow airfoil (which also has advantages in delaying Mach number effects at high speeds) has a sharper nose, a sharper and reflexed profile near the trailing edge, and has its point of maximum thickness at about 45% of chord compared with about 25% for the conventional airfoil. (4) The pressure distributions on the surfaces of the conventional airfoil are displayed and drawn for a modest angle of incidence. The point of minimum pressure on the upper surface is far forward, and is followed by a region of positive pressure gradient. The area between the two pressure curves is proportional to the lift. (5) Using the same tunnel speed, the laminar-flow airfoil is set at such incidence as to produce the same lift, as indicated by the area between the two pressure curves. (6) Repeat. The pressure distributions for the two profiles are very different. A large adverse pressure gradient does *not* appear on the upper surface of the laminar-flow airfoil until about 70% of chord, compared with about 30% of chord for the conventional airfoil. Thus transition to turbulence is delayed, and the laminar-flow airfoil has a greater fraction of its surface covered by a laminar boundary layer. (7) View of the wake rake and the manometer to which it is attached. (8) The momentum defect curve in the wake is observed and marked for the conventional airfoil. (9) The momentum-defect curve for the laminar-flow airfoil is then compared at the angle of incidence giving the same lift. The maximum momentum defect is not as great, the wake is narrower, and the area under the curve is smaller; thus the laminar-flow airfoil has less skin-friction drag.

CREDITS: Scenes from the film "Boundary Layer Control" by D. C. Hazen, Princeton University; Consultants, A. H. Shapiro, M.I.T., and D. C. Hazen, Princeton University; Producer, R. Bergman, E.D.C.; Editor, E. Carini, E.D.C. Produced by the National Committee for Fluid Mechanics Films and Education Development Center, Inc., with the support of the National Science Foundation.

FM-119
Reduction of Airfoil Friction Drag by Suction

PURPOSE: To show how a turbulent boundary layer on an airfoil can be rendered laminar through a thinning of the boundary layer by means of suction through the airfoil surface. Also, to show how this affects the momentum defect in the wake.

APPARATUS: A symmetrical, conventional airfoil of subsonic type is tested in a low-speed two-dimensional wind tunnel. Slots spaced along the upper surface allow the boundary layer to be sucked into the interior. Smoke injected through a series of fine tubes upstream makes the streak lines visible; these are also the streamlines in steady flow. An array of total-head tubes (a "rake") downstream of the airfoil shows the momentum defect in the wake by the rise in levels of the manometer tubes to which they are connected.

SEQUENCES: (1) A close-up view of a smoke streakline close to the upper surface. The boundary layer is laminar for some distance behind the leading edge, then becomes turbulent. Fluttering and breakup of the smoke line shows transition to turbulence. (2) The instability and transition of the laminar boundary layer is seen in slow-motion photography. (3) As the incidence of the airfoil is increased from zero, the point of boundary-layer transition moves forward. This comes about because the point of minimum pressure on the upper surface (which is followed by an adverse pressure gradient) moves forward as the incidence increases. (4) Suction is applied to two slots on the upper surface (at about 5% and 15% of chord length aft of the nose). The boundary layer is made thinner and thereby becomes more stable, with the result that transition to turbulence occurs further downstream. The flow is shown alternately with suction on and suction off, for comparison. (5) A similar experiment, with eight suction slots uniformly spaced along the upper surface. When suction is applied, the boundary layer becomes laminar over the entire upper surface. (6) A view of the rake of total head tubes and the manometer board to which it is attached. (7) The manometer board is shown alternately with no suction on the airfoil and with the eight suction slots operative. When suction is applied, the total-head defect curve becomes lower and narrower. This indicates that the skin-friction drag has been reduced. The point of maximum momentum defect is shifted slightly because suction is applied only to the upper surface of the airfoil. Whether there is a net gain in airplane performance depends on the amount of power required to suck in the boundary layer.

REFERENCE: Lachmann, G. V. (Ed.), *Boundary Layer Control* (2 vols.), Pergamon Press, 1961.

CREDITS: Scenes from the film "Boundary Layer Control" by D. C. Hazen, Princeton University; Consultants, A. H. Shapiro, M.I.T., and D. C. Hazen, Princeton University; Producer, R. Bergman, E.D.C.; Editor, E. Carini, E.D.C. Produced by the National Committee for Fluid Mechanics Films and Education Development Center, Inc., with the support of the National Science Foundation.

FM-120
Some Methods for Increasing Lift Coefficient

PURPOSE: To show a variety of boundary-layer-control techniques for increasing the maximum lift coefficient which an airfoil can achieve.

APPARATUS: Two dimensional airfoils are tested in a low-speed wind tunnel. Smoke is injected upstream of the airfoils through a series of fine tubes to render the streak-lines visible. These are also the streamlines when the flow is steady.

SEQUENCES: (1) An airfoil of conventional shape has its incidence increased. At a certain incidence, stall occurs: the boundary layer on the upper surface separates, the external flow detaches from the upper surface, and the lift is reduced below its maximum value. The succeeding experiments show methods by which the stall may be delayed to higher incidences, and the maximum lift increased. (2) Distributed suction is applied along the upper surface. This maintains an attached flow to very high incidence, with correspondingly high lift. With high incidence, the flow pattern changes dramatically as the suction is turned on and off. (3) The downward deflection of a nose flap also eliminates the separation on the upper surface. The deflected nose flap, which in effect decreases the incidence while increasing the camber of the airfoil, produces at high incidence a pressure distribution with a less severe adverse gradient. (4) A leading-edge slot brings high pressure air from the lower surface to the region of lower pressure on the upper surface. The resulting air jet energizes the boundary layer on the upper surface in the region of adverse pressure gradient and reduces the extent of separation. (5) In an attempt to gain additional lift, a trailing-edge flap may be deflected too far, resulting in a severe flow separation. However, if a slot is opened to bleed a jet of high pressure air from the lower surface onto the flap, separation is reduced somewhat, and the lift is increased. The use of two slots in series improves the situation even further. (6) With the trailing-edge flap deflected and producing a separated flow, the separation is decreased and the lift increased when the boundary layer ahead of the flap is sucked in through a slot. (7) An internally-powered jet is blown over the severely-deflected trailing-edge flap. This energizes the boundary layer and reduces the separation. (8) *Jet-flap effect.* This is not a means of boundary-layer control, but rather a means of generating lift without incidence and/or camber. The jet flap is a high-speed jet issuing at a downward angle from the trailing edge of an airfoil, in this case of symmetrical section. The jet curves and can thereby support a pressure difference. The potential flow field is thus modified so as to produce circulation and lift.

REFERENCE: Lachmann, G. V. (Ed.), *Boundary Layer Control* (2 vols.), Pergamon Press, 1961.

CREDITS: Scenes from the film "Boundary Layer Control" by D. C. Hazen, Princeton University; Consultants, A. H. Shapiro, M.I.T., and D. C. Hazen, Princeton University; Producer, R. Bergman, E.D.C.; Editor, E. Carini, E.D.C. Produced by the National Committee for Fluid Mechanics Films and Education Development Center, Inc., with the support of the National Science Foundation.

FM-122
Non-Linear Shear Stress Behavior in Steady Flows

PURPOSE: To show the relation of flow rate to pressure head and the velocity profile in steady laminar flow through tubes for fluids characterized by a non-linear stress-deformation relation.

APPARATUS AND PROCEDURE: (1) Two identical flow systems are used, each comprising a large cylindrical reservoir discharging through a vertical capillary tube. The total pressure head for the right-hand system is twice that for the left system. The same fluid is used in both systems. To begin the experiment, the closures at the ends of the capillaries are removed simultaneously. They are closed again when 10 ml. of fluid have flowed into a graduated cylinder from the left system. The volume collected by the right cylinder is then noted. This experiment is performed with three different fluids: (a) Glycerin: the flow rates, like the driving heads, are in the ratio of 2 to 1; (b) A particular polymer solution: the flow rates are in the ratio of 4.4 to 1; (c) A particular suspension: the flow rates are in the ratio 1.6 to 1. (2) A dyed fluid line is positioned across the diameter of a tube containing a fluid. A driving pressure is suddenly applied. The ensuing shape of the dyed line is an indication of the velocity profile. This experiment is performed with two fluids: (a) an oil and (b) a polymer solution.

DISCUSSION: (a) For a Newtonian fluid in steady laminar flow, the shear stress is proportional to the shear rate. If this flow occurs in a cylindrical tube, the velocity profile is parabolic and the rate of flow is proportional to the pressure head. The glycerin in experiment (1a) and the oil in experiment (2a) behave as Newtonian fluids. (b) When the stress-deformation relation is not linear, the velocity profile in a tube is not parabolic and the rate of flow is not proportional to the pressure head. If the "apparent viscosity" (the ratio of shear stress to shear rate) decreases as the shear rate increases, the fluid behavior is usually called *shear-thinning* or *pseudoplastic*. The velocity profile is more blunt than a parabola and the ratio of flow rate to pressure head increases as the head increases. Shear-thinning behavior is exhibited by the polymer solutions in experiments (1b) and (2b). (c) If the apparent viscosity increases as the shear rate decreases, the fluid behavior is usually called *shear-thickening* or *dilatant*. The velocity profile is sharper than a parabola and the ratio of flow rate to pressure head decreases as the flow rate increases. The latter phenomenon is exhibited by the suspension used in experiment (1c).

REFERENCES: (1) H. Markovitz in "Rheology: Theory and Applications" (F. R. Eirich, ed.), Vol. 4, Chapter 6, Academic Press, New York, 1967; (2) A. G. Frederickson, "Principles and Applications of Rheology", Prentice-Hall, Englewood Cliffs, New Jersey, 1964.

CREDITS: Scenes from the NCFMF film "Rheological Behavior of Fluids", by Hershel Markovitz, Mellon Institute, Carnegie-Mellon University. Scenes of velocity profiles by A. G. Frederickson and N. N. Kapoor, University of Minnesota. Consultants, C. Conn, E.D.C., A. H. Shapiro, M.I.T., and H. Markovitz; Producer, R. Bergman, E.D.C.; Editor, W. Tannebring, E.D.C.; Art Director, R. P. Larkin, E.D.C. Produced by the National Committee for Fluid Mechanics Films and Education Development Center (formerly Educational Services Inc.), with the support of the National Science Foundation.

FM-123
Normal-Stress Effects in Viscoelastic Fluids

PURPOSE: To show some effects associated with normal stresses in the flow of fluids characterized by a non-linear stress-deformation relation.

EXPERIMENTS: (1) The behavior of glycerin is contrasted with that of cake batter in a kitchen mixer. The cake batter (a non-Newtonian fluid) climbs up the beater shafts, whereas the surface of the glycerin (a Newtonian fluid) is concave upwards. (2) In the principal experiment, fluid is sheared in the 1/4-inch gap between two parallel discs, the lower of which is rotating about its vertical axis. The distribution of stress normal to the stationary upper disc is indicated by the fluid level in manometer tubes placed over holes spaced across a diameter. The experiment is performed first with glycerin, then with a polymer solution. (a) With *glycerin* (Newtonian), the normal stress distribution is parabolic, with the lowest stress at the center of rotation. (b) With the *polymer solution*, the normal stress is highest at the center of rotation. (3) A small tank with a vertical dividing wall has identical orifices in the floors of the two compartments, one of which contains glycerin while the other contains a polymer solution. The same driving pressure is applied to both compartments. The diameter of the emerging glycerin stream is the same as that of the orifice. The polymer stream upon emerging swells to two or three times the diameter of the orifice.

DISCUSSION: In a rotating Newtonian fluid the normal stress (pressure) increases with distance from the center of curvature, a consequence of centrifugal forces. In a Newtonian fluid undergoing the simple laminar flow of experiment (2), the normal stresses at any point are the same in the axial, radial, and circumferential directions. However, for a fluid with a non-linear stress-deformation relation, the normal stresses in these directions are no longer equal. Consequently the distribution of normal stresses is quite different from that of a Newtonian fluid. In particular, the normal stress on the upper plate is highest at the center for the non-linear fluid. This also explains why the cake batter climbs up the mixer shaft. In experiment (3) the expansion of the polymer stream upon emerging from the orifice is an example of normal stress effects in a more complicated flow situation. This effect must often be taken into account by designers of extrusion dies for high-polymer plastics.

REFERENCES: (1) H. Markovitz in "Rheology: Theory and Applications (F. R. Eirich, ed.), Vol. 4, Chapter 6, Academic Press, New York, 1967. (2) B. D. Coleman, H. Markovitz and W. Noll, "The Viscometric Flows of Non-Newtonian Fluids", Springer, New York, 1966. (3) A. G. Frederickson, "Principles and Applications of Rheology", Prentice-Hall, Englewood Cliffs, New Jersey, 1964.

CREDITS: Scenes from the NCFMF film "Rheological Behavior of Fluids", by Hershel Markovitz, Mellon Institute, Carnegie-Mellon University. Consultants, C. R. Conn, E.D.C., A. H. Shapiro, M.I.T. and H. Markovitz, Mellon Institute; Producer, R. Bergman, E.D.C.; Editor, W. Tannebring, E.D.C.; Art Director, R. P. Larkin, E.D.C. Produced by the National Committee for Fluid Mechanics Films and Education Development Center (formerly Educational Services Inc.), with the support of the National Science Foundation.

FM-124
Memory Effects in Viscoelastic Fluids

PURPOSE: To show experiments with fluids for which the flow behavior depends on the stress history.

EXPERIMENTS: (1) A high-polymer solution is poured out of a beaker, but when the beaker is righted the fluid snaps back into the container. (2) A steel ball first bounces when it is dropped into a high-polymer solution, then falls slowly. (3) A ball of silicone putty bounces and recovers its shape when it is thrown against a solid surface, yet it slowly flows into a puddle under the force of gravity. (4) A fluid is sheared in the annulus between two coaxial cylinders by suddenly rotating the inner cylinder. When the cylinder is released, it reverses its direction and rotates towards its original position. The original angular displacement is partially recovered. The amount of recovery decreases if a longer time elapses before the cylinder is released. (5) In another coaxial cylinder apparatus a torque is suddenly applied to the inner cylinder through the action of a weight on a string that is wrapped around a drum on the shaft of that cylinder. The torque remains constant until the weight is disengaged by a slip toggle. The angular rotation as a function of time is recorded vertically on a chart which moves horizontally at constant speed. This experiment is first performed with glycerin and then with a polymer solution.

DISCUSSION: In a viscoelastic (or memory) fluid, the current value of the stress depends in part on the deformation which the material has experienced in the recent past and not simply on the current rate of deformation. Conversely, the current stress does not determine the current state of deformation. In experiment (5) (which was designed so that bearing friction and inertial effects are small), glycerin acts like a Newtonian fluid; the cylinder rotates with constant angular velocity as long as a constant torque is applied, and stops rotating when the torque is released. With the polymer solution in that experiment, the rate of rotation is not constant while the torque is constant. Furthermore, the direction of rotation reverses when the torque is removed; i.e. the solution deforms when no stress is applied. Unlike the Newtonian fluid, not all the energy used to produce the flow was dissipated; some was stored and then recovered. Thus, the fluid exhibits some elastic characteristics. Experiments (1), (2), (3), and (4) illustrate this type of behavior qualitatively. The deformation of a viscoelastic fluid depends more strongly on the recent history of the stress than it does on the more ancient stresses; the fluid is said to have a "fading memory" of its earlier states of stress. Experiments (3) and (4) exhibit this characteristic.

REFERENCE: (1) H. Markovitz in "Rheology: Theory and Applications" (F. R. Eirich, ed.), Vol. 4, Chapter 6, Academic Press, New York, 1967. (2) A. S. Lodge, "Elastic Liquids", Academic Press, New York, 1964.

CREDITS: Scenes from the film "Rheological Behavior of Fluids", by Hershel Markovitz, Mellon Institute, Carnegie-Mellon University. Consultants, C. R. Conn, E.D.C.; A. H. Shapiro, M.I.T., and H. Markovitz, Mellon Institute; Producer, R. Bergman, E.D.C.; Editor, W. Tannebring, E.D.C.; Art Director, R. P. Larkin, E.D.C. Produced by the National Committee for Fluid Mechanics Films and Education Development Center (formerly Educational Services Inc.), with the support of the National Science Foundation.

FM-125
Examples of Cavitation

PURPOSE: To show a variety of cavitating flows.

SEQUENCES: (1) *A Venturi Throat*. The venturi discharges water at its right-hand end to the atmosphere. As the flow rate increases, the pressure at the throat decreases, according to Bernoulli's integral. When the throat pressure gets low enough, the water boils and a cloud of cavitation bubbles appears. The bubbles collapse as they move into the region of higher pressure in the diffuser. (2) *A Cavitating Hydrofoil of Symmetrical Cross-Section*. The foil is in a circulating water channel. The first experiment is at low channel speed, without cavitation; the next is at high speed, and cavitation occurs on the suction surface. The cavitation region is two-phase, with individual bubbles growing, by expansion of microscopic vapor or gas nuclei moving into or already present in the low pressure region, and then collapsing. (3) *Supercavitating Hydrofoil*. When the cavitated region extends beyond the body about which it occurs, the flow is said to be "supercavitating". A properly designed supercavitating hydrofoil (which has a sharp leading edge and a trailing edge from which the flow leaves cleanly) has the least loss of performance under conditions where a large amount of cavitation cannot be avoided. The first scene is at low tunnel speed, without cavitation. The second scene is at high speed, under "fully-cavitating" conditions: a large closed cavity exists above the upper surface and extends into the wake. This is very different from the two-phase bubbly condition of the previous sequence. (4) *Propeller in a Water Channel*. The ambient pressure and tunnel flow are maintained constant, and the propeller rpm is increased. Cavitation occurs first in the low pressure cores of the trailing vortices from the blade tips. At higher rpm, cavitation also occurs on the suction surfaces of the blades. (5) *Supercavitating Propeller*. The blades, which have sections like that of the supercavitating foil, are designed to operate effectively under conditions of strong cavitation. A large part of the suction surface is fully cavitating and there is a large helical cavity springing from each blade. (6) *Cavitation Damage*. An aluminum button is driven at 14,000 cps by a magnetostriction oscillator so that the face of the button moves perpendicular to itself. It is immersed in water. The pressure at the button face oscillates from high to low. The low pressures cause cavitation. When the cavitation bubbles collapse on the face of the button during the high-pressure phase they produce large stresses and cause damage to the material in a short time. In the next scene, a marine propeller shows pitting due to cavitation damage. (7) *Wake Cavitation*. Water flows past a flat disk whose axis is aligned with the flow. As the speed increases, cavitation bubbles first appear in the localized regions of low pressure in the turbulent separated wake. At higher speeds, a complete cavity forms in the wake.

CREDITS: Scenes from the film "Cavitation" by P. Eisenberg, Hydronautics, Inc.; Consultants, P. Eisenberg, and A. H. Shapiro, M.I.T. Producer/Director, R. Bergman, E.D.C.; Editor, E. Carini, E.D.C. Produced by the National Committee for Fluid Mechanics Films and Education Development Center, Inc., with the support of the National Science Foundation.

FM-126
Cavitation on Hydrofoils

PURPOSE: To show the effects of cavitation on the flow patterns and force characteristics of hydrofoils.

APPARATUS: All the experiments are performed in a water channel in which the flow speed and ambient pressure are adjustable. The hydrofoil models are mounted on a strut which can be rotated to change the incidence.

SEQUENCES: (1-3) A *conventional hydrofoil* with a rounded nose is tested, with the three variables (speed, incidence, ambient pressure) adjusted one at a time. (1) As the speed increases, with incidence and pressure held constant, cavitation first occurs at the junction of the hydrofoil with the strut. Then it occurs on the foil at the point of minimum pressure (approximately the point of maximum thickness). The cavitation region is two-phase and unsteady, involving the generation and collapse of bubbles, and extends for some distance toward the trailing edge. (2) With the speed and pressure constant and no cavitation on the foil, the incidence is increased from zero, resulting in the appearance of a similar cavitating zone. (3) With the speed and incidence constant, and no cavitation on the foil, the ambient pressure is reduced, and similar cavitation occurs. (4) A *supercavitating hydrofoil*, designed to operate effectively when the cavitating region extends beyond the foil, is tested. (a) First the speed is increased until a small region of cavity flow appears on the upper surface, springing from the sharp leading edge. (b) Then, with this same speed, the incidence is increased. The cavity zone on the upper surface grows larger and covers virtually the entire foil. (c) Finally, with this same speed and incidence, the ambient pressure is reduced. The cavity zone gets even larger and extends to a closed cavity zone in the wake. (5) The conventional hydrofoil is supported on a balance that measures lift and drag. The water speed and foil incidence are kept constant. At first the ambient pressure is high enough to prevent cavitation. As the ambient pressure is reduced, cavitation appears and grows on the suction surface of the airfoil. Simultaneously the lift decreases sharply and the drag increases. Reason: prior to the inception of cavitation, reducing the ambient pressure has no effect on the forces because the pressure at every point is decreased by exactly the same amount. However, the pressure on the foil cannot go below the vapor pressure. Once cavitation occurs on the suction surface, therefore, the pressure distribution changes, and in such manner as to decrease the lift and increase the drag.

CREDITS: Scenes from the film "Cavitation" by P. Eisenberg, Hydronautics, Inc.; Consultants, P. Eisenberg, and A. H. Shapiro, M.I.T. Producer/Director, R. Bergman, E.D.C.; Editor, E. Carini, E.D.C. Produced by the National Committee for Fluid Mechanics Films and Education Development Center, Inc., with the support of the National Science Foundation.

FM-127
Cavitation Bubble Dynamics

PURPOSE: To illustrate some phenomena that occur in the dynamics of expanding and contracting bubbles.

SEQUENCES: (1) The electrolysis of water forms simultaneous bubbles of H_2 and O_2 which rise from the cathode and anode, which are out of sight below the bottom of the frame. Since $2H_2O \rightarrow 2H_2 + O_2$, the H_2 bubble has twice the volume of the O_2 bubble. As the bubbles rise in the gravity field the ambient pressure in the water decreases. As they rise, the smaller O_2 bubbles increase only slightly in size. At a certain level the larger H_2 bubbles are unstable and grow very rapidly. Reason: the curve of equilibrium ambient pressure (necessary to balance surface tension and the internal pressure of vapor plus non-condensables) vs. bubble radius, has a minimum in it. When the ambient pressure reaches this minimum value, the equilibrium becomes unstable, and the bubble grows exponentially in time. (2) Water flows past a body of revolution on which the longitudinal pressure distribution has a minimum. Near the point of minimum pressure, cavitation bubbles appear; they grow as they are swept downstream to a region of lower pressure; and they subsequently collapse as they move further downstream into a region of higher pressure. (3) A similar experiment, with non-condensable gas in solution. The cavitation bubbles do not now collapse to zero radius. They rebound in size; the compressibility of the non-condensable gas and the inertia of the radially-moving water comprise a spring-mass system. (4) Cavitation bubbles appear in bulk fluid when the pressure is suddenly reduced, and then collapse when the pressure is rapidly increased. Close inspection of the collapsing bubble shows that the interface does not remain spherical, but distorts. This is the result of an instability which is similar to that of a heavy fluid lying on top of a light fluid. (5) When the collapsing bubble is on a wall the inward-moving liquid forms a high-speed jet of liquid directed toward the wall. This jet may be a major source of damage to the material of the wall.

CREDITS: Experiments and film by A. T. Ellis, University of California, San Diego, R. T. Knapp, California Institute of Technology, and A. Hollander, California Institute of Technology; Consultants, P. Eisenberg, Hydronautics, Inc., and A.H. Shapiro, M.I.T.; Producer, R. Bergman, E.D.C.; Editor, E. Carini, E.D.C. Produced by the National Committee for Fluid Mechanics Films and Education Development Center, Inc., with the support of the National Science Foundation.

FM-128
Cavity Flows

PURPOSE: To show examples of cavity-type flows which sometimes occur, usually due to cavitation, when a liquid flows past a body. In a cavity flow there is a region filled with the vapor of the liquid and/or with a non-condensable gas. The liquid-gas interface of the cavity is a free surface on which the pressure is nearly constant.

APPARATUS: A water tunnel is used in which cavitation on a body can be induced either by increasing the flow speed or by reducing the ambient pressure in the tunnel.

SEQUENCES: (1) A hydrofoil is operated under conditions of "supercavitation", i.e. a vapor cavity region exists above the entire suction surface and extends into the wake. At normal photographing speed, the cavity in the wake appears to be closed at its rearward end. In high-speed photography, however, the water flowing around the rearward end of the cavity is observed to form a jet that moves upstream within the cavity. This upstream-moving jet is characteristic of cavity flows. (2) Water flows normal to a thin circular disk. After startup, the ambient pressure is held constant. As the speed increases, cavitation bubbles appear in the low-pressure vortex cores of the turbulent shear layer in the separated wake. Then, as the speed increases further, a closed cavity region, attached to the disk, forms in the wake behind. (3) The circular disk is tested at an ambient pressure and speed such that cavitation does not occur. However, the injection of air into the wake behind the disk develops a cavity flow which is essentially the same as that formed by the vapor in the preceding experiment. (4) A sphere is shot vertically into an open tank of water at high speed. The camera is oriented so that the vertical entry appears horizontal. Initially, an open cavity is formed extending from the sphere to the water surface. When the sphere is at sufficient depth, the cavity pinches off and becomes closed. Once the cavity is closed, a re-entrant jet is formed at the rear of the cavity. At first the gas in the cavity is air, but ultimately (if the sphere did not slow down) the air would be entrained out of the cavity and the cavity would be filled with vapor.

CREDITS: Scenes from the film "Cavitation" by P. Eisenberg, Hydronautics, Inc.; high-speed water entry scene from the U.S. Naval Ordnance Laboratory; Consultants, P. Eisenberg, and A. H. Shapiro, M.I.T.; Producer/Director, R. Bergman, E.D.C.; Editor, E. Carini, E.D.C. Produced by the National Committee for Fluid Mechanics Films and Education Development Center, Inc., with the support of the National Science Foundation.

FM-129
Compressible Flow Through Convergent-Divergent Nozzle

PURPOSE: To show pressure distributions and wave phenomena for a convergent-divergent nozzle of fixed geometry when it is operated over a wide range of pressure ratios.

APPARATUS: A convergent-divergent half-nozzle of rectangular cross-section receives air from the atmosphere and discharges to an exhauster pump. The back pressure downstream of the nozzle is controlled by a means of a valve ahead of the exhauster. Below the nozzle is a manometer board whose tubes are lined up with the corresponding pressure taps spaced along the straight lower wall of the nozzle. Schlieren optics are used in later scenes to show wave patterns in the nozzle flow. The inlet reservoir pressure (atmospheric, at left) is held constant, while the back pressure (at right) is slowly lowered.

SEQUENCES: (1) The channel and manometer board. (2) Opening of the exhaust valve starts flow through the system; the pressures and the manometer levels fall. (3) The back pressure is slightly below atmospheric. The channel is a purely subsonic venturi, with the pressure falling to the throat and then rising to the exit. (4) Further reduction in back pressure increases the speed until the flow is exactly sonic at the throat. The rising pressure beyond the throat shows that the flow decelerates and is thus subsonic in the divergent section. (5, 6) Further reduction in back pressure has no effect on the pressure distribution up to the throat. The flow is choked. Beyond the throat the pressure continues to fall for some distance, indicating that the flow has become supersonic. The supersonic region is terminated by a region of abrupt pressure rise followed by a region of slower pressure rise, signifying a shock system followed by subsonic deceleration. (7) Schlieren observation shows that Mach waves appear downstream of the throat when the flow becomes supersonic, and that the region of rapid pressure rise coincides with the shock location. (8) Strips of tape on the lower wall (which produce small steps transverse to the flow) generate Mach waves that clearly delineate the supersonic region. (9) As the back pressure is lowered further, the shock system moves downstream, until the whole nozzle is filled with supersonic flow.

CREDITS: Scenes from the NCFMF film "Channel Flow of a Compressible Fluid" by D. Coles, California Institute of Technology; Consultants, D. Coles, and A.H. Shapiro, M.I.T.; Producer/Director, A.H. Pesetsky, E.D.C.; Editor, W. Gaddis, E.D.C. Produced by the National Committee for Fluid Mechanics Films and Education Development Center, Inc., with the support of the National Science Foundation.

FM-130
Starting of Supersonic Wind Tunnel with Variable-Throat Diffuser

PURPOSE: To display pressure distributions and wave patterns for a convergent-divergent nozzle followed by a variable-area diffuser when the two main parameters of the system are varied: (i) the area of the second throat as a fraction of the area of the first throat, and (ii) the pressure ratio across the system. Also, to show how these parameters influence the establishment of supersonic flow in the test section (i.e. *starting*).

APPARATUS: The nozzle, test section and diffuser are all of rectangular cross-section. Air enters from the atmosphere and is drawn through the system by an exhauster pump. A valve ahead of the pump controls the back pressure of the system. Below the flow channel is a manometer board whose tubes are lined up with the corresponding pressure taps spaced along the channel. A schlieren system makes waves in the channel visible.

SEQUENCES: (1) Close-up of the supersonic nozzle, and of the variable diffuser throat while it is being reduced. (2) Opening the valve reduces the back pressure; flow commences. (3) The diffuser throat is made smaller than the nozzle throat, and the back pressure is gradually reduced. At first the flow increases, all pressures fall and all speeds increase. The diffuser throat chokes first, while the nozzle flow is still wholly subsonic. Thereafter, the flow rate remains constant and there is no further change in speed or pressure ahead of the diffuser throat. Only the flow beyond the diffuser throat becomes supersonic. (4) The diffuser throat is made slightly larger than the nozzle throat, and the back pressure is gradually reduced. At first all pressures fall and all speeds increase, as before. However, now the nozzle throat chokes first, and a region of supersonic flow, terminated by a starting shock, appears in the nozzle. Before the starting shock reaches the end of the first nozzle, the diffuser throat also chokes. The tunnel is now *blocked*, in that further reduction of back pressure cannot influence the flow in the nozzle. (5) The diffuser throat is made large enough to pass the starting shock, and the back pressure is gradually reduced. The nozzle throat chokes but the diffuser throat does not; instead, the starting shock passes through the test section and is then swallowed through the diffuser throat. The channel is *started*, with supersonic flow existing from the nozzle throat to beyond the diffuser throat. (6) With the tunnel started, and with the diffuser throat area fixed, the back pressure is slowly increased until the shock jumps through the diffuser throat back into the nozzle, thus *unstarting* the tunnel. By lowering the back pressure, the shock is swallowed again, and the tunnel re-starts. The back pressure for starting is lower than that for unstarting. (7) With a low, constant back pressure the diffuser throat is deliberately made too small to pass the starting shock, so that the channel is blocked. As the diffuser throat is slowly enlarged, a point is reached at which the shock is swallowed and the tunnel starts.

CREDITS: Scenes from the NCFMF film "Channel Flow of a Compressible Fluid" by D. Coles, California Institute of Technology; Consultants, D. Coles, and A.H. Shapiro, M.I.T.; Producer/Director, A.H. Pesetsky, E.D.C.; Editor, W. Gaddis, E.D.C. Produced by the National Committee for Fluid Mechanics Films and Education Development Center, Inc., with the support of the National Science Foundation.

FM-134
Laminar and Turbulent Pipe Flow

PURPOSE: To show how laminar and turbulent pipe flows differ with respect to fluid mixing, jet formation, and frictional pressure drop.

APPARATUS: A long plastic pipe, open to the atmosphere at the right-hand end, is fitted with a trumpet-shaped inlet at the left-hand end. The inlet lies within a large, closed reservoir. A positive-displacement pump feeds the reservoir. The pump draws liquid through a mixing valve from two reservoirs, one containing water and the other containing a glycerine-water solution. By adjusting the valve, the viscosity of the liquid passing through the pipe may be varied while the volume flow rate is maintained constant. The pipe Reynolds number, VD/ν, is thus changed through variations in the kinematic viscosity (ν) while the velocity (V) and the diameter (D) are fixed.

SEQUENCES: (1) A piece of drawn-out glass tubing is arranged to inject a thin stream of dye into the center of the inlet. When the flow is laminar, the dye streakline persists in a clear thread-like form all the way to the exit of the long pipe. Diffusive mixing is negligible. (2) The viscosity is reduced sufficiently for the flow to be turbulent. For some distance the dye streakline is stable and unmixed. Then the streakline exhibits transverse waves of growing amplitude and irregularity, although the dye remains distinct. Ultimately a condition of fully-developed turbulence is reached, and the dye mixes so rapidly throughout the cross-section as to become invisible. (3) A close-up of the exit jet when laminar. It is steady and has a smooth surface. A "wattle" hangs below the main part of the jet. Its shape is governed by gravity, by the parabolic velocity profile, and by surface tension. (4) A close-up of the exit jet when turbulent. The surface appears blurred. (5) Slow-motion of the turbulent exit flow, showing that the blurring was due to unsteadiness and surface roughness produced by transverse turbulent motions. (6) A pressure tap midway along the pipe is connected by a horizontal tube to a vertical manometer leg positioned near the pipe exit. The manometer level measures the frictional pressure drop between the pressure tap and the exit. (7) At first the viscosity is high, the Reynolds number is low, and the flow is laminar. The pressure drop is 10 inches of liquid. With the flow held constant, the viscosity is reduced by a certain amount. At the resulting Reynolds number, the flow is still laminar. After some interval of time required for the less viscous fluid to fill the tube, the pressure drop is reduced to 7 inches. (8) The viscosity is further reduced, producing a Reynolds number at which the flow becomes turbulent. Now the *reduction* of viscosity, with speed held constant, produces an *increase* of pressure drop to 8 inches, thus showing dramatically how turbulence increases frictional losses.

CREDITS: Scenes from the NCFMF film "Turbulence" by R.W. Stewart, University of British Columbia; Consultants, A.H. Shapiro, M.I.T. and R.W. Stewart; Producer/Director, J. Friedman, E.D.C.; Editor, W. Hansard, E.D.C. Produced by the National Committee for Fluid Mechanics Films and Education Development Center, Inc., with the support of the National Science Foundation.

FM-135
Averages and Transport in Turbulence

PURPOSE: To illustrate the unsteadiness of turbulent flows, to suggest how meaningful time averages may be defined, and to show how turbulence augments mass and momentum mixing through convective transport.

APPARATUS: Water flows under turbulent conditions in a rectangular duct. One method of visualizing the flow is by means of hydrogen bubbles that are produced by electrolysis at a wire normal to the flow. Various shapes of bubble-marked fluid lines are released into the flow by pulsing the current and by insulating parts of the wire. A second visualization method involves injection of two differently colored dyes.

SEQUENCES: (1) A continuous bubble sheet is released. It deforms rapidly into disorderly, changing patterns. (2) Thin bubble lines are released periodically. Each is distorted in a complicated way, and successive lines are distorted in very different ways. This shows the inherent unsteadiness of turbulent flows, and that turbulent flows are irreproducible in detail. (3) Still pictures of the previous sequence, each taken at a short fixed time interval following the release of a bubble line, are additively superposed by optical means. As the number of samples increases, an average shape takes form. This shows how average velocity profiles may be meaningfully defined in a flow that is inherently unsteady. (4) A checkerboard pattern of bubble squares is released. These deform in three dimensions and exhibit transverse mixing motions normal to the mean flow. (5, 6, 7) The apparatus is modified to have very rough walls. Blue dye is injected on the axis of the duct, and red dye near the lower wall. This makes it possible to see the transverse motions and the mixing. As compared with the smooth wall experiment, roughening of the wall increases the ratio of turbulent to mean velocities: compare the scales of the unsteady and time-averaged motions. The downward-moving blue dye and upward-moving red dye reveal the turbulent transport normal to the mean flow. Examination shows that, on the average, the downward-moving blue dye (which comes from the high-speed region near the center) has, at the same distance from the wall, a higher forward speed than upward-moving red dye (which comes from the low-speed region near the wall). This accounts for a turbulent transport of momentum which appears in the equations of mean motion as a Reynolds stress.

CREDITS: Scenes from the NCFMF film "Turbulence" by R.W. Stewart, University of British Columbia; Consultants, A.H. Shapiro, M.I.T. and R.W. Stewart; Producer/Director, J. Friedman, E.D.C.; Editor, W. Hansard, E.D.C.; Art Director, R.P. Larkin, E.D.C. Produced by the National Committee for Fluid Mechanics Films and Education Development Center, Inc., with the support of the National Science Foundation.

FM-136
Structure of Turbulence

PURPOSE: To illustrate significant features of the large-scale and small-scale structures of turbulent flows.

APPARATUS: In the first half of the film, two free jets of the same size and speed, but having different Reynolds numbers, are compared. Each jet is made by injecting a dyed liquid of a certain viscosity into a reservoir of the same liquid free of dye. In the second half, a dyed jet flow of liquid into like liquid is compared with a channel flow of liquid.

SEQUENCES: (1) The upper half of a split screen shows a jet with Re = 20,000, and the lower half a jet of the same size and speed but with a higher viscosity such that Re = 400. From direct comparison it appears that the large-scale turbulent structure is the same for the two jets. This observation confirms that the large-scale eddies of a turbulent fluid motion are dominated by inertial forces and are virtually independent of Reynolds number. (2) The same two flows are now seen by means of shadowgraph visualization. Since the shadowgraph method reveals the second derivatives of the optical index of refraction, the small-scale structure is emphasized. The small-scale structure in the high-Reynolds-number jet is similar to but much finer-grained than in the low-Reynolds-number jet. This shows that, if the scale of speed (and thus of ultimate energy dissipation) is held constant, the smallest eddies in the turbulent-energy cascade that can survive without being dissipated by viscosity decrease in size as the viscosity decreases. (3) A close-up of sequence (2), again showing the similar character of the small structure, but the difference in scale at different Reynolds numbers. (4) A still picture of the shadowgraph of the jet at Re = 20,000. A circular portion of the photograph, showing a part of the jet interior, has been cut out and mounted so that it may be rotated. As it is turned, it seems to fit the rest of the pattern well in any orientation, showing that, in a turbulent flow at high Reynolds number, the small-scale structure is both homogeneous and isotropic. (5, 6) The large-scale structure of the dyed jet is seen again, followed by a view of a channel flow in which the distortions of a bubble sheet show the large-scale structure. The large-scale motions of these flows, which have different geometries, are very different. (7) When the same flows are compared by shadowgraph, however, the small-scale structures as seen here on a split screen are very similar, the main difference being in size. (8) A still of the preceding sequence. When a magnifying glass is placed over a part of the picture of the channel flow, thus bringing it to the same scale of small-structure size as the picture of the jet flow, it becomes clear that the only real difference in small-scale structure is in scale of size. In general, there is a locally isotropic regime, at the small-scale end of the energy cascade, which is similar for all kinds of high-Reynolds-number turbulence.

CREDITS: Scenes from the NCFMF film "Turbulence" by R.W. Stewart, University of British Columbia; Consultants, A.H. Shapiro, M.I.T. and R.W. Stewart; Producer/Director, J. Friedman, E.D.C.; Editor, W. Hansard, E.D.C.; Art Director, R.P. Larkin, E.D.C. Produced by the National Committee for Fluid Mechanics Films and Education Development Center, Inc., with the support of the National Science Foundation.

FM-137
Effects of Density Stratification on Turbulence

PURPOSE: To show how stable and unstable vertical gradients of density affect the development and level of turbulence.

APPARATUS: A horizontal duct of rectangular cross-section is divided in half by a thin partition for a certain distance beyond the inlets. The latter are arranged so that each half of the channel can be fed with liquid from a different source. The two liquid streams join beyond the end of the partition. The inlet flange is rotatable about the longitudinal axis of the duct, so that the upper half of the channel (above the partition) and the lower half (below the partition), each carrying its own liquid, can be interchanged.

SEQUENCES: (1) The flows in the two halves are identical in density, viscosity, speed, etc. They differ only in color: the upper fluid is dyed red and the lower fluid gold. The two turbulent flows mingle fairly quickly under these conditions of equal density, to produce a single turbulent channel flow. (2a) Now the upper liquid, colored gold, is hot, and is less dense than the lower cold liquid, colored blue. The mixing is much less rapid, and the turbulence level much lower, than in the first sequence where the densities were equal. The rise in the center of gravity associated with turbulent mixing in this case draws on the kinetic energy of the turbulence. Thus a stable density stratification tends to suppress turbulence. (2b) The channel is rotated by hand about its longitudinal axis so that the more dense liquid (blue) lies above the less dense liquid (gold). The turbulent activity is greatly increased and mixing is much more rapid than when the densities were equal. Unstable density gradients introduce buoyancy forces that promote turbulence. (3) A close-up of the preceding scene showing the high turbulence level associated with an unstable density gradient. (4) Smog over a city, showing that stable stratification sometimes occurs in the atmosphere. When air close to the ground is colder and heavier than the air above it, a so-called inversion is present. Inversions produce very stable cloud and smog configurations which, by preventing upwards mixing, keep polluted air trapped over a city. (5) Unstable conditions also occur in the atmosphere, as shown by time lapse movies of convective clouds which exhibit large verticle motions owing to buoyancy forces. (6) In an oil fire at a refinery, huge clouds of black smoke billow upwards. The configuration of hot gas surrounded by cold air is unstable, and promotes high turbulence levels.

CREDITS: Scenes from the NCFMF film "Turbulence" by R.W. Stewart, University of British Columbia. Oil fire courtesy Los Angeles County Fire Department. Consultants, A.H. Shapiro, M.I.T. and R.W. Stewart; Producer/Director, J. Friedman, E.D.C., Editor, W. Hansard, E.D.C. Produced by the National Committee for Fluid Mechanics Films and Education Development Center, Inc., with the support of the National Science Foundation.

FM-138
Taylor Columns in Rotating Flows
(At Low Rossby Number)

PURPOSE: To demonstrate the tendency of a slightly-perturbed rotating flow in a homogeneous fluid to be two-dimensional when the Rossby number is small. The two-dimensional constraint produces so-called "Taylor columns" or "walls."

APPARATUS: A circular tank of water is mounted on a turntable which rotates at constant speed until the water acquires solid-body rotation. The water is viewed by a camera mounted on the turntable; thus the motions are observed with respect to the rotating reference frame. Before the pure solid-body rotation is perturbed in each experiment, the water appears stationary. For each experiment, the perturbation magnitude is scaled by the Rossby number, $V/L\Omega$, where V is the scale of the perturbation velocity in the rotating frame, L is an appropriate length scale, and Ω is the angular rotation speed. When the Rossby number is small, the inviscid Helmholtz vorticity equation reduces to a requirement that the velocity field not vary in the direction of the rotation axis — that is the Taylor-Proudman theorem of two-dimensionality. The corresponding requirement from Kelvin's circulation theorem is that the projected areas of fluid circuits on planes perpendicular to the rotation axis do not change. Both effects somewhat resemble those in elastic bodies.

SEQUENCES: (1) Ink is first injected into the tank *without* rotation. It spreads like a turbulent jet. (2) When the water is in solid-body rotation, the injected ink (after it has been slowed to low Rossby number by mixing) arranges itself into tall vertical sheets, showing two-dimensionalization. (3) A sphere is towed horizontally through the rotating water. The water above the sphere is colored by falling dye crystals. The column of water above (and also below) the sphere moves with it; the sphere carries with it a circular cylinder of water. (4) When the sphere is speeded up, the Rossby number becomes too large for the Taylor-Proudman theorem to hold, and the column of water above (and below) the sphere is left behind. (5) The sphere is towed vertically upward at low Rossby number. Dye shows that the sphere pushes a circular column of water ahead of it, and draws a circular column behind. The contraction of the vertical vortex lines above, and the stretching below, produces vorticity of opposite signs as seen in the rotating frame. (6) A buoyant ping-pong ball released at the bottom in *non-rotating* water rises rapidly. (7) The same experiment in rotating water. The ball first accelerates, then slows almost to a stop, illustrating the quasi-elastic character of the Coriolis effects at low Rossby number. Then the ball rises steadily at a much lower speed than without rotation. The flow vorticity changes lead to changes in the pressure differences between the top and bottom sides which increase the pressure drag on the sphere.

CREDITS: Scenes from the NCFMF film "Rotating Flows" by Dave Fultz, University of Chicago; Consultants, D. Fultz, A.H. Shapiro, M.I.T.; Producer, C. Conn, E.D.C.; Editor, E. Carini, E.D.C. Produced by the National Committee for Fluid Mechanics Films and Education Development Center, with the support of the National Science Foundation.

FM-139
Small-Amplitude Gravity Waves
in an Open Channel

PURPOSE: To show the fluid motions produced by small-amplitude gravity waves. Also to demonstrate the dependence of wave speed on fluid depth for long gravity waves.

APPARATUS: Waves are produced by moving a piston located in the left-hand end of an open water channel 40 feet long, 12 inches high and 8 inches wide. The velocity amplitude of the piston is small enough for the waves to be of small amplitude, i.e., the height change in the wave is small compared with the water depth. Fluid motions are visualized with particles that have effectively neutral buoyancy.

SEQUENCES: (1) The piston oscillates horizontally at high frequency. Deep-water (or short) waves propagate to the right in the channel. The length of these waves is comparable to (as here) or less than the depth of fluid. (2) The fluid particles move in approximately circular paths for these short waves. The amplitude of the motion decreases with depth and the particles near the bottom hardly move at all. (3) The piston oscillates horizontally at low frequency. Shallow water (or long) waves propagate to the right. Here and in the remaining sequences, the wave length is large compared to the depth. (4) The particles move in flat ellipses, the horizontal amplitudes of which are nearly the same at all depths. The flow is nearly one-dimensional since the fluid motion depends only on time and horizontal distance from a reference point. (5) The piston is given a rightward velocity, and produces a depth-increasing step wave. Fluid particles accelerate only when the wave is passing. After the wave has passed, the particles have constant "drift" velocity in the direction of propagation of the wave. (6) The displacement of a particle is recorded at equal time intervals after passage of the step wave. Two superimposed arrows follow the leading and trailing edges of the wave. The vertical acceleration of the particle is much smaller than the gravitational acceleration; thus, the vertical pressure gradient is nearly hydrostatic. (7) Small-amplitude step waves are produced in the channel with three different depths in the ratio 1:4:9. Moving arrows over the leading-edge of the wave illustrate that the wave speeds are in the ratio 1:2:3, that is, proportional to the square root of the depth.

DISCUSSION: Shallow water waves differ from deep water waves in that vertical accelerations may be neglected when compared to the gravitational acceleration. The vertical pressure gradient is thus hydrostatic. For shallow water waves, the fluid accelerates longitudinally because of longitudinal pressure gradients associated with the slope of the free surface. The pressure differences in the horizontal direction are nearly the same at all depths and the fluid motions are correspondingly nearly one dimensional. As a depth-increasing wave passes, the pressure on a fluid particle increases, and for this reason depth-increasing waves are also called 'compression waves'.

CREDITS: Scenes from the film "Waves in Fluids" by A. E. Bryson, Harvard University. Consultants, A. H. Shapiro, M.I.T., and A. E. Bryson, Harvard University; Producer, R. Bergman, E.D.C.; Editor, E. Carini, E.D.C. Produced by the National Committee for Fluid Mechanics Films and Education Development Center, Inc. (formerly Educational Services Inc.), with the support of the National Science Foundation.

FM-140
Flattening and Steepening of Large-Amplitude Gravity Waves

PURPOSE: To show how large-amplitude, shallow-water expansion waves flatten and compression waves steepen as they propagate.

APPARATUS: An open water channel, 40 feet long, 8 inches wide is used. Depth-decreasing (expansion) waves are produced by the lifting of a sluice gate at the left end. Depth-increasing (compression) waves are produced by accelerating a piston to the right at the left end.

SEQUENCES: (1) The sluice gate at the left is opened. The fluid accelerates to the left while an expansion wave propagates to the right. The camera is moved with the leading edge of the expansion wave. The trailing edge of the wave falls behind the leading edge and the wave profile spreads out, or flattens. (2) One reason for the flattening of an expansion wave is that the wave speed relative to the fluid is proportional to $\sqrt{gh}$. This speed is less at the shallower trailing edge than that at the deeper leading edge of the wave. (3) Another reason is that the speed of the trailing edge relative to the channel is further reduced by the leftward velocity acquired by the fluid there. The wave speed relative to the channel is $u + c$, where u is the fluid speed, and c is the wave speed relative to the fluid. Relative to the leading edge, the speed c is less at the trailing edge; furthermore u is negative at the trailing edge, whereas $u = 0$ at the leading edge. (4) The piston at the left end is accelerated to the right. This accelerates the fluid to the right, and the depth-increasing (compression) wave also propagates to the right. The camera moves with the leading edge of the depth-increasing wave. The wave steepens as it moves. The local wave velocity is higher at the deeper trailing edge of the wave than at the shallower leading edge because of the $\sqrt{gh}$ dependence. (5) Furthermore, passage of the compression wave accelerates the fluid in the direction of the wave motion and the fluid velocity u at the trailing edge adds to the increased local wave speed, $\sqrt{gh}$, thus allowing the trailing edge to overtake the leading edge even more quickly.

CREDITS: Scenes from the film "Waves in Fluids" by A. E. Bryson, Harvard University. Consultants, A. H. Shapiro, M.I.T. and A. E. Bryson, Harvard University; Producer, R. Bergman, E.D.C.; Editor, E. Carini, E.D.C. Produced by the National Committee for Fluid Mechanics Films and Education Development Center, Inc. (formerly Educational Services Inc.), with the support of the National Science Foundation.

FM-141
The Hydraulic Surge Wave

PURPOSE: To illustrate the characteristics of a positive surge wave.

APPARATUS: (a) An open water channel, 40 feet long, 8 inches wide is used. A piston at the left end is pushed to the right to produce the waves. (b) Another channel 2 feet wide with water flowing steadily from right to left. Water at a high head behind a sluice gate issues supercritically beneath the gate. A second sluice gate downstream of the test section regulates the back pressure (head).

SEQUENCES: (1) The piston at the left end is moved with a carefully-programmed speed to the right and produces a depth-increasing (compression) wave of large amplitude. This depth-increasing (compression) wave steepens until the slope of a portion of its leading edge becomes vertical. (2) Continuation of (1). The wave topples and develops into a positive surge wave, i.e. a wave of stationary shape and uniform speed. Dissipation of mechanical energy into internal energy occurs continually within the wave through turbulence and viscosity. (3) The wave speeds of surge waves of different heights are compared in three channels with the same depth, h_1, in the undisturbed water ahead of the wave. The speed of a positive surge wave depends on both h_1 and the depth behind the wave, h_2. It is given by $V_{surge} = \sqrt{g (h_1 + h_2) (h_2/h_1)/2}$

As predicted by this equation, the surge-wave speed is seen to increase with the depth ratio h_2/h_1. (4) The piston produces a small-amplitude wave and is then programmed to produce a surge wave of stationary form which overtakes the small-amplitude wave ahead of it. A small-amplitude wave ahead of the surge wave moves at the speed, $\sqrt{gh_1}$, relative to the fluid; this is always less than the speed of a surge wave behind the small-amplitude wave. (5) Having produced a surge wave of stationary form, the piston speed is changed slightly to produce a small-amplitude wave behind the surge wave which catches up to it. The small amplitude wave behind travels at a speed $V = \sqrt{gh_2} + V_{surge} [1-(h_1/h_2)]$ where the first term is the local wave speed relative to the fluid, and the second term is the drift speed of the fluid behind the surge wave. The drift speed plus the local wave speed is always greater than the speed of the surge wave. Thus, a small-amplitude wave behind the surge wave, and moving in the same direction as the surge wave, always overtakes it. (6) Moving the camera at the speed of the surge wave, the fluid ahead appears to be moving into the wave front. The wave appears stationary and the fluid behind has a speed less than that of the fluid ahead. The flow is steady in this reference frame. The surge wave is compared in split screen to a hydraulic jump in a steady channel flow.

CREDITS: Scenes from the film "Waves in Fluids" by A. E. Bryson, Harvard University. Consultants, A. H. Shapiro, M.I.T., and A. E. Bryson, Harvard University; Producer, R. Bergman, E.D.C.; Editor, E. Carini, E.D.C. Produced by the National Committee for Fluid Mechanics Films and Education Development Center (formerly Educational Services Inc.), with the support of the National Science Foundation.

FM-142
The Hydraulic Jump

PURPOSE: To show some hydraulic jumps and investigate their characteristics.

APPARATUS: The laboratory experiments are done in an open channel 2 feet wide with flow from right to left. Water at a high head behind a sluice gate at the right issues supercritically beneath the gate. A second sluice gate downstream of the test section regulates the back pressure (head).

SEQUENCES: (1) A large hydraulic jump below a dam. (2) a circular hydraulic jump on a plate below a kitchen water tap. (3) A hydraulic jump in the laboratory channel. The jump is quasi-stationary in laboratory coordinates and is the region where the fluid velocity rapidly decreases and the depth of the fluid rapidly increases. Although the flow in the jump region is steady in the average, the conditions there are locally unsteady in time. (4) A paddle disturbs the flow upstream of the hydraulic jump. Both upstream-and downstream-propagating disturbance waves are washed into the jump because the fluid speed upstream is greater than the wave speed relative to the fluid. The ratio of fluid speed u to the local wave speed $\sqrt{gh}$ is called the Froude number, $F = u/\sqrt{gh}$, where h is the depth of the fluid. Upstream of the jump, F is always greater than unity. This is called supercritical (or "shooting") flow. (5) (Repeat) (6) A paddle disturbs the flow downstream of the hydraulic jump. The upstream-moving disturbance wave works its way up to the jump because the fluid speed behind the jump is smaller than the wave speed relative to the fluid, i.e. the Froude number is less than unity there. This is called subcritical (or "tranquil" flow. (7) Repeat. (8) A pitot tube measuring the total head is traversed through the hydraulic jump. The total head decreases because of turbulent energy production and viscous dissipation. This total head loss in the jump is given by $H_1 - H_2 = \dfrac{(h_2 - h_1)^3}{4h_1h_2}$ where h_1, and h_2 are the depths ahead of and behind the jump, respectively.

CREDITS: Scenes from the film "Waves in Fluids" by A. E. Bryson, Harvard University. Consultants, A. H. Shapiro, M.I.T., and A. E. Bryson, Harvard University; Producer, R. Bergman, E.D.C.; Editor, E. Carini, E.D.C. Produced by the National Committee for Fluid Mechanics Films and Education Development Center, Inc. (formerly Educational Services Inc.), with the support of the National Science Foundation.

FM-143
Free-Surface Flow Over a Towed Obstacle

PURPOSE: To show a variety of subcritical and supercritical free-surface flows over a two-dimensional submerged obstacle.

APPARATUS: A long horizontal flume of rectangular cross-section contains water initially at rest. A two-dimensional obstacle with a flat bottom and a lenticular upper surface is towed along the floor of the channel at various speeds. The camera moves with the obstacle; in this coordinate system, we see a streaming flow past a stationary submerged obstacle.

SEQUENCES: (1) A preliminary scene: the obstacle is placed in the tank and observed with a stationary camera. (2) The same experiment is filmed with a moving camera, which shows the flow relative to the obstacle. (3) A defining sketch for the approach Froude number, $F = U/\sqrt{gh}$, where U is the obstacle speed relative to the undisturbed flow far upstream, h is the water depth, and g is the acceleration of gravity. The local ratio of flow speed to speed of long waves varies from place to place in the flow. (4) *Low subcritical speed, F = 0.05* The flow is everywhere subcritical. The free surface dips as it approaches the crest of the obstacle, then rises again to its undisturbed value. From Bernoulli's integral the water speed is a maximum at the crest of the obstacle. (5) *High subcritical speed, F = 0.4.* The flow upstream of the obstacle is subcritical. The water accelerates and the free-surface drops over the obstacle. At about the crest the flow becomes critical; beyond that point the flow accelerates further to supercritical speed, and the free-surface drops to a low level in the lee of the obstacle. The supercritical flow in the wake changes abruptly to subcritical as it passes through a hydraulic jump in which the speed decreases and the water depth increases. (6) *Supercritical speed, F = 2.3.* A hydraulic jump occurs ahead of the obstacle, making the flow subcritical just upstream of the obstacle. However, the flow accelerates over the obstacle and becomes supercritical again in the lee. (7) *Higher supercritical speed, F = 2.6.* Now the upstream hydraulic jump is washed downstream during a transient phase and the flow in the neighborhood of the obstacle becomes everywhere supercritical. As the flow approaches the crest of the obstacle, the level rises. Beyond the crest, the level falls to its upstream undisturbed value. (8) Repeat of experiments at F = 0.05, 0.4, 2.3, 2.6. (9) The obstacle accelerates from rest to a subcritical speed. The upstream-propagating transient waves generated during the acceleration move ahead of the obstacle, since the wave speed $\sqrt{gh}$ exceeds the obstacle speed U. (10) The obstacle accelerates from rest to a supercritical speed. The transient waves generated during the acceleration are washed downstream and do not influence the flow ahead of the obstacle.

NOTE: Compare these free-surface flows with one-dimensional flows of a gas through a duct of varying area.

CREDITS: Scenes from the film "Stratified Flow" by R. R. Long, The Johns Hopkins University; Consultants, A. H. Shapiro, M.I.T., and R. R. Long; Producer/Director, R. Bergman, E.D.C.; Editor, E. Carini, E.D.C. Produced by the National Committee for Fluid Mechanics Films and Education Development Center, with the support of the National Science Foundation.

FM-144
Flow of a Two-Liquid Stratified Fluid
Past an Obstacle

PURPOSE: To show the great variety of flow and wave patterns possible when a two-liquid stratified system, with a free air surface at the top, is disturbed by a moving submerged obstacle.

PARAMETERS: For a given shape of obstacle, the parameters include (a) the internal Froude number of the flow approaching the model $(U/\sqrt{(\Delta\rho/\rho)\,gh})$ where U is the speed of the obstacle relative to the undisturbed flow far upstream, g is the acceleration of gravity, h is the total fluid depth, ρ is the mean density, and $\Delta\rho$ is the difference in density between the two liquids; (b) the height ratio of the two liquid layers; (c) the ratio of obstacle height to liquid height. Different combinations of these parameters yield different flows and wave patterns.

APPARATUS: A long horizontal flume with a rectangular cross-section is filled with two immiscible liquids that differ in density by about 3%. In most of the experiments the obstacle has a flat-bottomed lenticular shape and is towed along the bottom of the channel. Usually the camera moves with the model; in this coordinate system we see a streaming flow past a stationary obstacle.

SEQUENCES: (1) A paddle is drawn through the channel; it develops waves in the liquid-liquid interface and in the liquid-air interface. (2) A large obstacle is towed at high speed and produces disturbances in both interfaces. (3) A schematic sketch of the two-liquid system and the obstacle. (4) In an experiment at very low internal Froude number, the liquid-air interface is undisturbed, and the liquid-liquid interface dips down slightly over the obstacle, showing that the flow is subcritical everywhere. (5) At the same low Froude number, but with a shorter obstacle of the same height, a train of lee waves is generated. (6) With the obstacle of (5), but at a higher internal Froude number, the lee waves are much stronger. (7) With a higher obstacle, an internal hydraulic jump occurs in the lee. (8) At a higher internal Froude number, the liquid-liquid interface swells symmetrically over the obstacle. but the free surface is undisturbed. The flow is supercritical with respect to internal waves and subcritical with respect to free-surface waves. (9) When the upper layer is thin, an internal hydraulic *drop* occurs in the lower liquid in the lee of the obstacle. (10) At a very high internal Froude number, disturbances are generated in both interfaces. (11) A ship model produces very little disturbance in the free surface, but very large waves in the interface. This produces high wave drag with virtually no surface disturbances. Such situations are sometimes encountered by ocean-going vessels in localities where fresh water flows from rivers into the sea, or where warm water lies above cold water.

CREDITS: Scenes from the film "Stratified Flow" by R. R. Long, The Johns Hopkins University; Consultants, A. H. Shapiro, M.I.T., and R. R. Long; Producer/Director, R. Bergman, E.D.C.; Editor, E. Carini, E.D.C. Produced by the National Committee for Fluid Mechanics Films and Education Development Center, with the support of the National Science Foundation.

FM-145
Flow of a Continuously-Stratified
Fluid Past an Obstacle

PURPOSE: To show several types of flow patterns that can occur when a continuously-stratified fluid flows over a submerged obstacle.

APPARATUS: A long horizontal flume with a rectangular cross-section is filled with a salt solution of varying concentration such that the density increases continuously from the free surface to the bottom of the channel. Small plastic spheres, approximately neutrally-buoyant, make the fluid motion visible. Although the beads are unevenly distributed, the liquid density distribution is approximately linear from top to bottom. An obstacle, flat on the bottom and lenticular on top, is towed along the bottom of the channel. In some runs the camera moves with the obstacle; in this coordinate system we see a streaming flow past a stationary obstacle. The internal Froude number is $U/\sqrt{(\Delta\rho/\rho)\,gh}$, where U is the fluid speed, g is the acceleration of gravity, h is the total depth, ρ is the mean density and $\Delta\rho$ is the density difference from top to bottom.

SEQUENCES: (1) A sketch showing the density distribution. (2) At a high internal Froude number, the flow is supercritical, and is symmetrical fore and aft of the obstacle. (3) At a lower internal Froude number, lee waves are present. They are associated with vertical oscillations of inertial parcels of liquid in a restoring buoyancy-force field. (4) A sketch to show that vorticity is generated because surfaces of constant pressure and surfaces of constant density are not coincident. (5) An experiment with strong lee waves showing the relation between the generated vorticity and the wave oscillations. (6) With a large obstacle and a low internal Froude number, there is a large upstream influence. The flow ultimately is characterized by a series of horizontal jets interlayered with horizontal stagnant regions. (7) A drawing of observed streamlines (from weather balloon data on January 30, 1952) in the lee of the Sierra Nevada Ridge in California. There are about three lee waves present. (8) A laboratory experiment to model the Sierra Nevada Ridge at the same internal Froude number. The agreement between the model and the meteorological observations is good.

CREDITS: Scenes from the film "Stratified Flow" by R. R. Long, The Johns Hopkins University; Consultants, A. H. Shapiro, M.I.T. and R. R. Long; Producer/Director, R. Bergman, E.D.C.; Editor, E. Carini, E.D.C. Produced by the National Committee for Fluid Mechanics Films and Education Development Center, with the support of the National Science Foundation.

FM-146
Examples of Flow Instability (Part I)

PURPOSE: To show examples of the great variety of fluid-mechanical instabilities.

SEQUENCES: (1) A drop of milk falls into a beaker of water. The drop rapidly forms into a vortex ring. The ring is unstable and breaks up into a few smaller vortex rings. This process continues in cascade fashion until there are many tiny vortex rings. This instability involves gravity, inertial, and surface tension forces. (2) A slow laminar stream of water exits downward from a tap. When a finger is brought from below into the jet a buckling instability appears. This instability involves inertial, pressure, and surface tension forces. (3) A plume of smoke rises from a cigarette in a still room. The smoke jet is at first laminar, then more-or-less regular waves appear at a certain level, and ultimately the jet becomes turbulent. This instability involves gravity, inertial, and viscous forces. (4) A thin layer of liquid is heated from below in a frying pan. Hexagonal cells appear, in which there is circulation upwards in the middle of each cell and circulation downwards near the vertices. This instability involves gravity, inertial and viscous forces, with heat conductivity playing an important role. (5) Time-lapse movies show the appearance and disappearance of clouds. (6) Time-lapse movies show turbulence in a cloud bank. (7) The flame from the wick of a kerosene lamp. A low flame is stable and steady. A high flame is unstable and unsteady. (8) Air is released into a fish tank through a vertical tube. An instability involving surface tension causes the air to rise in discrete bubbles rather than in a continuous stream.

NOTE: Additional examples are shown in the continuation loop: FM-147, EXAMPLES OF FLOW INSTABILITY (Part II).

CREDITS: Scenes from the NCFMF film "Flow Instabilities" by E.L. Mollo-Christensen, M.I.T.; Consultants, E.L. Mollo-Christensen and A.H. Shapiro, M.I.T.; Producer/Director, A.H. Pesetsky, E.D.C.; Editor, M. Chalufour, E.D.C., Produced by the National Committee for Fluid Mechanics Films and Education Development Center, Inc., with the support of the National Science Foundation.

FM-147
Examples of Flow Instability (Part II)

NOTE: This is a continuation of FM-146, EXAMPLES OF FLOW INSTABILITY (Part I).

PURPOSE: To show examples of the great variety of fluid mechanical instabilities.

SEQUENCES: (9) A circular cylinder is moved on a curved path through a liquid with a Reynolds number such that the boundary layer and wake are unstable. Periodic oscillations of growing amplitude and beautiful form appear in the wake. (10) In a similar experiment, the flow is unstable but the instabilities do not appear until a bubble bursts through the liquid surface and produces enough disturbance to trigger the instability. (11) A long, horizontal box contains two immiscible liquids of slightly different density. When the box is tilted, a gravity wave is generated in the interface (this is not an instability). The passage of the gravity wave causes the fluids above and below the interface to be displaced in opposite directions. This in turn leads to a viscous shear layer in the neighborhood of the interface. At high-enough Reynolds number, this shear flow is unstable (Kelvin-Helmholtz instability), and a periodic instability grows, as shown by smaller waves in the interface. (12) A jet of water is discharged from a nozzle into water. Streaklines of dyed water mark the upper and lower edges of the jet. The shear layer at the jet boundary is unstable. Waves appear, grow, roll up into ring vortices, and higher-order instabilities ultimately make the jet tubulent. (13) A tank contains two immiscible liquids, the red liquid at the bottom being heavier than the blue liquid at the top. An open container is lowered and filled with the red (heavy) liquid. The container, filled with heavy liquid, is very carefully raised into the blue (light) liquid with the open end of the container held downward. Because of the stabilizing action of surface tension, the heavy liquid does not flow out of the container even though it is on top of the light liquid. When a detergent is injected near the interface, the surface tension is reduced, and a one-wave-length instability in the interface grows to the point where the heavy liquid flows out one side while the light liquid flows into the other. Holding water in an inverted tumbler is a well-known parlour trick.

CREDITS: Scenes from the NCFMF film "Flow Instabilities" by E.L. Mollo-Christensen, M.I.T., and from "Übergang von Laminarer zu Turbulenter Strömung" by R. Wille, Technische Universtät, Berlin. Consultants, E.L. Mollo-Christensen and A.H. Shapiro, M.I.T.; Producer/Director, A.H. Pesetsky, E.D.C.; Editor, M. Chalufour, E.D.C. Produced by the National Committee for Fluid Mechanics Films and Education Development Center, Inc., with the support of the National Science Foundation.

FM-148
Experimental Study of a Flow Instability

PURPOSE: To show the experimental determination of the neutral stability curve by means of a particular example: the generation of surface waves by wind.

APPARATUS: A long, horizontal channel of rectangular cross-section is filled with liquid up to about half its depth. Above the liquid, which is stationary, air is drawn through the channel by a fan. Near the upstream end, a cylinder which spans the channel lies just below the liquid surface. When oscillated up and down by mechanical means, the cylinder serves as a wavemaker that introduces disturbances into the interface at controlled frequency.

SEQUENCES: (1) A schematic of the apparatus. (2) When the wind speed is low (below critical), waves generated by the wavemaker die out as they propagate downstream. (3) When the wind speed is high (above critical) waves generated by the wavemaker amplify as they propagate downstream. (4) In a more systematic experiment, the wavemaker oscillates at a fixed frequency while the wind speed is gradually increased in a manner approximately linear with time. A strip recorder shows two traces: (a) the wind speed, as measured by an anemometer (b) the wave height at a certain location downstream of the wavemaker, as measured by a capacitance gage. At low wind speeds, the recorded wave height is essentially zero. Beyond a certain wind speed (the critical speed), the recorded wave height begins to increase and becomes larger as the speed increases. (5) The strip chart is pasted at the appropriate level on a large graph on which frequency is the ordinate and wind speed is the abscissa. (6) The experiment is repeated at other frequencies and the resulting strip charts are placed at their appropriate levels on the large graph. A line is then drawn through the points on the strip charts where wave amplification first becomes appreciable. This is the *neutral stability curve* which separates the stable frequency-speed domain at the left from the unstable frequency-speed domain at the right. (7) At high wind speeds (beyond critical), waves are seen to grow even with the wavemaker removed. This occurs because accidental disturbances of many frequencies are usually present in an experiment unless great pains are taken to eliminate them. The most prominent wave length seen is usually that one having the highest amplification rate.

CREDITS: Scenes from the NCFMF film "Flow Instabilities" by E.L. Mollo-Christensen, M.I.T. Consultants, E.L. Mollo-Christensen and A.H. Shapiro, M.I.T.; Producer/Director, A.H. Pesetsky, E.D.C.; Editor, M. Chalufour, E.D.C.; Art Director, R.P. Larkin, E.D.C. Produced by the National Committee for Fluid Mechanics Films and Education Development Center, Inc., with the support of the National Science Foundation.

Appendix

**The National Committee
for Fluid Mechanics Films**

Frederick H. Abernathy, Harvard University
(Chairman, 1970–1971)

Arthur E. Bryson, Jr., Harvard University
(Chairman, 1965–1968) Currently at Stanford
University

Donald Coles, California Institute of Technology

Stanley Corrsin, The Johns Hopkins University

Dave Fultz, University of Chicago

Robert A. Gross, Columbia University

Stephen J. Kline, Stanford University

Erik L. Mollo-Christensen, Massachusetts Institute
of Technology (Chairman, 1968–1970)

Walter L. Moore, University of Texas

Ascher H. Shapiro, Massachusetts Institute of
Technology (Chairman, 1961–1965, and 1971–)

Hsuan Yeh, University of Pennsylvania

Charles R. Conn, II, Associate Director

Bruce Egan, Associate Director

**Principals and Members
of the Advisory Committees
for the Sound Films**

The Chairman of each Advisory Committee is denoted by (C).

The Fluid Dynamics of Drag, Parts I, II, III, IV (1960)
Principal:
Ascher H. Shapiro, Massachusetts Institute of Technology

Vorticity (1961)
Principal:
Ascher H. Shapiro, Massachusetts Institute of Technology

Deformation of Continuous Media (1963)
Principal:
John L. Lumley, The Pennsylvania State University
Advisory Committee:
Stanley Corrsin, The Johns Hopkins University; Dave Fultz, University of Chicago; Ascher H. Shapiro (C), Massachusetts Institute of Technology

Flow Visualization (1963)
Principal:
Stephen J. Kline, Stanford University
Advisory Committee:
Frederick H. Abernathy, Harvard University; Robert C. Dean, Jr., Dartmouth College; Ascher H. Shapiro (C), Massachusetts Institute of Technology

Pressure Fields and Fluid Acceleration (1963)
Principal:
Ascher H. Shapiro, Massachusetts Institute of Technology
Advisory Committee:
Arthur E. Bryson, Jr. (C), Harvard University; Stephen J. Kline, Stanford University

Surface Tension in Fluid Mechanics (1964)
Principal:
Lloyd Trefethen, Tufts University
Advisory Committee:
Myron J. Block, Block Associates; Peter Griffith, Massachusetts Institute of Technology; Alan S. Michaels, Massachusetts Institute of Technology; William C. Reynolds, Stanford University; Ascher H. Shapiro (C), Massachusetts Institute of Technology

Waves in Fluids (1964)
Principal:
Arthur E. Bryson, Jr., Harvard University
Advisory Committee:
Robert A. Gross, Columbia University; Arthur T. Ippen, Massachusetts Institute of Technology; Walter L. Moore, University of Texas; Ascher H. Shapiro (C), Massachusetts Institute of Technology

Boundary Layer Control (1965)
Principal:
David C. Hazen, Princeton University

Advisory Committee:
Joseph J. Cornish, Mississippi State College; Arnold M. Kuethe, University of Michigan; George B. Matthews, University of Virginia; Ascher H. Shapiro (C), Massachusetts Institute of Technology

Rheological Behavior of Fluids (1965)
Principal:
Hershel Markovitz, Mellon Institute
Advisory Committee:
J. L. Ericksen, The Johns Hopkins University; Herbert Leaderman, National Bureau of Standards; J. L. Lumley, The Pennsylvania State University; Edward W. Merrill, Massachusetts Institute of Technology; A. B. Metzner, University of Delaware; Ascher H. Shapiro (C), Massachusetts Institute of Technology

Secondary Flow (1965)
Principal:
Edward S. Taylor, Massachusetts Institute of Technology
Advisory Committee:
Howard W. Emmons, Harvard University; Dave Fultz, University of Chicago; William R. Hawthorne, Cambridge University; Ascher H. Shapiro (C), Massachusetts Institute of Technology

Channel Flow of a Compressible Fluid (1967)
Principal:
Donald Coles, California Institute of Technology
Advisory Committee:
Arthur E. Bryson, Jr., Harvard University; Stephen J. Kline, Stanford University; Anatol Roshko, California Institute of Technology; Ascher H. Shapiro (C), Massachusetts Institute of Technology; Edward S. Taylor, Massachusetts Institute of Technology

Low-Reynolds-Number Flows (1967)
Principal:
Geoffrey I. Taylor, Cambridge University
Advisory Committee:
Arthur E. Bryson, Jr., Harvard University; Stanley Corrsin (C), The Johns Hopkins University; S. G. Mason, Pulp and Paper Research Institute of Canada; Erik L. Mollo-Christensen, Massachusetts Institute of Technology; Ascher H. Shapiro, Massachusetts Institute of Technology

Magnetohydrodynamics (1967)
Principal:
J. Arthur Shercliff, University of Warwick
Advisory Committee:
Arthur E. Bryson, Jr., Harvard University; James A. Fay, Massachusetts Institute of Technology; Robert A. Gross (C), Columbia University; Ascher H. Shapiro, Massachusetts Institute of Technology; Herbert H. Woodson, Massachusetts Institute of Technology

Cavitation (1968)
Principal:
Phillip Eisenberg, Hydronautics Incorporated
Advisory Committee:
Arthur E. Bryson, Jr., Harvard University; Ralph D. Cooper, Office of Naval Research; Albert T. Ellis, University of California at San Diego; Virgil E. Johnson, Jr., Hydronautics Incorporated; Walter L. Moore, University of Texas; Ascher H. Shapiro (C), Massachusetts Institute of Technology; Murray Strasberg, Naval Ship Research and Development Center

Eulerian and Lagrangian Descriptions in Fluid Mechanics (1968)
Principal:
John L. Lumley, The Pennsylvania State University
Advisory Committee:
Frederick H. Abernathy, Harvard University; Arthur E. Bryson, Jr., Harvard University; Stanley Corrsin, The Johns Hopkins University; Clarence A. Kemper, Joseph Kaye & Co.; Stephen J. Kline, Stanford University; Hans A. Panofsky, The Pennsylvania State University; Ascher H. Shapiro (C), Massachusetts Institute of Technology

Flow Instabilities (1968)
Principal:
Erik L. Mollo-Christensen, Massachusetts Institute of Technology
Advisory Committee:
Arthur E. Bryson, Jr., Harvard University; Donald Coles, California Institute of Technology; Howard W. Emmons, Harvard University; Louis N. Howard, Massachusetts Institute of Technology; Philip S. Klebanoff, National Bureau of Standards; Stephen J. Kline (C), Stanford University; Marten T. Landahl, Massachusetts Institute of Technology; Ascher H. Shapiro, Massachusetts Institute of Technology

Fundamentals of Boundary Layers (1968)
Principal:
Frederick Abernathy, Harvard University
Advisory Committee:
Arthur E. Bryson, Jr., Harvard University; Francis H. Clauser, University of California; James A. Fay, Massachusetts Institute of Technology; Sydney Goldstein, Harvard University; Robert A. Gross (C), Columbia University; Phillip S. Klebanoff, National Bureau of Standards; Edward L. Resler, Jr., Cornell University; Ascher H. Shapiro, Massachusetts Institute of Technology

Rarefied Gas Dynamics (1968)
Principals:
Frederick S. Sherman, University of California, Berkeley; Franklin C. Hurlbut, University of California, Berkeley
Advisory Committee:
Arthur E. Bryson, Jr., Harvard University; Donald

Coles, California Institute of Technology; Robert
A. Gross, Columbia University; Charles H. Kruger,
Stanford University; Hans W. Liepmann, California
Institute of Technology; Ascher H. Shapiro (C),
Massachusetts Institute of Technology

Stratified Flow (1968)
Principal:
Robert R. Long, The Johns Hopkins University
Advisory Committee:
Norman H. Brooks, California Institute of Tech-
nology; Arthur E. Bryson, Jr., Harvard University;
Dave Fultz, University of Chicago; Walter L. Moore
(C), University of Texas; Chia-Shun Yih, University
of Michigan

Aerodynamic Generation of Sound (1969)
Principals:
James Lighthill, Imperial College of Science and
Technology; John Ffowcs-Williams, Imperial Col-
lege of Science and Technology
Advisory Committee:
Arthur E. Bryson, Jr., Stanford University; Ira Dyer,
Bolt, Beranek and Newman; John Laufer, Univer-
sity of Southern California; Erik L. Mollo-Christen-
sen (C), Massachusetts Institute of Technology;
Ascher H. Shapiro, Massachusetts Institute of
Technology

Rotating Flows (1969)
Principal:
Dave Fultz, University of Chicago
Advisory Committee:
Arthur E. Bryson, Jr. (C), Stanford University;
Robert C. Dean, Jr., Dartmouth College; Harvey P.
Greenspan, Massachusetts Institute of Technol-
ogy; Raymond Hide, Meteorological Office, Brack-
nell, England; Yale Mintz, University of California,
Los Angeles; Erik L. Mollo-Christensen, Massa-
chusetts Institute of Technology; Ascher H. Sha-
piro, Massachusetts Institute of Technology

Turbulence (1969)
Principal:
Robert W. Stewart, University of British Columbia
Advisory Committee:
Arthur E. Bryson, Jr., Stanford University; Donald
Coles, California Institute of Technology; Stanley
Corrsin (C), The Johns Hopkins University; Robert
H. Kraichnan; John L. Lumley, The Pennsylvania
State University; Ascher H. Shapiro, Massachu-
setts Institute of Technology

**Staff Members of
Education Development Center, Inc.
Who Contributed to the Production of
the Fluid Mechanics Films**

Executive Producer
Kevin Smith

Directors
Richard Bergman
Quentin Brown
Jack Churchill
John Friedman
Victor Komow
Alan H. Pesetsky

Editors
Richard Bergman
Elvin Carini
Michel Chalufour
William Hansard
Jack Hirschfeld
Jack Kaufman
Alan H. Pesetsky
Bill Tannebring

Cameraman
Abraham Morochnik

Unit Manager
Francis L. Meagher

Graphics
R. Paul Larkin

Technicians
Martha A. Booker
Kevin Cameron
George F. Fardy
James M. Henry
R. Kenneth Jones
Andrew Littell

Production Manager
John J. Barta

1497
174?